Oxygen Complexes and Oxygen Activation by Transition Metals

Oxygen Complexes and Oxygen Activation by Transition Metals

Edited by
Arthur E. Martell and
Donald T. Sawyer
Texas A&M University
College Station, Texas

Plenum Press • New York and London

Library of Congress Cataloging in Publication Data

IUCCP Symposium on Oxygen Complexes and Oxygen Activation by Metal Complexes (1987: Texas A&M University)
Oxygen complexes and oxygen activation by transition metal / edited by Arthur E. Martell and Donald T. Sawyer.
p. cm.
"Proceedings of the Fifth Annual IUCCP Symposium, Oxygen Complexes and Oxygen Activation by Metal Complexes, held March 23-26, 1987, at Texas A&M University, College Station, Texas"—T.p. verso.
Includes bibliographies and index.
ISBN 0-306-42789-3
1. Oxygen—Congresses. 2. Catalysis—Congresses. 3. Transition metals—Congresses. 4. Martell, Arthur Earl, 1916- —Congresses. I. Martell, Arthur Earl, 1916- . II. Sawyer, Donald T. III. Texas A&M University. Industry-University Cooperative Chemistry Program. IV. Title. V. Title: Oxygen activation by transition metals.
QD181.01I93 1987
546′.72125—dc19 87-32170
CIP

Proceedings of the Fifth Annual IUCCP Symposium, Oxygen Complexes and Oxygen Activation by Metal Complexes, held March 23-26, 1987, at Texas A&M University, College Station, Texas

A Division of Plenum Publishing Corporation
233 Spring Street, New York, N.Y. 10013

Printed in the United States of America

PREFACE

This monograph consists of manuscripts, summary statements, and poster abstracts submitted by invited speakers and poster contributors who participated in the symposium "Oxygen Complexes and Oxygen Activation by Transition Metals," held March 23-26, 1987, at Texas A&M University. This meeting was the fifth annual international symposium sponsored by the Texas A&M Industry-University Cooperative Chemistry Program (IUCCP). The co-chairmen of the conference were Professors Arthur E. Martell and Donald T. Sawyer of the Texas A&M University Chemistry Department. The program was developed by an academic-industrial steering committee consisting of the co-chairmen and members appointed by the sponsoring chemical companies Dr. James F. Bradzil, The Standard Oil Company, Ohio; Dr. Jerry R. Ebner, Monsanto Company; Dr. Craig Murchison, Dow Chemical Company; Dr. Donald C. Olsen, Shell Development Company; Dr. Tim R. Ryan, Celanese Chemical Company; and Dr. Ron Sanderson, Texaco Chemical Company.

The subject of this conference reflects the intense interest that has developed in academic institutions and industry on several aspects of dioxygen chemistry. These include the formation of dioxygen complexes and their applications in facilitated transport and oxygen separation; homogeneous and heterogeneous catalysis of oxidation; and oxygenation of organic substrates by molecular oxygen.

The conference differs in two respects from several other symposia on dioxygen chemistry held during the past few years. First, there is extensive industrial participation, especially with respect to oxygen activation. Secondly, the conference purview involves the broadest possible scope of the general subject, from oxygen complex formation and degradation to oxygen transport and activation of dioxygen in catalytic processes.

Eighteen invited papers were presented in seven sessions, in addition to the presentation and discussion of twenty contributed poster papers.

A highlight of the meeting was a lecture entitled "The Nature of the Reactivity of Activated Oxygen Species" by Sir Derek Barton at the symposium banquet.

We thank Liz Porter for assistance with logistics of the symposium and Mary Martell for help with the organization and preparation of the final manuscript.

Arthur E. Martell
Donald T. Sawyer

CONTENTS

ABSTRACTS OF POSTERS

APPENDIX

OPENING REMARKS BY F. BASOLO AT THE SYMPOSIUM HONORING A. E. MARTELL

I've been around a long time doing coordination chemistry, as you people know, and for about twenty years we did synthetic oxygen carrier type work. This was started in 1964 by a graduate student by the the name of Al Crumbliss who is now a professor at Duke University. Crumbliss and I decided we would study solution chemistry of some of the cobalt chelates that Martell and Calvin and others had looked at primarily in solid-state gas-phase interactions. Fortunately Al chose a system that gave for the first time monomeric dioxygen complexes of a series of cobalt compounds. We were never able to get a good suitable single crystal to give Jim Ibers to do the X-ray structure, but we were able to wave our hands and speculate about the structure on the basis of IR and of EPR spectra with the help of Brian Hoffman. We suggested that the dioxygen cobalt chelates have an end-on bent structure which later Ward Robinson showed by X-ray structure to be correct. As you know several of these structures have been found not only for cobalt, but also for iron.

Perhaps the most exciting thing at Northwestern University was the work of one of my postdoctorals and Hoffman. Dave Pettering and Hoffman decided to make what they called cobaglobin which they showed to be an excellent model for the natural protein. The only difference being that the iron in hemoglobin is replaced with cobalt, everything else is pretty much the same. Cobaglobin is similar to hemoglobin in its cooperative uptake of dioxygen, and in its pH effect on the uptake of dioxygen, but the cobalt system has one thing the iron system natural protein does not have and that is the cobalt dioxygen adduct is EPR active. Since oxyhemoglobin is EPR silent one can not probe it with EPR, but Hoffman was able to get valuable information from EPR studies of cobaglobin.

We then went on to study iron complexes and we were not clever enough - we are not good enough organic chemists - to put big bulky groups on Schiff bases, which is what we were using at the time, to prevent the irreversible formation of the μ-oxo bridge (Fe(III)-O-Fe(III)). My graduate student, Dave Anderson, was attempting this when we began to read where Jim Collman had made the "picket-fence" porphyrin and a little later I ran into Jack Baldwin who had made the "capped porphyrin". We then abandoned this steric approach and along with others showed that at low temperature one gets reversible dioxygen uptake in synthetic iron complexes. We studied the kinetics and mechanism of oxygen uptake and release at low temperature. We also attached iron porphyrin onto an imidazole modified silica gel and showed that this works as an oxygen carrier at room temperature, because the irons can not come together to form the stable μ-oxo bridge.

Charlie Weschler, my postdoctorate, studied manganese porphyrin and he discovered the first example of such a compound that reversibly adds dioxygen. During all this work we got letters from theoreticians in

Strasbourg who do ab initio calculations. They were very happy when we published our cobalt dioxygen paper suggesting the end-on bent structure, saying their ab initio calculations agreed very nicely with this. Although I know nothing about ab initio calculations, I too was happy that they were happy and everything went along smoothly until we published our work on the manganese-dioxygen complex. The situation there is quite different, instead of an end-one bent structure we think the $Mn-O_2$ is a T shape peroxy type manganese(IV) structure.The structure was based on our IR and EPR spectral results, as was also used to assign the $Co-O_2$ structure. This time the letter from Strasbourg said they were very unhappy with our structure, because their calculations favored a $Mn(II)-O_2$ end-on bent structure. I guess that they wanted me to be very unhappy, but since I do not understand ab initio calculations I could not be too troubled by their calculations. Thanks to one of our hosts and speakers here, Professor Michael Hall, we now have calculations which agree with our proposed structure.

In 1983, as you know, I was president of the American Chemical Society, and during that same year a renewal was necessary for my NIH grant. I must not have given my renewal request the loving care that I had always given my proposals previously, because it got bounced and did not get funded. I was too busy to try and put it back together again, so I have not done any of this kind of work since 1983, but we did have a lot of fun during those twenty years of working on these kinds of systems.

I appreciate being asked to participate in this symposium honoring my friend Art Martell. I am particularly looking forward to hearing all of these fine talks which will bring me up to date on what has been happening in the field the past few years.

BONDING OF DIOXYGEN TO TRANSITION METALS

Michael B. Hall

Department of Chemistry
Texas A&M University
College Station, Texas 77843

INTRODUCTION

In this chapter I would like to provide the reader with an introduction to the nature of the bond between molecular oxygen and transition metals. As examples of these systems, I will focus on models for cobalt, iron and manganese porphyrins.[1,2] Before discussing these fairly complicated metal systems, I would like to briefly review some basic bonding models for small molecules.

HYDROGEN

Dihydrogen is a deceptively simple molecule. If one thinks about the usual molecular orbital (MO) representation, one begins with an orbital on each hydrogen, let us call them orbital a and orbital b, then one makes linear combinations of these two orbitals,

$$\phi_+ = (a+b), \qquad \phi_- = (a-b) \tag{1}$$

One finds the in-phase combination, ϕ_+, is lower in energy than the out-of-phase combination, ϕ_-, and that the electron density of ϕ_+ is larger between the nuclei than the atomic density while that of ϕ_- is smaller. Thus, we identify ϕ_+ as the bonding molecular orbital and ϕ_- as the antibonding molecular orbital. The ground state of H_2 is described as ϕ_+^2, two electrons in the bonding molecular orbital. In order to satisfy the Pauli Exclusion principle these must have opposite spin, one with $m_s = +½$ or α spin, the other with $m_s = -½$ or β spin.

In the usual valence bond (VB) description of molecules one draws the primary resonance structures as Lewis dot diagrams. For dihydrogen the primary structure is, of course, H:H or H-H where the two dots or the straight line indicates a pairing of the two 1s H electrons one with α spin the other with β spin. In a qualitative sense we are describing the same bond here as in molecular orbital theory. However, when one develops the mathematical expressions for the wave functions one finds that the expressions are quite different.

In molecular orbital theory the total wavefunction, ψ_{MO}, is a Slater determinent (to satisfy Pauli Exclusion Principle) of the molecular orbitals $\phi_+\alpha$ and $\phi_+\beta$.

$$\psi_{MO} = \begin{vmatrix} \phi_+\alpha\ (1) & \phi_+\alpha\ (2) \\ \phi_+\beta\ (1) & \phi_+\beta\ (2) \end{vmatrix} \tag{2}$$

where the numbers (1) and (2) label the electrons. In the remaining equations we will not write these electron labels, but will always assume that functions are in the order electron (1) then electron (2). If one expands this determinant, ignoring any normalization constants, one find that

$$\psi_{MO} = \phi_+\phi_+\ (\alpha\beta - \beta\alpha) \tag{3}$$

where the spin component represents a singlet state. Now we substitute equation (1) for ϕ_+ and we find that

$$\psi_{MO} = (aa + ab + ba + bb)\ (\alpha\beta - \beta\alpha) \tag{4}$$

This wavefunction implies that the two electrons are equally likely to be found on the same atom (aa) or (bb) as they are to be found on different atoms (ab + ba). When the nuclei are very close together this is a good approximation, but as the nuclei move further apart it becomes an increasingly poor approximation. The overall behavior of the potential energy curve is illustrated in Figure 1, where one can see that the (MO) result parallels the experimental (exp) curve at short distances but fails at large distances. The failure to properly dissociate is a common feature of simple molecular orbital wavefunctions.

The VB function typically has the opposite problem. Ignoring normalization constants, one can write the VB wavefunction as

$$\psi_{VB} = (ab + ba)\ (\alpha\beta - \beta\alpha) \tag{5}$$

Here again one finds the singlet spin function multiplying a spacial function, but now the terms with both electrons on the same atom are missing. Thus, the VB wavefunction dissociates properly, but is missing some components which are important at short distances. The typical VB behavior is illustrated in Figure 1. One can see that a more accurate

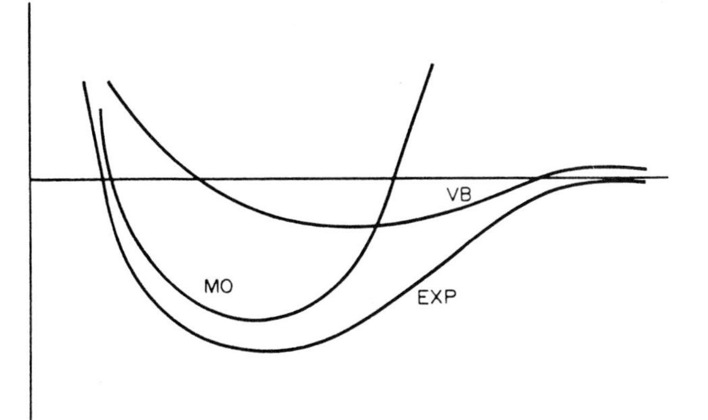

Figure 1. Qualitative potential energy curves for molecular orbital (MO) and valence bond (VB) wavefunction.

description of the potential energies curve could be made by joining the MO and VB curves.

One can do this mathematically by considering the wavefunction formed by putting both electrons into the antibonding MO, ϕ_-, and expanding the determinant. The result is shown in equation (6).

$$\psi'_{MO} = (aa - ab - ba + bb)(\alpha\beta - \beta\alpha) \qquad (6)$$

If one compares this to ψ_{MO} in equation (4), one sees that the terms (ab) and (ba) enter with the opposite sign. Thus, if one forms a new wavefunction by subtracting some fraction of ψ'_{MO} from ψ_{MO},

$$\psi = \psi_{MO} - \lambda\psi_{MO'},$$

one can produce a wavefunction ψ which at short internuclear distances resembles ψ_{MO} ($\lambda = 0$) but at large internuclear distances resembles ψ_{VB} ($\lambda = 1$). One can use λ as a variational parameter to obtain a wavefunction which is more accurate over the entire potential energy curve than either the MO or the VB treatment. The process of adding a variable amount of a wavefunction which is doubly excited with the respect to the usual MO function is called configuration interaction (CI). Although CI may not be necessary for a qualitative description all molecules especially when they are near their equilibrium geometry, it appears to be important in the correct description of metal-dioxygen bonds.

OXYGEN

In O_2 the two oxygen atoms, whose atomic configuration is $1s^2 2s^2 2p^4$, interact with each other to produce a molecular orbital configuration $1\sigma_g^2 1\sigma_u^2 2\sigma_g^2 2\sigma_u^2 3\sigma_g^2 1\pi_u^4 1\pi_g^2$, where the 1σ orbitals and the

2r orbitals are linear combinations of the 1s and 2s atomic orbitals, respectively. The primary O_2 bond is manifest in the remaining MO's. The $3\sigma_g$ is the in-phase combination of the 2p atomic orbitals in the σ direction, while the $1\pi_u$ and $1\pi_g$ are the in-phase and out-of-phase combinations, respectively, of the 2p atomic orbitals with π symmetry. This ground state electronic configuration gives rise to three states of different energy which are given in Table I. By applying Hund's Rule, one easily sees that the ground state is the $^3\Sigma_g^+$ state.

Table I. Relative Energy and Valence Bond Structures for O_2 States

State	E (cm^{-1})	VB
$^1\Sigma_g^+$	13,195	$\vert\overset{\uparrow}{\underline{O}}-\underset{\downarrow}{\bar{O}}\vert \leftrightarrow \vert\underset{\downarrow}{O}\equiv\overset{\uparrow}{O}\vert$
$^1\Delta_g$	7,918	$\underline{\bar{O}}=\underline{\bar{O}}$
$^3\Sigma_g^+$	0	$\vert\overset{\uparrow}{\underline{O}}-\underset{\uparrow}{\bar{O}}\vert \leftrightarrow \vert\underset{\uparrow}{O}\equiv\overset{\uparrow}{O}\vert$

However, when students are asked to draw the VB structure or Lewis-dot diagram for O_2, they often draw $\underline{\bar{O}}=\underline{\bar{O}}$, which is correct for the first excited state but not for the ground state. Although this is often cited as a failure of VB theory, it is not. It is simply a failure of the practitioner to write the correct VB structures. The correct one (see Table I) is a resonance hybrid between singly and triply bonded structures both of which have two unpaired electrons. Thus with the correct resonance structures VB theory also produces a O_2 molecule with a double bond and two unpaired electrons.

The bond order and bond distances for the molecule and negative ions are given in Table II.

Table II. Bond Order and Bond Distance for O_2, O_2^- and O_2^{2-}.

Species	B.O.	R_e (Å)
O_2	2	1.21
O_2^-	1½	1.34
O_2^{2-}	1	1.49

As electrons are added to the $1\pi_g$ orbital, the bond lengths increases to 1.34 Å for the superoxide ion and to 1.49 Å for the peroxide ion. Some earlier workers reported 1.28 Å as the bond length for superoxide and this has caused some confusion in making comparisons with the dioxygen in transition metal complexes.

Another "simple" system that is sometimes used as a standard of comparison to describe the metal-dioxygen bond is the ozone molecule. The molecule is not well described by the standard valence bond structures, 1, but is well deserted as a biradical, 2.[3]

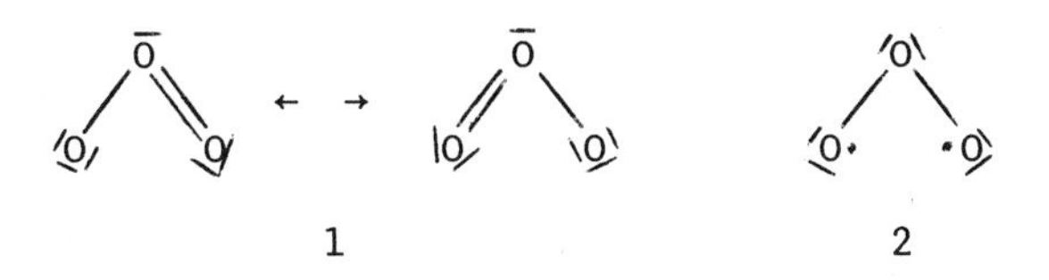

The correct description in MO theory requires the use of a configuration interaction wavefunction for the four π electrons. The three orbitals involved are shown in Figure 2. The usual MO description would be $1b_1^2$ $1a_2^2$ $2b_1^0$. However, this $^1A_1$ state is predicted to be higher in energy than the $^3B_2$ state described by the configuration $1b_1^2$ $1a_2^1$ $2b_1^1$. Experimentally the ground state is $^1A_1$ but only by including the doubly-excited configuration $1b_1^2$ $1a_2^0$ $2b_1^2$ in a configuration interaction wavefunction can one achieve the correct description and energy. Thus, when one refers

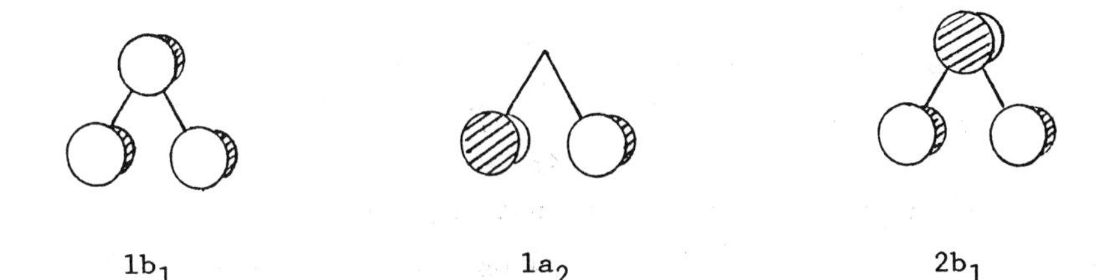

Figure 2. Qualitative π MO's of ozone.

to a MO_2 system as having ozone like character[4] one is referring to a system where the M and O atomic orbitals have similar energies and small overlaps such that CI is necessary for a correct description.

METAL-DIOXYGEN BONDING[5]

Among the most interesting metal complexes which show the binding of dioxygen are the porphyrin complexes. Shown below is the porphyrin ring, 3, and the model ligand, 4, used in our calculations.[1,2] The accuracy of this model ligand has been discussed previously and will not be repeated here. A simple ligand system is necessary in order to reduce the computational time to manageable levels.

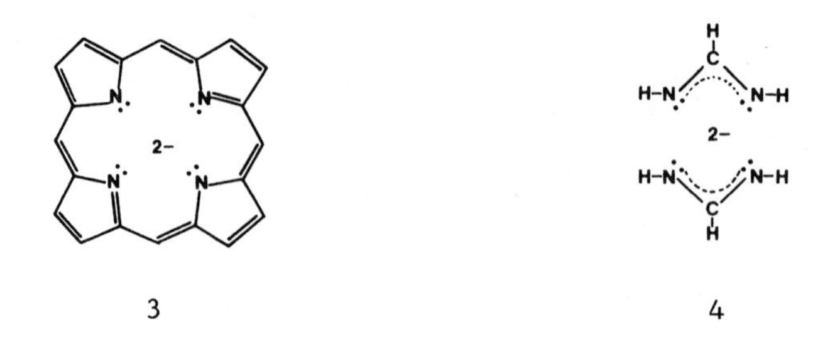

3 4

The two prevalent geometries for the metal dioxygen bond are shown in 5, the end-on or Pauling geometry, and 6, the side-on or Griffith geometry.

One area which often generates considerable argument, some semantic, is the question of the dioxygen oxidation state, i.e.

5 6

dioxygen, superoxo or peroxo. One can take a very formal view and insist that all electron transfer is from ligand to metal and all bonds to the metal should be viewed as dative.[6] This view usually results in a high oxidation state metal. One can also take a purely structural view; naming all end-on systems as superoxo and side-on systems as peroxo.[7] In this paper, I will try to clarify the nature of the argument about the electron distribution, and to offer a complete theoretical description of the bonding. I will emphasize the actual electron distribution as opposed to the formal oxidation state.

Cobalt

In all cobalt porphyrin and related complexes the dioxygen is found to be end-on as in 5. Figure 3 shows a qualitative molecular orbital diagram for the interaction of the upper O_2 valence orbitals with the metal 3d orbitals. The primary interactions are between the $1\pi_g{}^s$, the in-plane π^* O_2 orbitals, and the $3d_{z^2}$ or a hybrid containing substantial d character, and the $1\pi_g{}^a$, the out-of-plane π^* O_2 orbital,

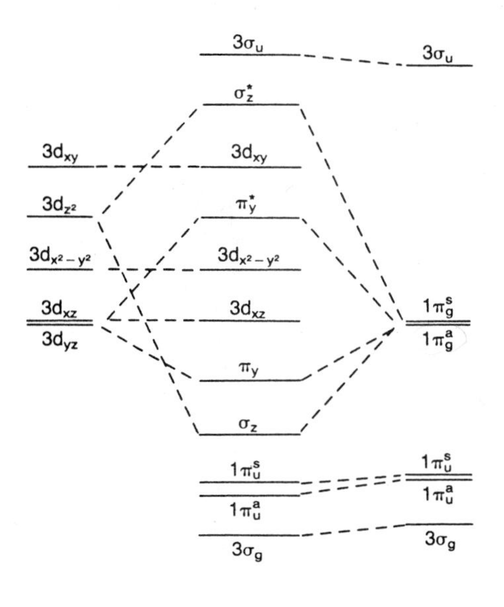

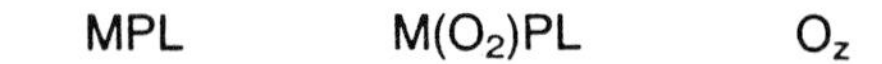

Figure 3. MO diagram for end-on dioxygen system. The left side shows the qualitative energy ordering of the metal d for the M(porphyrin model)(terminal ligand) system, while the right side shows the order of the O_2 orbitals.

and the $3d_{yz}$. The former interaction produces the σ_z and σ_z*, the bonding and antibonding M-O_2 σ orbital, the latter interaction produces the π_y and π_y*, the bonding and antibonding M-O_2 π orbitals. For $CoO_2(P)L$ system there are a total of 15 electrons to place in orbitals shown in Figure 3. Thus, this molecule has 1 electron in the π_y* and 2 electrons in the orbitals below the π_y*.

The main arguments about the charge distribution in these complexes involves the distribution of metal and ligand in the σ_z, π_y, and π_y*, the question of whether the molecular orbital representation is accurate or CI is needed, and whether the lower energy σ and π orbitals become involved to a significant extent. In the case of the CoO_2 systems ESR spectroscopy has shown that the unpaired electron is almost entirely on the O_2 ligand. Thus, the π_y* orbital must be almost pure $\pi_g{}^a$, and therefore, the π_y must be nearly pure $3d_{yz}$. With that potential problem solved the main argument in the CoO_2 systems resides in the nature of the σ_z molecular orbital. If the σ_z is mainly $\pi_g{}^s$ with only a small amount of metal character than the system should be viewed as $Co^{3+} \leftarrow O_2^-$ with a dative σ bond. This electron distribution could occur with or without the involvement of the lower energy π_u and σ_g orbitals. If these lower energy orbitals become involved that would indicate a strong enough Co-O_2 interaction to cause rehybridization of the O_2 orbitals. If the σ_z is nearly an equal mixture of metal and π_g^s the system should be viewed as Co^{2+}-O_2^0 with a covalent σ bond between Co and O_2. For the latter electron distribution there are two bond types. If the Co-O_2 σ interaction were so weak that CI was needed to properly describe it then the spin-coupling model of Co^{2+}-O_2^0 would the best view, but if the interaction was strong such that rehydridization of the in-plane orbitals of O_2 occurs then the spin-coupling model would be inappropriate. Thus, for the CoO_2 system we have four possibilities

$Co^{3+}\leftarrow O_2^-$	Co^{2+}-O_2^0
simple dative bond	spin coupled
or	or
rehybridized dative bond	rehybridized covalent bond

The results of our calculations strongly support a $Co^{3+}\leftarrow O_2^-$ electron distribution with rehybridized O_2 orbitals. The key molecular orbitals are shown in Figure 4. The calculations confirm the interpretation of the ESR;[8,9] the doubly occupied π_y orbital is pure $3d_{yz}$ and the π_y* is pure π_g^a. As can be seen from the number of contours in Figure 4. The calculations even predict the fact that more of the unpaired electron resides on the distal oxygen.

The σ bond is comprised of two important interactions displayed on the right of Figure 4, donation from the doubly occupied π_g^s and donation from the doubly occupied π_u^s. Because both the π_g and π_u become involved in this bond, it is best viewed as involving a rehybridization of the O_2. The distortion in the π orbitals are due to the mixing of some $3\sigma_g$ character into them.

Figure 4. MO of cobalt-dioxygen system. Plots on left are for π system, while those on the right are for the σ system.

An isolated Co-O_2 σ bond and an O lone pair can be formed from a linear combinations of the two σ orbitals in Figure 4. The sum of these two orbitals would be an isolated Co-O bond, while the difference would represent a lone pair on the distal oxygen. Examination of the CI results suggests that the MO representation described above is accurate and that CI makes no important qualitative changes in this description. The electron distribution in this complex closely resembles a rehybridized superoxide ion forming a dative bond to a cobalt (+3) ion.

Iron

Figure 3 can again be used to describe qualitatively the problems inherent in resolving the argument about the electron distribution in the iron-porphyrin-dioxygen system. The FeO_2 system has one less electron than the CoO_2 system, and, hence, the $\pi_y{}^*$ is the lowest unoccupied molecular orbital. Since the iron system is a closed-shell system we do not have a convenient probe such as ESR that allows us to determine unequivocally the distribution of charge in the π system. Hence, the description of FeO_2 system is complicated because one may argue about the distribution in both the σ and π system.

For the π system the key is the nature of the interaction and charge distribution in the π_y molecular orbital. The two likely extremes for this orbital are either a d_{yz} orbital with some small amount of donation into the $\pi_g{}^a$ O_2 orbital (back-bonding), or a strong mixture of d_{yz} and $\pi_g{}^a$ which would best be described as one electron in each orbital spin-coupled to form a covalent bond. The π system could also be complicated somewhat by some mixing of the π_y with the $\pi_u{}^a$. The σ system has the same four choices that the σ system of cobalt had. Even if we leave out any $\pi_u{}^a$ involvement in the π_y and any complications due to CI, we already have 8 possible descriptions. Two of these correspond to Fe^{1+}-$O_2{}^+$ descriptions and need not be considered further. The remaining six are:

$Fe^{2+}(S=0) \leftarrow O_2{}^0(S=0)$	$Fe^{2+}(S=1)$-$O_2{}^0(S=1)$	$Fe^{3+} \leftarrow O_2{}^-$
simple dative or rehybridized dative	spin-coupled or rehybridized covalent	spin-coupled or rehybridized covalent

Our results support $Fe^{2+}(S=0)$-$O_2{}^0(S=0)$ with a rehybridized dative bond as the major contributor to the Fe-O_2 bond. The key molecular orbitals are shown in Figure 5. The π system consists of a π_u orbital

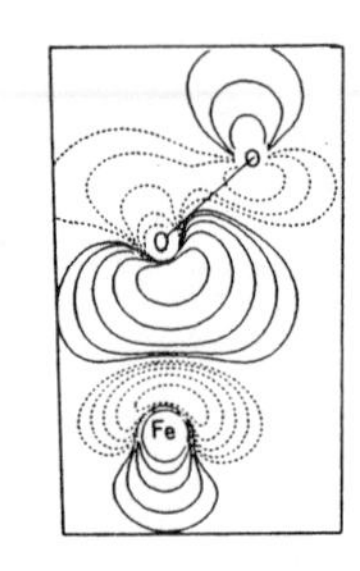

Figure 5. MO of iron-dioxygen system. The π orbitals are on the left and the σ orbital is on the right.

with a small amount of metal character and an unequal distribution of O character with more density on the distal O and of a Fe $3d_{yz}$ orbital delocalized on to the O_2, but with substantially more character on the bonded O. This orbital picture seems to correspond closely to a localized versions of the typical 3-orbital, 4-electron interaction such as found in ozone. If we have 3 equal energy atomic orbitals (a, b, and c) the delocalized MOs will be

$$\phi_1 = a+b+c, \qquad \phi_2 = a-c, \qquad \phi_3 = a-b+c. \tag{8}$$

In ozone and FeO_2 both ϕ_1 and ϕ_2 are filled. The orbitals in Figure 4 roughly correspond to the linear combinations

$$\phi_1 + \phi_2 = 2a+b, \qquad \phi_1 - \phi_2 = b+2c \tag{9}$$

Thus, the first has most of the Fe character (a) while the second has most of the distal O character (c). What makes ozone different from other 3-orbital, 4-electron systems is that the energy of the triplet state $\phi_1^2\phi_2^1\phi_3^1$ is lower in energy than the singlet $\phi_1^2\phi_2^2$, unless CI is included by subtracting a small amount of $\phi_1^2\phi_3^2$ from $\phi_1^2\phi_2^2$. It turns out that the FeO_2 system has the same behavior. From this behavior one might be tempted to say that the FeO_2 has substantial $Fe^{2+}(S=1)$-$O_2^0(S=1)$ or ozone-like character. However, an examination of ϕ_3 (Figure 5) which in ozone would have slightly more central atom (b) character and equal terminal atom (a and c) character, shows that in FeO_2 ϕ_3 has substantially more π_g^a character than Fe character and that the distribution is nearly equal between the two oxygen (b and c). Thus, in spite of some resemblance to ozone a better description is a Fe $3d_{yz}$ orbital strongly nearly equal between the two oxygen (b and c). Thus, in spite of some resemblance to ozone a better description is a Fe $3d_{yz}$ orbital strongly delocalized into the π_g^a. This view returns us to thinking of O_2 as a neutral ligand with strong π acceptor properties in the plane perpendicular to the molecular plane of FeO_2.

The Fe-O_2 σ bond, right side of Figure 5, shows rehybridization in the molecular plane to produce an O_2 lone pair which then donates electron density to the metal. The final Fe-O_2 bonding MO appears to contain more O_2 than Fe character, but doesn't preclude some contributions from a covalent Fe-O bond used in the ozone-like model. Still, $Fe^{2+}(S=0) \leftarrow O_2^0(S=0)$ is the major contributor to the bonding. In this model there is donation from an O_2 lone pair to the metal to form the σ bond and back-donation from the metal to the O_2 π_g^a orbital to add partial double bond character to the Fe-O bond and reduce the bond order of the O-O bond. This description does not preclude some contribution to the bonding from the other two models nor does it preclude discussing the bonding starting with one of these models and then modifying the description of the electron flow.

All of the experimental results on FeO_2 porphyrin can be accommodated within this model. Beginning with the system as $Fe^{2+}(S=0) \leftarrow O_2^0(S=0)$ one has sufficient back-donation from the d_{yz} to the π_g^a to reduce the O-O bond order and explain the reduction in the O-O stretching frequency and the increase in the O-O bond length. Although the stretching frequency approachs that of the superoxide ion so does the O-O stretch in ozone and HO_2.[9] Furthermore, where the O-O bond distance is accurately known it is significantly shorter than that found for the superoxide ion.[9] The Mössbauer spectrum of these system is often interpreted to favor Fe^{3+} with a configuration $d_{yz}^1\, d_{xz}^2$.[10] However, our calculations suggest that the d_{yz} occupation is larger. The remaining orbital asymmetry observed in the Mössbauer spectrum arises from an expansion of the d_{yz} toward the O_2 ligand and a contraction of the d_{xz} away from the O_2 ligand.

Manganese

Although earlier ab-initio calculations without CI suggested that the MnO_2 porphyrin geometry was also end-on,[11] our CI results support a side-on structure, 7 as do the ESR and IR results. The ESR results suggest three possible electronic configurations for the three unpaired electrons: $d_{x^2-y^2}^1 d_{yz}^1 d_{xz}^1$, $d_{x^2-y^2}^1 d_{yz}^1 d_{z^2}^1$, or $d_{x^2-y^2}^1 d_{yz}^1 d_{xy}^1$. In addition to the question of the distribution of the unpaired electrons

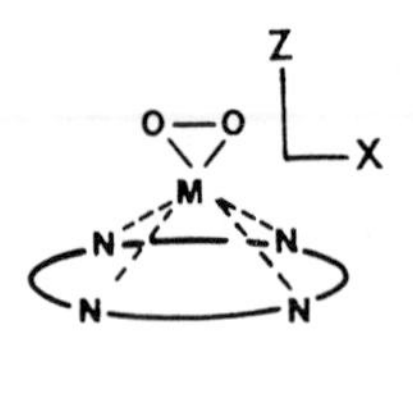

7

the $Mn\text{-}O_2$ bonding electrons could be accommodated by several electron distributions. Beginning with a Mn^{2+} high-spin d^5 system, the bonding could occur by a spin coupling mechanism which in structure 7 would involve the d_{xy} and d_{xz} orbitals. Thus, only the second configuration would be acceptable for this $Mn^{2+}(S=5/2)\text{-}O_2^{0}(S=1)$ model. A second alternative would be a $Mn^{2+}(S=3/2)\text{-}O_2^{0}(S=0)$ scheme. Here we would use the $^1\Delta_g$ state of O_2 and use the empty $d_z{}^2$ as the acceptor orbital for the $\pi_u{}^s$ pair and the d_{xz} as the donor orbital to the $\pi_g{}^s$. This model would suggest the third configuration for the unpaired electrons; however, if the donation to the $d_z{}^2$ were weak, the second configuration model would also be acceptable. These two configurations would also be invoked by a peroxide model $Mn^{4+}(S=3/2)$ - $O_2{}^{2-}(S=0)$. One could also postulate a superoxide model $Mn^{3+}(S=2)\text{-}O_2{}^{-}(S=1/2)$. In this model the $\pi_g{}^{s1}$ electron would spin couple with $d_{xz}{}^1$ to form one bond. In addition there would be donation from the $\pi_u{}^{s2}$ orbital into the d_z2 orbital. Again either configuration two or three would be appropriate.

We examined several geometries and a number of possible electronic configurations, and concluded that the second configuration was the lowest in energy.[2] The three singly occupied orbitals are shown in Figure 6. Furthermore, the calculations totally eliminated the spin-coupling model $M^{2+}(S=5/2)\text{-}O_2^{0}(S=1)$ as a possible description.

The doubly occupied orbitals in the MnO_2 plan are shown in Figure 7; in addition to these the $\pi_g{}^a$ orbital, which is perpendicular to this plane, is also doubly occupied. The orbital on the left side represents a $\pi_u{}^s$ orbital donating some density to the metal. Most of the interaction must be with the s and p orbitals on the metal since the d_z2 remains low enough in energy to be singly occupied. The orbital on right side represents a strong covalent bond between the $\pi_g{}^s$ and the d_{xz} orbital. Notice that it is almost an equal mixture of the two components. Configuration interaction also plays an important role in this interaction, such that a reasonable representation would be to

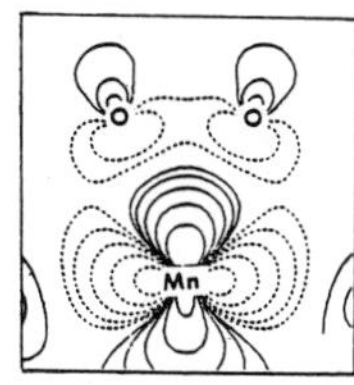

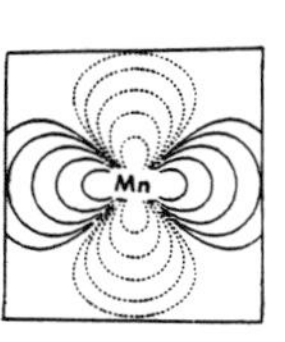

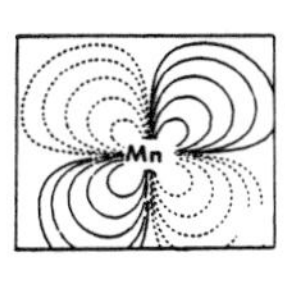

Figure 6. Singly occupied MO of manganese-dioxygen system: the d_{z^2}, the $d_{x^2-y^2}$ and the d_{yz}.

think of a valence bond description where one electron in the d_{xz} orbital formed a covalent bond with one electron in the π_g^s. Thus, the dioxygen configuration corresponds to O_2^- $3\sigma_g^2 \pi_u^{a2} \pi_u^{s2} \pi_g^{a2} \pi_g^{s1}$, and the manganese corresponds to Mn^{3+} $d_{x^2-y^2}^1 d_{yz}^1 d_{z^2}^1 d_{xz}^1 d_{xy}^0$.

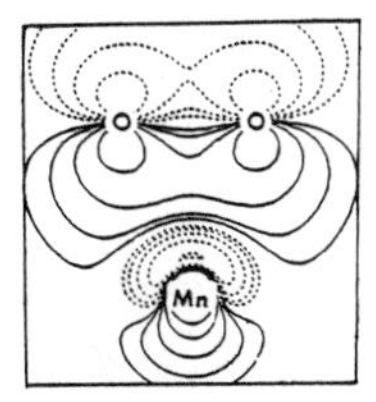

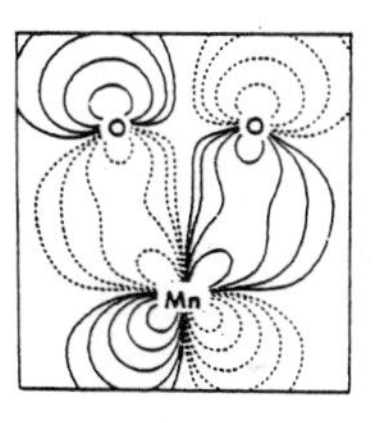

Figure 7. Doubly occupied MO of manganese-dioxygen system: the π_u^s to metal donation and the π_g^s-d_{xz} interaction.

Summary

In this paper we have attempted to summarize our work on MO_2 porphyrin systems and provide enough background to allow the reader to appreciate some of the difficulties in arriving at a simple description of the charge distribution. Our calculations and analysis suggest that the final charge distribution in both the cobalt and manganese porphyrine is closest to M^{3+}-O_2^-, while that in the iron system is closest to Fe^{2+}-O_2^0. One may start with a variety of formal oxidation states and arrive at this final charge distribution.

Acknowledgement

The author would like to thank the National Science Foundation (CHE83-09936 and CHE86-19420) and the Robert A. Welch Foundation (A-648) for their support.

References

1. J. E. Newton and M. B. Hall, Inorg. Chem. 23:4627 (1984).
2. J. E. Newton and M. B. Hall, Inorg. Chem. 24:2573 (1985).
3. W. A. Goddard III, T. H. Dunning, Jr., W. J. Hunt, and P. J. Hay, Acc. Chem. Res. 6:368 (1975).
4. W. A. Goddard III and B. D. Olafson, Proc. Natl. Acad. Sci. U.S.A. 72:2335 (1975); B. D. Olafson and W. A. Goddard III, Proc. Natl. Acad. Sci. U.S.A. 74:1315 (1977).
5. For recent reviews see: M.-M. Rohmer in "Quantum Chemistry: The Challenge of Transition Metals and Coordination Chemistry," A. Veillard ed., NATO ASI Series C, Vol. 176, D. Reidel Pub. Co., Dordrecht, 1986, p. 377; M. H. Gubelmann and A. F. Williams, Structure and Bonding, 55:1 (1983); and references therein.
6. D. A. Summerville, R. D. Jones, B. M. Hoffman, and F. Basolo, J. Chem. Educ. 56:157 (1979).
7. L. Vaska, Acc. Chem. Res. 9:175 (1976).
8. G. McLendon and A. E. Martell, Coord. Chem. Rev. 19:1 (1976); R. D. Jones, D. A. Summerville, and F. Basolo, Chem. Rev. 79:139 (1979); and references therein.
9. R. S. Drago, Coord. Chem. Rev. 32:97 (1980); and references therein.
10. Y. Maeda, J. Phys. Colloq. C2, 40:C2-514 (1979).
11. A. Dedieu, M.-M. Rohmer, and A. Veillard, Adv. Quantum Chem. 16:43 (1982) and references therein.
12. C. J. Weschler, B. M. Hoffman and F. Basolo, J. Am. Chem. Soc. 97:5278 (1975); B. M. Hoffman, C. J. Weschler, and F. Basolo, J. Am. Chem. Soc. 100:7253 (1978); and M. W. Urban, K. Nakamoto, and F. Basolo, Inorg. Chem. 21:3406 (1982).

OXYGEN BINDING BY THE METALLOPROTEINS HEMERYTHRIN, HEMOCYANIN, AND HEMOGLOBIN

Thomas M. Loehr

Oregon Graduate Center

19600 N. W. Von Neumann Drive, Beaverton, Oregon 97006-1999

INTRODUCTION

Respiring organisms have evolved three principal oxygen transport proteins, hemoglobins, hemerythrins, and hemocyanins, that possess radically different polypeptide structures, subunit aggregates, and active site structures. Hemoglobins are by far the most widespread, occurring in all mammals and vertebrates, many invertebrates, selected eukaryotic micro-organisms, and even some leguminous plants. Hemoglobins are largely tetrameric proteins consisting of $\alpha_2\beta_2$ subunits each of molecular weight ≈16,000; however, some invertebrate hemoglobins consist of huge aggregates with molecular weights into the millions. Vertebrate muscle tissue also contains a monomeric oxygen storage/transport protein, myoglobin, that is very similar to a hemoglobin monomer. Hemoglobin and myoglobin contain a "heme" prosthetic group: an iron complex of a macrocyclic tetrapyrrole, such as protoporphyrin IX. Crystal structures of these proteins in various states of ligation have been reported and form a thorough basis for the elucidation of oxygen coordination, protein allosteric control, cooperativity of oxygen binding, and macromolecular assembly [1].

Hemerythrin (the red blood pigment) is a nonheme, binuclear iron protein that occurs among several phyla of marine invertebrates. The protein lacks a distinct prosthetic group, but has a solvent derived oxo bridge between the two iron atoms in oxidized forms of the protein. The dominant form of hemerythrin contains eight polypeptides of identical 13,500-dalton subunits, and in general, shows no cooperativity in O_2 binding. As with hemoglobin, muscle tissues contain a myohemerythrin. Crystallographic studies of both structures have been reported for various physiological and non-physiological forms of these metalloproteins [2, 3].

Hemocyanin (the blue blood pigment) is the respiratory protein of molluscs and arthropods. It also contains a nonheme, binuclear metal center, but has copper rather than iron atoms at its active site. Until recently, least was known about the structures of the highly aggregated hemocyanins (MW $>10^6$), however, the publication of the crystal structure of a deoxy or met hemocyanin hexamer from the spiny lobster finally provides considerable insight into an area of longstanding speculation and controversy [4-6]. The minimal functional subunit molecular weights of molluscan hemocyanins are ≈50 kD and those of arthropodan species ≈75 kD.

In each instance, oxygen binding by these diverse proteins appears to be describable as an oxidative addition reaction (Table I), whereby a reduced metal center in the deoxy protein becomes oxidized with the concomitant reduction of the incoming dioxygen molecule. (For an extensive discussion on the description of oxidation states of M and O_2 in dioxygen complexes see Niederhoffer, Timmons, and Martell [7]).

TABLE I. ACTIVE SITES AND OXYGENATION REACTIONS OF O_2-TRANSPORT PROTEINS

Protein	Active Site	Reaction
Hemoglobin	Iron Porphyrin	$Fe^{II} + O_2 = Fe^{III}—O_2^-$
Hemerythrin	Binuclear Iron	$(Fe^{II})_2 + O_2 = (Fe^{III})_2—O_2^{2-}$
Hemocyanin	Binuclear Copper	$(Cu^{I})_2 + O_2 = (Cu^{II})_2—O_2^{2-}$

Structural studies have shown that the oxygen binding sites are generally hydrophobic, devoid of polar side chains from the protein. The principal exception is the conspicuous presence of the distal histidine group on the O_2-binding side of the heme pocket in hemoglobin and myoglobin that is strongly implicated in hydrogen-bonding interactions with the bound dioxygen. (Traylor refers to these sites as "polar pockets" [8]). There is currently much support for the importance of hydrogen bonding in the stabilization of the oxygenated protein and model porphyrin dioxygen adducts [8-13]. One may think of the oxygenated species as an intermediate that can dissociate via two competing processes, i.e., the release of the dioxygen molecule from the carrier or oxidative dissociation leading to an oxidized metal center and reduced oxygen species:

$$M + O_2 \rightleftharpoons [M^+—O_2^{n-}\cdots X^+] \longrightarrow M^+ + O_2^{n-}$$

Transport   Oxidation

For reversible oxygen binding to be favored over the oxidation pathway, a hydrophobic binding site will facilitate the release of the neutral dioxygen molecule, thus enhancing the off-rate over the oxidative rate. Additionally, interactions such as hydrogen bonding (X = H) or coordination to a second metal ion (X = M') that "secure" both ends of the bound dioxygen, will serve to stabilize the oxygenated intermediate by lowering the off-rate relative to the on-rate.

This review will focus on the known structural details of the active sites of these three protein classes and discuss the evidence that has been presented, largely from vibrational spectroscopy, for the molecular and electronic structures of the coordinated dioxygen in the oxygenated forms of these proteins.

VIBRATIONAL SPECTROSCOPY OF DIOXYGEN COMPLEXES IN PROTEINS AND MODELS

The vibrational frequency of the dioxygen moiety is an excellent reporter for its electronic structure by reflecting its bond strength. Table II lists frequencies typical of the four dioxygen species that span >1000 cm^{-1} as the bond order changes from 2.5 for the oxygenyl cation to 1.0 for peroxide. Reference to these characteristic frequencies aids in the identification of the nature of the bound oxygen in metal-dioxygen complexes. A representative list of such complexes, including the spectroscopic data for the three respiratory proteins, is given in Figure 1 (patterned after Suzuki et al. [14]). The O-O vibrational frequencies are observed to fall into three ranges: a high (1160-1100 cm^{-1}), analogous to superoxides; a low (740-880 cm^{-1}), analogous to peroxides; and a middle range (940-1010 cm^{-1}) that straddles these descriptive boundaries. Oxyhemoglobin and oxymyoglobin exhibit vibrational frequencies among the superoxo

TABLE II. BOND PROPERTIES, π-ELECTRONS, AND FREQUENCIES OF O_2 SPECIES

Species	O-O Distance	Bond Order	π^* e^-	ν(O-O)
$O_2^{+\cdot}$	1.12 A	2.5	↑ __	1860 cm^{-1}
O_2	1.21	2.0	↑ ↑	1550
$O_2^{-\cdot}$	1.33	1.5	↑↓ ↑	1130
O_2^{2-}	1.49	1.0	↑↓ ↑↓	815

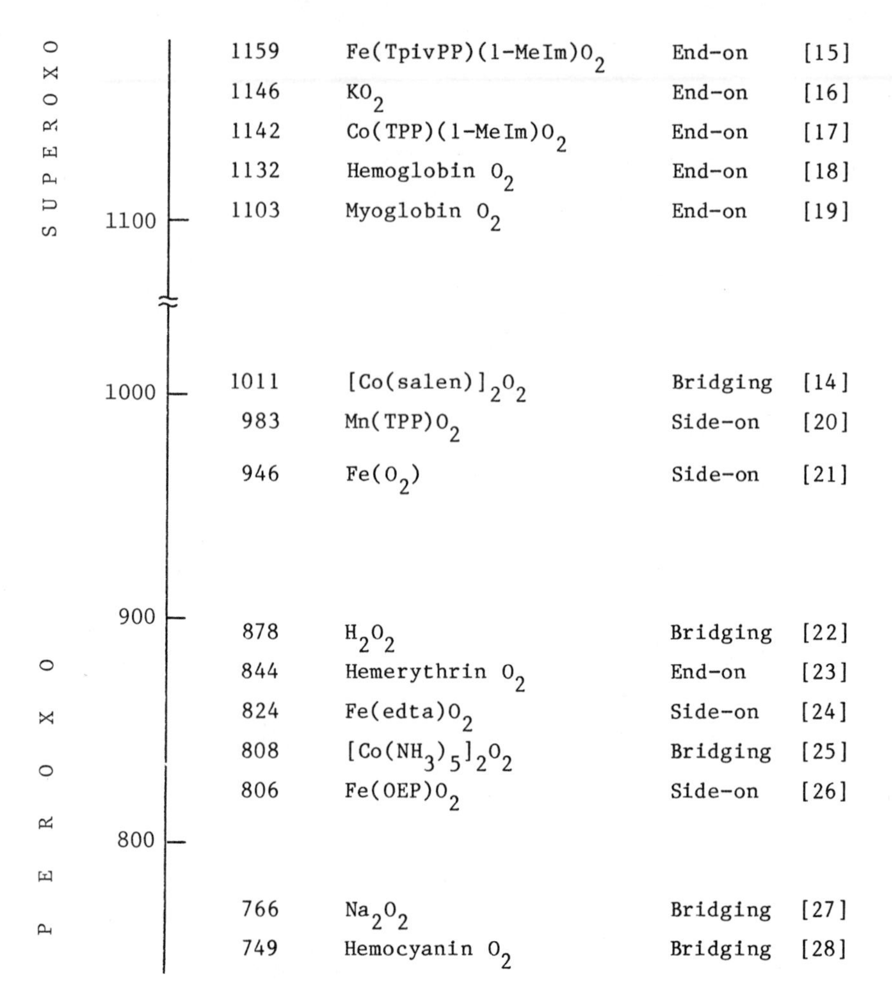

Figure 1. O-O vibrational frequencies of selected dioxygen complexes.

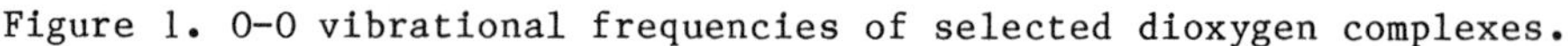

group, and have end-on coordination as is typical of mononuclear superoxo complexes [7]. This distinct classification from vibrational spectroscopy is one of the reasons that these oxyhemeproteins may be described as $Fe(III)—O_2^-$ complexes [7]. Dioxygen adducts of picket-fence porphyrin [15], Co(TPP) [17], as well as Co-substituted hemoglobin and myoglobin [29] fall within this range.

For the peroxo (low range) and peroxo-like (intermediate range) complexes (Figure 1), the coordination geometry of the dioxygen in mononuclear species switches from end-on to side-on binding. The only known example of end-on peroxo coordination to a single metal occurs with oxyhemerythrin. However, in this instance, the peroxo ligand approaches a bridging configuration by virtue of being protonated (see below). In binuclear model complexes, a bridging disposition of the peroxide predominates. This appears to be the structure in oxyhemocyanin.

Aqueous hydrogen peroxide has a ν(O-O) frequency at 878 cm^{-1} [22], whereas metal-peroxo complexes exhibit lower values (Figure 1). For example, the binuclear, peroxo-bridged pentaamminecobalt(III) complex [25] and the side-on bonded Fe(octaethylporphyrin) peroxo complex [26] have nearly identical frequencies at ≃807 cm^{-1}. A similar side-on bonded complex is the peroxo Fe(edta) species with a value of 824 cm^{-1} [24]. Both oxyhemerythrin and oxyhemocyanin have ν(O-O) within the peroxo complex range, even though their absolute values differ by 100 cm^{-1}.

USE OF OXYGEN AND HYDROGEN ISOTOPES IN THE STUDY OF DIOXYGEN COMPLEXES

The positive identification of ν(O-O) and ν($Fe-O_2$) in infrared or Raman spectra rests on the observation of the shift of the respective bands upon isotopic substitution. For most of the examples cited in Figure 1, O-18 was used to verify ν(O-O) from its substantial isotope shift. Typical values are 50-70 cm^{-1} in superoxide complexes and 40-50 cm^{-1} in peroxo complexes.

Considerable information on the coordination geometry of the dioxygen ligand has been derived from the use of mixed isotopes of oxygen, since the number of vibrational bands and their intensities contain information on the symmetry of the coordinated ligand. The interpretation of the experimental results are illustrated in Figure 2. When the mixed isotope

Binuclear Site:		M–O—O*–M	M₂>O—O*
Mononuclear Site:		M<(O–O*)	M–O—O*
ν(M-O)	500 cm^{-1}	1	2
ν(O-O)	800 cm^{-1} or 1100 cm^{-1}	1	(2)

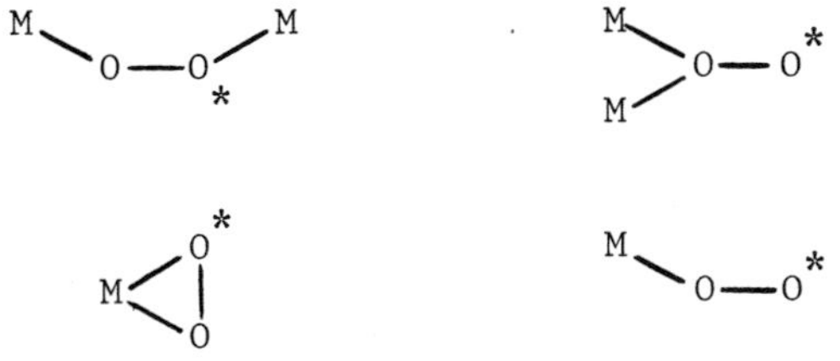

Figure 2. Expected Bands for ν(O-O) and ν(M-O) using Mixed Isotope $O-O^*$

binds symmetrically with respect to the metal center, either in a bridging fashion across a binuclear site or side-on to a single metal ion, then the two oxygens of the molecular ligand are equivalent and only single (O-O) and ν(M-O) bands would be expected. However, if the coordinated dioxygen binds through a single oxygen, either to one metal of a binuclear pair or to the metal of a mononuclear complex, then the oxygen atoms are inequivalent and should give rise to two O-O and M-O bands of equal intensity representing the statistically equal populations of the two forms. In practice, such experiments have provided much insight into the bonding in the respiratory proteins. However, for some model complexes with end-on coordination, the ν(O-O) has sometimes failed to show the expected two bands; the reason for this is not well understood [30].

If the bound dioxygen ligands are protonated or involved in hydrogen-bonding interactions, then deuterium exchange experiments may reveal sensitivity of vibrational modes to the replacement of exchangeable protons with deuterium. For M-O-O-H and M-O-O···H, the protons may exert sufficient influence on the O-O and even M-O vibrations to yield isotope shifts as large as 2 to 5 cm^{-1} in D_2O. Again, such experiments have shed considerable light on the effects of hydrogen bonding of the coordinated dioxygen ligands in model complexes as well as the respiratory proteins.

OXYHEMOGLOBIN

The resonance Raman spectrum of hemoglobin oxygenated with mixed-isotope oxygen (>88% O-18) obtained by Duff et al. [31] is shown in Figure 3. The result of this experiment, showing two distinct $Fe-O_2$ vibrations at 567 and 540 cm^{-1}, indicates the presence of two inequivalent oxygen atoms in oxyhemoglobin, and is only consistent with end-on binding of the superoxo ligand to the heme iron. The two Fe-O vibrations from the mixed

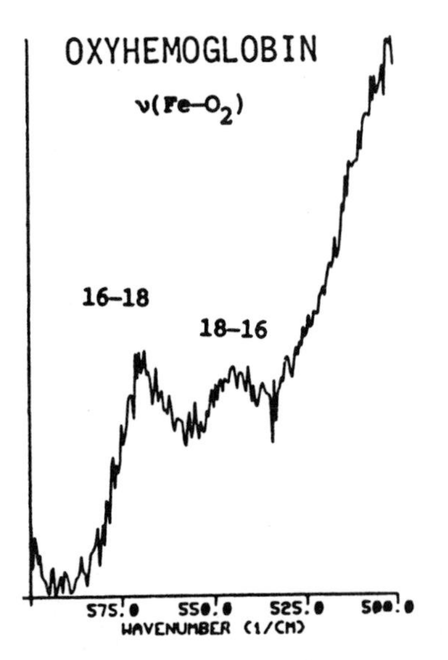

Figure 3. Raman spectrum of hemoglobin + 89% $^{16}O-^{18}O$ [31].

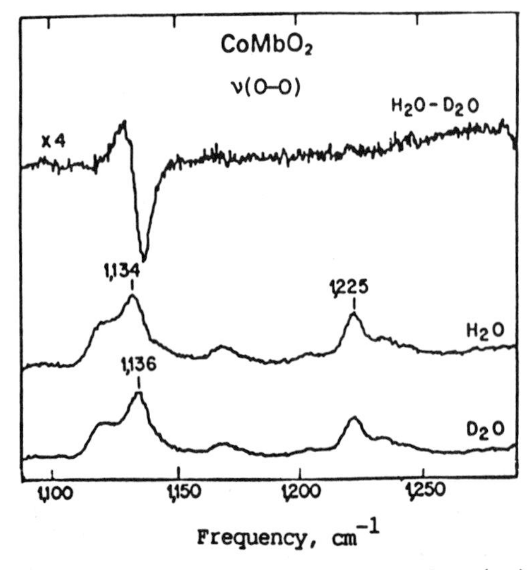

Figure 4. Raman spectrum of Co(II)-myoglobin in H_2O and D_2O [33].

isotope experiment (assigned to Fe-^{16}O and Fe-^{18}O, respectively) are in identical positions to those observed when hemoglobin is reacted with pure $^{16}O_2$ and $^{18}O_2$, respectively [32]. Hence, the isotopic identity of the terminal oxygen has little or no influence on the Fe-O vibrational frequencies of the mixed isotopes. A summary of the frequencies and isotopic behavior for both ν(M-O) and ν(O-O) is given in Table III.

TABLE III. FREQUENCIES OF O-O AND M-O_2 VIBRATIONS AND THEIR ^{18}O AND D-ISOTOPE EFFECTS IN OXY FORMS OF RESPIRATORY PROTEINS.

	ν(O-O)				ν(M-O_2)		
Species	16/16	16/18	18/18	D_2O	16/16	16/18	18/18
$FeHbO_2$	1132*		1066		567	567,540	540
$CoHbO_2$	1122*		1063	+5	537		514
HrO_2	844	825,819	796	+4	503	501,485	482
HcO_2	749	728	708			n.o.	

* centroid of Fermi resonance pair; n.o. = not observed

Although the metal-dioxygen vibrations of oxyhemoglobin have been investigated by resonance Raman spectroscopy, the corresponding O-O vibrations have never been clearly observed and this information is thus far only available for the iron protein from infrared spectroscopy [18, 19]. It must be presumed that the axially coordinated dioxygen ligands in FeHb and FeMb make smaller contributions to the electronic states, and consequently, their O-O vibrations are ineffectively resonance enhanced. However, for cobalt-substituted hemoglobin and myoglobin, ν(O-O) values have been reported [29, 33]. The resonance Raman spectrum of Co-oxymyoglobin ($CoMbO_2$) is shown in Figure 4 in both H_2O and D_2O solutions. The ^{18}O-substitution studies of Tsubaki & Yu [29] have proven that the O-O vibration of the coordinated superoxo ligand occurs at 1122 cm^{-1} but appears as peaks at 1107 and 1137 cm^{-1} due to Fermi resonance with a porphyrin mode. The data in Figure 4 demonstrate that this vibration is also sensitive to deuterium exchange. Although the upshift is small, the difference spectrum makes this point unambiguously. Similar data were obtained for $CoHbO_2$. The deuterium sensitivity of these O-O vibrations supports the view that the dioxo ligand is hydrogen bonded in HbO_2 and MbO_2.

The spectroscopic data discussed above for the heme respiratory proteins are totally consistent with recent high-resolution crystal data

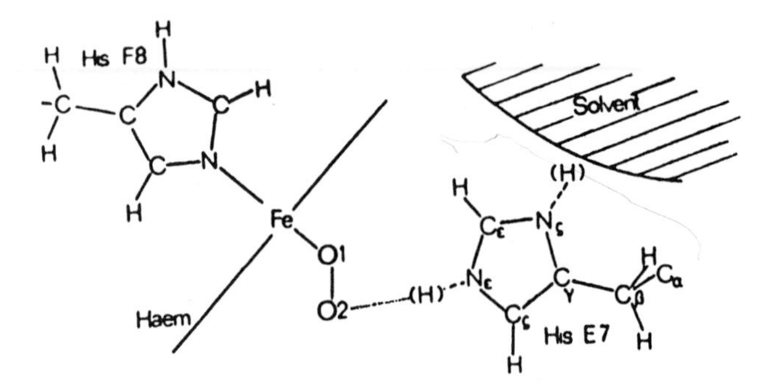

Figure 5. Hydrogen bonding of O_2 in oxymyoglobin [9].

available for hemoglobin [34], myoglobin [9], and the insect hemoglobin, erythrocruorin [35]. In all cases, the O-O is bonded end-on in a non-linear fashion. In erythrocruorin, an immobilized water molecule is hydrogen bonded to the terminal oxygen atom of the ligand. The hydrogen bond in MbO_2 is revealed distinctly from the neutron diffraction study of Phillips and Schoenborn [9], and the structural data are reproduced in Figure 5. This illustration shows the end-on bonded dioxygen with a 2.97 hydrogen bond to the protonated N_ε of the distal histidine E7. A final example of a terminal, end-on superoxo ligand that is intramolecularly hydrogen bonded is shown in Figure 6 for the oxygenated "basket-handle" iron porphyrin of Lavalette and coworkers [10]. The hydrogen bond extends to the amide proton of the meso substituent. When the amide is replaced by an ether group that is incapable of hydrogen bonding, the stability constant of the oxygenated complex drops by nearly an order of magnitude.

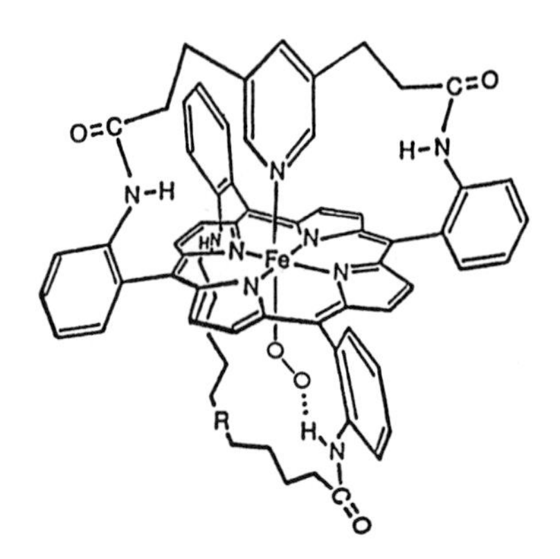

Figure 6. Hydrogen bonding of O_2 in basket-handle porphyrin [10].

OXYHEMERYTHRIN

Crystallographic studies [2, 3] of hemerythrin have shown that the active site of this protein consists of two octahedrally coordinated iron centers that are triply bridged by two bidentate, protein carboxylates (Asp 106 and Glu 58) and a solvent-derived μ-oxo group. Of the remaining six terminal coordination sites, five are occupied by histidines (His 73, His 77, and His 101 on the coordinatively saturated iron, and His 25 and His 54 on the ligand binding iron). Although the crystal structure of oxyhemerythrin is not yet available at high resolution, details regarding the coordination of dioxygen and its hydrogen-bonding interactions are available from resonance Raman spectroscopy. Oxyhemerythrin (HrO_2) was the first respiratory protein to have its O–O vibrational frequency determined [23]. The observation of the resonance-enhanced ν(O–O) at 844 cm^{-1} that shifted to 796 cm^{-1} in $^{18}O_2$ established that the dioxygen ligates as a peroxo group and that both metal atoms are involved in the oxidative addition of oxygen.

The resonance Raman spectrum in the O–O stretching region of hemerythrin oxygenated with mixed-isotope dioxygen (58% O–18) is shown in Figure 7 [36]. The spectrum is made up of three main components. The two flanking peaks arise from HrO_2 made from pure isotopes (16–16 and 18–18, respectively) and appear in the same relative intensity ratio as in the Raman spectrum of the gas mixture. The middle component, however, is not only smaller in height than in the original gas mixture, but is broader

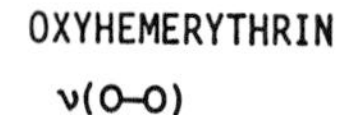

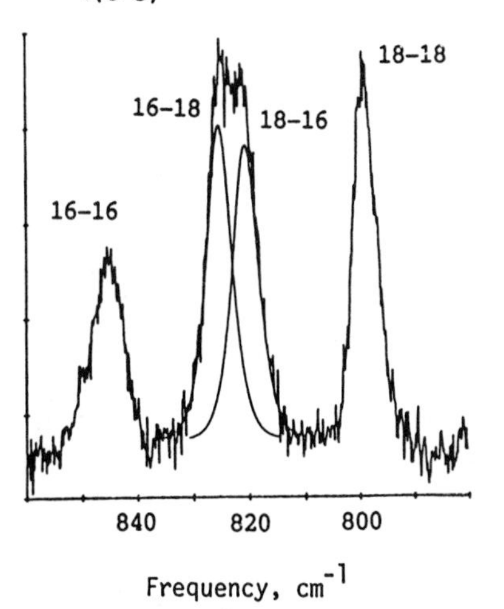

Figure 7. Raman spectrum of hemerythrin + 52% $^{16}O-^{18}O$ [36].

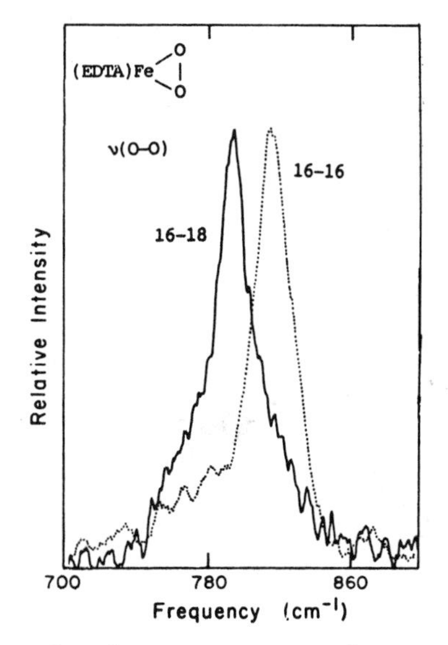

Figure 8. Raman spectrum of Fe(edta) + 90% $[^{16}O-^{18}O]^{2-}$.

and shows clear evidence of splitting (Table III). The sum of the two computer-generated curves drawn under the spectral envelope faithfully reproduces the observed spectral component. This splitting indicates that the oxygen atoms of the ligated peroxo group are inequivalent, as would be consistent for a terminally bonded, end-on geometry. Such binding is expected from the crystallographic results with azidomethemerythrin [37] and from preliminary results with oxyhemerythrin [38].

End-on coordination of a peroxide ligand to a mononuclear metal is unknown from model complexes (Figure 1). Recent results from our laboratory [24] on the vibrational spectrum of the $[Fe(edta)O_2]^{3-}$ complex using mixed isotopes of oxygen in hydrogen peroxide are shown in Figure 8 for comparison with the HrO_2 results. The peak shape of the mixed-isotope sample shows no evidence of broadening or splitting and is similar to that for the reference compound prepared from natural abundance H_2O_2. Furthermore, the absence of any measurable deuterium isotope effect on the O-O vibration in the Fe(edta)peroxo complex supports the expected symmetrical, side-on bonding of the peroxo group.

The unusual coordination geometry observed for oxyhemerythrin was explored further by deuterium exchange experiments. As shown in Table III, the O-O vibration shifts up by 4 cm^{-1} in D_2O solution. For aqueous hydrogen peroxide, ν(O-O) shows a similar shift of +2 cm^{-1} in D_2O [24], indicating that electronic effects associated with proton binding can outweigh mass effects. These observations support the description of oxyhemerythrin as a hydroperoxide species. Simple hydrogen bonding interactions of the peroxo group with a suitable protein donor group (similar to that shown in Figure 5 for oxyhemoglobin) are unlikely to account for the deuterium effect in this case. The oxygen binding pocket in hemerythrin is very hydrophobic and lacks any amino acid residues capable of acting as hydrogen bond donors in the vicinity of the bound peroxide.

Extensive studies on the resonance Raman spectra of hemerythrins [39] have established assignments for other active-site vibrational modes. The dominant feature with near-uv excitation is the strongly enhanced symmetric stretching mode of the Fe-O-Fe cluster (Figure 9). In oxyhemerythrin, ν(Fe-O-Fe) is at 486 cm^{-1} and shifts 14 cm^{-1} to lower energy when the μ-oxo bridge is replaced by O-18 [40]. The shoulder at 503 cm-1 is only weakly enhanced in the near-uv, but is a strong band with visible excitation. As expected, it shows no shift upon exchange of the μ-oxo group. However, when oxyhemerythrin is prepared in D_2O, both of these vibrations are affected. The symmetric Fe-O-Fe vibrational mode increases by 4 cm^{-1}, whereas the $Fe-O_2$ mode at 503 cm^{-1} decreases by 3 cm^{-1}.

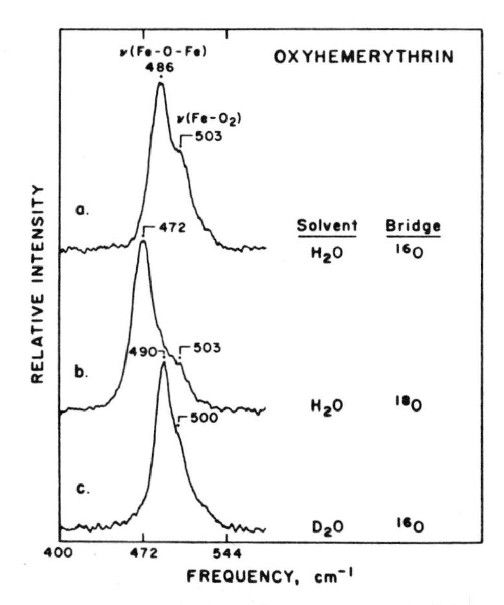

Figure 9. Raman scattering from μ-oxo bridge vibration with a) Fe-^{16}O-Fe, b) Fe-^{18}O-Fe, and c) protein in D_2O [40].

The isotopic replacement results in oxyhemerythrin may be explained as follows. The decrease in the 503-cm^{-1} band in D_2O is primarily a mass effect arising from the Fe-OOH(D) exchange. The increase in ν(Fe-O-Fe) is a result of hydrogen bonding of the μ-oxo group. The only available proton in the very hydrophobic oxygen binding pocket is from the hydroperoxide itself. The effect of hydrogen bonding is actually more extensive than the 4-cm^{-1} increase illustrated by the data in Figure 9. Methemerythrins with a variety of aprotic ligands exhibit ν(Fe-O-Fe) within the narrow range 510 ± 4 cm^{-1} [39, 40]. However, hydroxomet- and oxyhemerythrin both exhibit their symmetric Fe-O-Fe stretches at ≃490 cm^{-1}. This 20-cm^{-1} decrease is attributed to hydrogen-bonding interactions of these protic ligands with the μ-oxo bridge. In the case of deuterium substitution, these data indicate that the hydrogen bond is actually weaker in D_2O (by

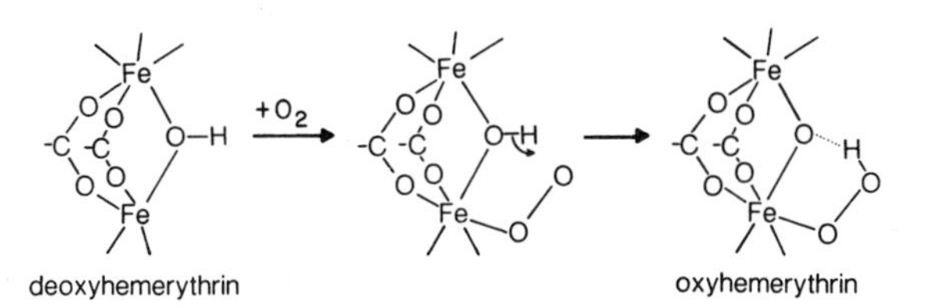

Figure 10. Proposed mechanism for reversible oxygenation [38]

an effective 4 cm^{-1} toward the value of the methemerythrins). We have interpreted these results with the proposal [40] that the hydroperoxide ligand in oxyhemerythrin is hydrogen bonded to the μ-oxo bridge as shown in Figure 10. Such a model is also in agreement with preliminary x-ray structural data on oxyhemerythrin [38].

A reaction sequence for the reversible oxygenation of deoxyhemerythrin is also presented in Figure 10. Although no high resolution x-ray structural data are yet available for the deoxy protein, the spectroscopic and magnetic properties of deoxyhemerythrin favor a hydroxo-bridged binuclear iron site. Such a structure is consistent with the weak antiferromagnetic coupling ($-J \simeq 10\ cm^{-1}$) observed in this form of the protein [41, 42], as well as the loss of a short (1.8 Å) Fe-O distance characteristic of the oxo-bridged oxy and met forms [43]. The proposed pathway is in agreement with the rapid kinetics and the known pH-independence of the oxygenation reaction. Thus, as the weakly coupled ferrous protein reacts with dioxygen, the developing charge on the peroxo ligand is neutralized by proton transfer. The resulting oxo-bridged binuclear ferric site exhibits strong antiferromagnetic coupling ($-J \simeq 100\ cm^{-1}$) characteristic of an Fe-O-Fe cluster [42]. As in the case of dioxygen addition to hemoglobin and porphyrin model compounds, the charge delocalization and hydrogen bonding in oxyhemerythrin must help to stabilize the oxygenated product and thereby promote high oxygen affinity.

OXYHEMOCYANIN

The recent crystal structure of *Panulirus interruptus* (spiny lobster) hemocyanin shows that each copper atom of the binuclear site is coordinated to three histidines, and these are the only amino acid residues sufficiently close to act as protein ligands [4, 5]. The Cu···Cu separation is estimated at 3.7 ± 0.25 Å which is in reasonable agreement with results derived from x-ray absorption spectroscopy [39]. The present crystal structure at a resolution of 3.2 Å is insufficiently clear to identify the bridging ligand(s) for this antiferromagnetically coupled copper pair. However, since the crystals used for structure determination were colorless, and lacked the intense 345-nm absorption band characteristic of oxyhemocyanin (HcO_2), they may well have been in the 2Cu(I)- (deoxy), 2Cu(II)- (met), or mixed valence Cu(II)Cu(I)- (halfmet) forms. A likely possibility for a bridging ligand would be a solvent-derived oxo or hydroxo group [39].

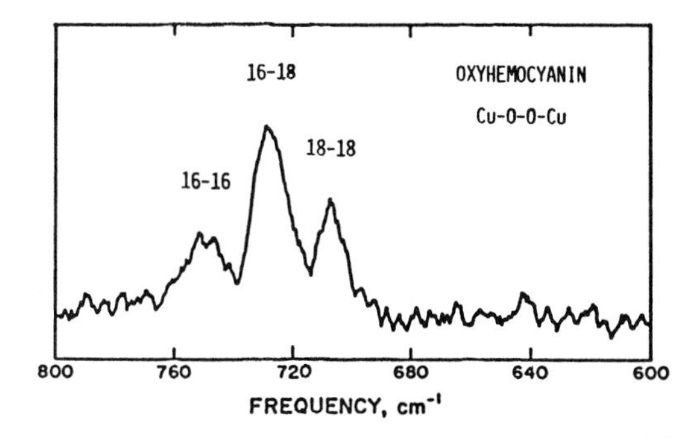

Figure 11. Raman spectrum of hemocyanin + 49% $^{16}O-^{18}O$ [45].

Vibrational spectroscopic data for oxyhemocyanin have been restricted to observations of the O–O vibration by visible excitation or to low frequency modes associated with Cu-imidazole ligation that are especially enhanced in the near uv [39, 44]. Metal-dioxygen vibrations have not been unambiguously detected. Early results from our laboratory defined the nature of the dioxygen ligand in oxyhemocyanin as a peroxo adduct from the observation of a resonance-enhanced mode at 749 cm^{-1} from <u>Busycon canaliculatum</u> (channeled whelk) hemocyanin that shifted to 708 cm^{-1} with O-18 gas [28].

The disposition of the peroxo ligand vis-a-vis the binuclear copper site was explored by the experiment using mixed-isotope dioxygen. The resonance Raman spectral data we obtained with 55 atom-% O-18 are illustrated in Figure 11 [45]. The three principal components are again due to the three types of isotopically labeled HcO_2 generated from the gas mixture as was seen for oxyhemerythrin (Figure 7). In this case, however, as distinct from the results for hemerythrin, the overall peak intensities and shapes repeat those of the spectrum of the free gas mixture. The central component remains intense and shows no sign of splitting. These results were interpreted as showing that the two oxygen atoms of the peroxo ligand are spectroscopically indistinguishable and led to the proposal that the dioxygen ligand is symmetrically bridged across the binuclear copper pair [45]. Based on these data alone it is actually not possible to distinguish between a 1,2-bridging geometry or a side-on peroxo group coordinated to a single metal ion. However, the symmetric distribution of three histidine ligands per copper as well as the $\simeq 3.7$ Å copper-copper distance that have been observed by x-ray crystallography [4, 5] greatly weigh in favor of a 1,2-bound peroxide.

A proposed structure of the active site in oxyhemocyanin which fits most of the available data [46] is shown in Figure 12. Further refinements of the crystal structure are of considerable interest to help to

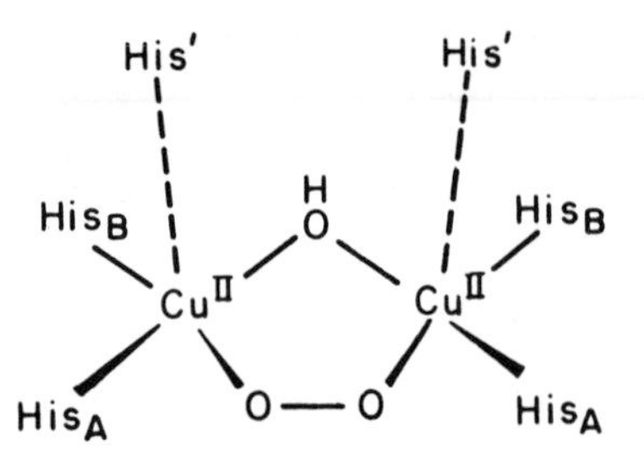

Figure 12. Model for dioxygen coordination in hemocyanin [46].

define the active sites in both the deoxy and oxy forms. With the absence of a protein bridge, the only likely pathway for magnetic coupling of the two copper ions is via a solvent-derived bridge. Although precise structural details of the identity and orientation of amino acid side chains in the vicinity of the dioxygen binding site are not yet clear, the site does contain a large number of aromatic amino acids with several fully conserved tryptophans [5]. Thus it appears very likely that hemocyanin also has a hydrophobic pocket as in hemoglobin and hemerythrin that would favor reversible oxygenation over oxidation.

ACKNOWLEDGMENTS

The author thanks Joann Sanders-Loehr for helpful discussion and criticisms during the preparation of this manuscript. He acknowledges the valuable contributions of his students and coworkers whose work is cited below. This research has been supported by grants from the National Institutes of Health (GM 18865).

REFERENCES

1. Dickerson, R. E.; Geis, I. (1983) Hemoglobin, Benjamin/Cummings, Menlo Park, CA.
2. Hendrickson, W. A.; Smith, J. L. (1981) in Invertebrate Oxygen Binding Proteins: Structure, Active Site and Function, Lamy, J.; Lamy, J.; Eds., Marcel Dekker, New York, pp. 343-352.
3. Sieker, L. C.; Stenkamp, R. E.; Jensen, L. H. (1982) in The Biological Chemistry of Iron, Dunford, H. B.; Dolphin, D.; Raymond, K. N.; Sieker, L. C., Eds., Reidel, Boston, p. 161.
4. Gaykema, W. P. J.; Volbeda, A.; Hol, W. G. J. (1986) J. Mol. Biol. 187, 255.
5. Linzen, B.; Soeter, N. M.; Riggs, A. F.; Schneider, H.-J.; Schartau, W.; Moore, M. D.; Yokota, E.; Behrens, P. Q.; Nakashima, H.; Takagi, T.; Nemoto, T.; Vereijken, J. M.; Bak, H. J.; Beintema, J. J.; Volbeda, A.; Gaykema, W. P. J.; Hol, W. G. J. (1985) Science 229, 519.

6. Solomon, E. I.; Eickman, N. C.; Gay, R. R.; Penfield, K. W.; Himmelwright, R. S.; Loomis, L. D. (1981) in Invertebrate Oxygen Binding Proteins: Structure, Active Site and Function, Lamy, J.; Lamy, J., Eds., Marcel Dekker, New York, pp. 487-502.
7. Niederhoffer, E. C.; Timmons, J. H.; Martell, A. E. (1984) Chem. Rev. 84, 137.
8. Traylor, T. G.; Koga, N.; Deardurff, L. A. (1985) J. Am. Chem. Soc. 107, 6504.
9. Phillips, S. E. V.; Schoenborn, B. P. (1981) Nature 292, 81.
10. (a) Lavalette, D.; Tetreau, C.; Mispelter, J.; Momenteau, M.; Lhoste, J.-M. (1984) Eur. J. Biochem. 145, 555; (b) Mispelter, J.; Momenteau, M.; Lavalette, D.; Lhoste, J.-M. (1983) J. Am. Chem. Soc. 105, 5165.
11. Jameson, G. B.; Drago, R. S. (1985) J. Am. Chem. Soc. 107, 3017.
12. Walker, F. A.; Bowen, J. (1985) J. Am. Chem. Soc. 107, 7632.
13. Chang, C. K.; Kondylis, M. P. (1986) J. Chem. Soc., Chem. Commun., 316.
14. Suzuki, M.; Ishiguro, T.; Kozuka, M.; Nakamoto, K. (1981) Inorg. Chem. 20, 1993.
15. Collman, J. P.; Brauman, J. I.; Halbert, T. R.; Suslick, K. S. (1976) Proc. Natl. Acad. Sci. USA 73, 3333.
16. Blunt, F. J.; Hendra, P. J.; Mackenzie, J. R. (1969) J. Chem. Soc. D, 278.
17. Jones, R. D.; Budge, J. R.; Ellis, P. E., Jr.; Linnard, J. E.; Summerville, D. A.; Basolo, F. (1979) J. Organomet. Chem. 181, 151.
18. Alben, J. O. (1978) The Porphyrins 3, 334.
19. Maxwell, J. C.; Volpe, J. A.; Barlow, C. H.; Caughey, W. S. (1974) Biochem. Biophys. Res. Commun. 58, 166.
20. Urban, M. W.; Nakamoto, K.; Basolo, F. (1982) Inorg. Chem. 21, 3406.
21. Abramowitz, S.; Acquista, N.; Levin, I. W. (1977) Chem. Phys. Lett. 50, 423.
22. Bain, O.; Giguere, P. A. (1955) Can. J. Chem. 33, 527.
23. Dunn, J. B. R.; Shriver, D. F.; Klotz, I. M. (1973) Proc. Natl. Acad. Sci. USA 70, 2582.
24. McCallum, J.; Ahmad, S.; Shiemke, A. K.; Appelman, E. H.; Loehr, T. M.; Sanders-Loehr, J., unpublished results.
25. Freedman, T. B.; Yoshida, C. M.; Loehr, T. M. (1974) J. Chem. Soc., Chem. Commun., 1016.
26. McCandlish, E.; Miksztal, A. R.; Nappa, M.; Sprenger, A, Q.; Valentine, J. S.; Stong, J. D.; Spiro, T. G. (1980) J. Am. Chem. Soc. 102, 4268.
27. Evans, J. C. (1969) J. Chem. Soc. D, 682.
28. Freedman, T. B.; Loehr, J. S.; Loehr, T. M. (1976) J. Am. Chem. Soc. 98, 2809.
29. Tsubaki, M.; Yu, N.-T. (1981) Proc. Natl. Acad. Sci. USA 78, 3581.
30. Watanabe, T.; Ama, T.; Nakamoto, K. (1984) J. Phys. Chem. 88, 440.
31. Duff, L. L.; Appelman, E. H.; Shriver, D. F.; Klotz, I. M. (1979) Biochem. Biophys. Res. Commun. 90, 1098.
32. Brunner, H. (1974) Naturwissenschaften 61, 129.
33. Kitagawa, T.; Ondrias, M. R.; Rousseau, D.; Ikeda-Saito, M.; Yonetani, T. (1982) Nature 298, 869.
34. Shaanan, B. (1982) Nature 296, 683.
35. Steigemann, W.; Weber, E. (1982) in Hemoglobin and Oxygen Binding, Ho, C., Ed., Elsevier, New York, 20.
36. Kurtz, D. M., Jr.; Shriver, D. F.; Klotz, I. M. (1976) J. Am. Chem. Soc. 98, 5033.
37. Stenkamp, R. E.; Sieker, L. C.; Jensen, L.H. (1984) J. Am. Chem. Soc. 106, 618.
38. Stenkamp, R. E.; Sieker, L. C.; Jensen, L. H.; McCallum, J. D.; Sanders-Loehr, J. (1985) Proc. Natl. Acad. Sci. USA 82, 713.
39. Loehr, T. M.; Shiemke, A. K. (1988) in Biological Applications of Raman Spectroscopy, Spiro, T. G., Ed., Vol. IV, Wiley, New York, in press.

40. Shiemke, A. K.; Loehr, T. M.; Sanders-Loehr, J. (1986) J. Am. Chem. Soc. 108, 2437.
41. Reem, R. C.; and Solomon, E. I. (1987) J. Am. Chem. Soc. 109, 1216.
42. Maroney, M. J.; Kurtz, D. M., Jr.; Nocek, J. M.; Pearce, L. L.; Que, L., Jr. (1986) J. Am. Chem. Soc. 108, 6871.
43. Elam, W. T.; Stern, E. A.; McCallum, J. D.; Sanders-Loehr, J. (1983) J. Am. Chem. Soc. 105, 1919.
44. Larrabee, J.A.; Spiro, T. G. (1980) J. Am. Chem. Soc. 102, 4217.
45. Thamann, T. J.; Loehr, J. S.; Loehr, T. M. (1977) J. Am. Chem. Soc. 99, 4187.
46. Reed, C. A. (1985) in Biological Inorganic Copper Chemistry, Karlin, K.D.; Zubieta, J., Eds., Adenine, New York, pp. 61-73.

METAL OXO COMPLEXES AND OXYGEN ACTIVATION

Thomas J. Meyer

Department of Chemistry
The University of North Carolina
Chapel Hill, North Carolina 27514

Dioxygen is at the same time both readily abundant and a powerful oxidizing agent in the thermodynamic sense. The utilization of its oxidizing equivalents in an efficient and controlled way is a matter of considerable interest in biology and in a number of technologically important issues including the oxidative activation of hydrocarbons, reactions 1 and 2, and in potential fuel cell applications.

$$O_2 + 2\ \text{>C=C<} \longrightarrow 2\ \text{>C(O)C<} \quad (1)$$

$$O_2 + 2\ \text{R-}\overset{|}{\underset{|}{C}}\text{-H} \longrightarrow 2\ \text{R-}\overset{|}{\underset{|}{C}}\text{-OH} \quad (2)$$

Although the epoxidation of olefins or the oxidation of hydrocarbons by O_2 in reactions 1 and 2 are highly spontaneous thermodynamically, if uncatalyzed they are immeasurably slow under mild conditions. The basis for the kinetic inertia lies both in the mechanistic limitations of dioxygen as an oxidant and in the mechanistic requirements of the hydrocarbons. In order to carry out such reactions at reasonable rates they must be catalyzed and as illustrated in Scheme 1 the catalyst or catalysts must both activate O_2 and have the mechanistic capability of transferring the oxidative equivalents to the organic reductant.

SCHEME 1

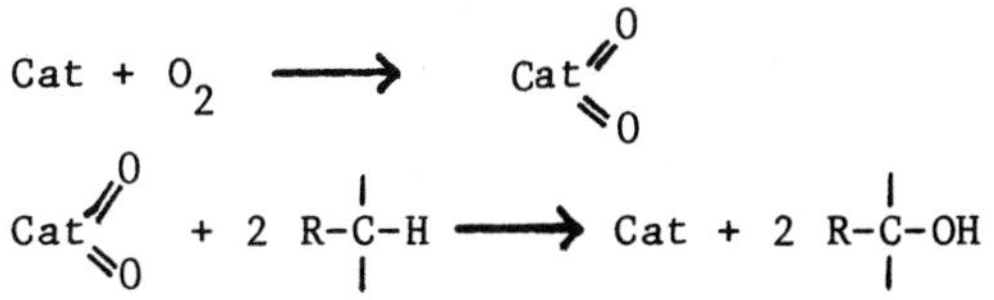

$$\text{Cat} + O_2 \longrightarrow \text{Cat}{\genfrac{}{}{0pt}{}{\diagup\!\!\!\diagup O}{\diagdown\!\!\!\diagdown O}}$$

$$\text{Cat}{\genfrac{}{}{0pt}{}{\diagup\!\!\!\diagup O}{\diagdown\!\!\!\diagdown O}} + 2\ \text{R-}\overset{|}{\underset{|}{C}}\text{-H} \longrightarrow \text{Cat} + 2\ \text{R-}\overset{|}{\underset{|}{C}}\text{-OH}$$

There is another, indirect route for the oxidation of hydrocarbons by dioxygen which takes recognition of the fact that reactions like 1 and 2 are net oxidation-reduction reactions, e.g.,

$$O_2 + 4\ H^+ + 4\ e^- \longrightarrow 2H_2O \quad (3)$$

$$\text{R-}\overset{|}{\underset{|}{C}}\text{-OH} + 2H^+ + 2\ e^- \longrightarrow \text{R-}\overset{|}{\underset{|}{C}}\text{-H} + H_2O \quad (4)$$

In principle, reactions 1 or 2 could be carried out in an electrochemical cell with the reduction of dioxygen (reaction 3) and the oxidation of the hydrocarbon (reaction 4) occuring at separate electrodes. The resulting electrochemical cell would be a fuel cell or an "electrochemical synthesis cell" in which the potential and current flow characteristics would be dictated by the differences in redox potentials between the half-cell reactions for reactions 3 and 4 and the kinetics of the redox interconversions at the two electrodes. To date, the problem of developing inexpensive high current density electrodes which operate near the potentials for the O_2/H_2O couple or oxidized hydrocarbon/reduced hydrocarbon couple remains unsolved.

One of the virtues of the indirect, electrochemical activation of dioxygen is that it simplifies the demands on the catalytic system.

SCHEME 2

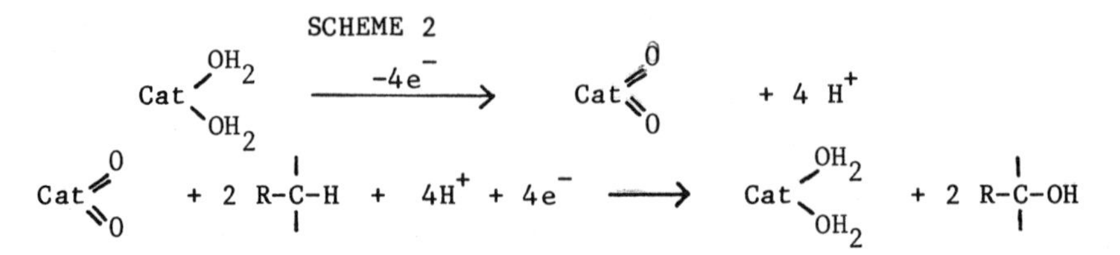

As shown in Scheme 2 the only requirement of the catalyst is that it undergo oxidation at an electrode to give an intermediate oxo reagent having the intrinsic chemical reactivity properties required to carry out the oxidation of the hydrocarbon.

Common electrode materials are no more skilled at reducing O_2 than they are at oxidizing organic compounds and it is necessary that the O_2/H_2O reaction be catalyzed if it is to occur at reasonable current densities close to the thermodynamic potential of the couple.

SCHEME 3

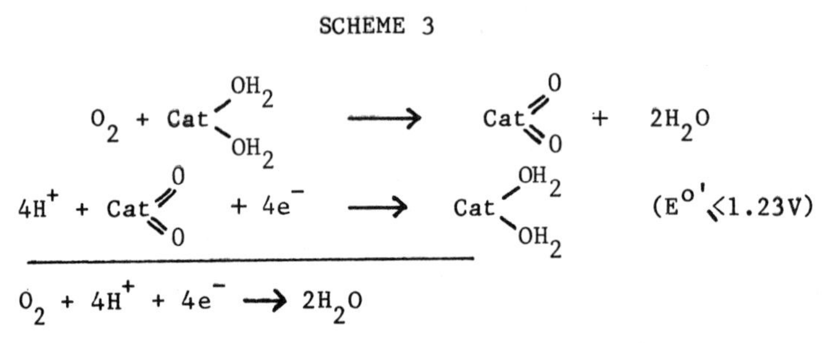

As suggested in Scheme 3, in a catalyzed reaction the potential available from the oxygen cathode becomes that of the catalyst couple and not that of the O_2/H_2O couple.

The reverse of dioxygen reduction, the oxidation of water to dioxygen is also a process of interest both as a mimic for photosynthesis and as a potentially important component in many artifical photosynthetic schemes. As shown in Scheme 4,

SCHEME 4

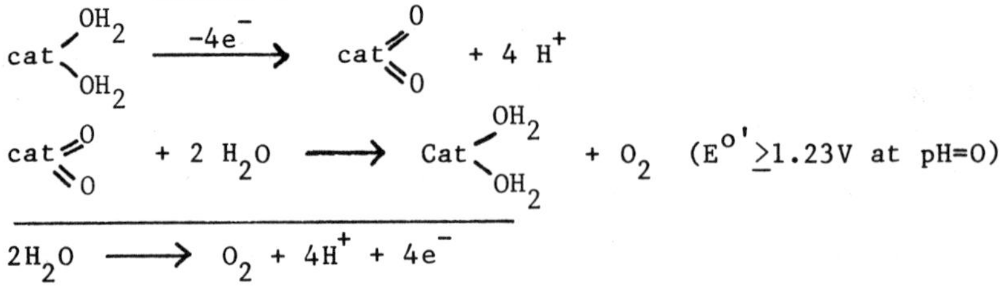

$$\mathrm{cat}(OH_2)_2 \xrightarrow{-4e^-} \mathrm{cat}(O)_2 + 4\ H^+$$

$$\mathrm{cat}(O)_2 + 2\ H_2O \longrightarrow \mathrm{Cat}(OH_2)_2 + O_2 \quad (E^{o\prime} \geq 1.23V \text{ at pH=0})$$

$$2H_2O \longrightarrow O_2 + 4H^+ + 4e^-$$

the mechanistic issues are the same as for the reduction of O_2 but in microscopic reverse.

The critical issue in either the direct or indirect activation of dioxygen lies in devising successful catalysts for the various steps in Schemes 1-4. The energetics involved in the redution of dioxygen in acidic solution are summarized in the Latimer diagram in Scheme 5,

SCHEME 5

(potentials vs. NHE; $a_{H^+} = 1$)

$$O_2 \xrightarrow{-0.10} HO_2 \xrightarrow{+1.50} H_2O_2 \xrightarrow{+0.71} OH + H_2O \xrightarrow{+2.85} H_2O$$

$$O_2 \xrightarrow{1.23} H_2O$$

Just a listing of the redox potentials is revealing in a mechanistic sense. For example, any mechanism which proceeds by stepwise one-electron transfer steps must necessarily involve the perhydroxyl radical (HO_2) which demands a reasonably good reducing agent if the reaction is to proceed at a reasonable rate. The situation is equally bad in the reverse direction where 1-electron oxidation would necessarily involve the intermediate hydroxyl radical which is a powerful oxidizing agent.

Similar problems exist in the 1-electron oxidation of organics because of the thermodynamic instability of the intermediate radicals. Seemingly, a successful catalyst for either O_2 reduction or hydrocarbon oxidation must incorporate a multiple electron capability so as to avoid high energy 1-electron intermediates. The point is further illustrated in Scheme 6 where the energetic consequences of 1-electron and multielectronic pathways are compared schematically for hydrocarbon and water oxidations.

SCHEME 6

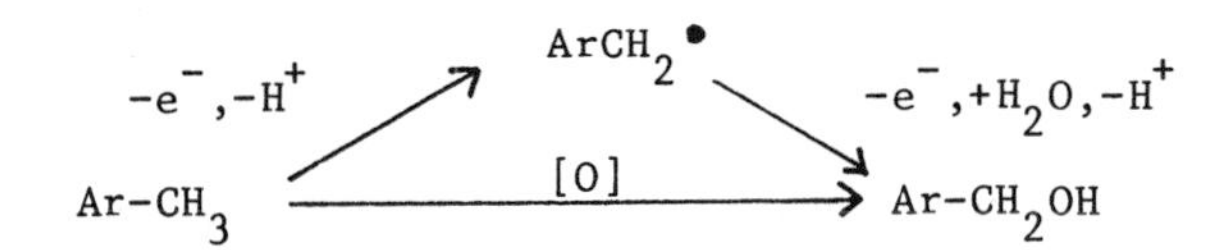

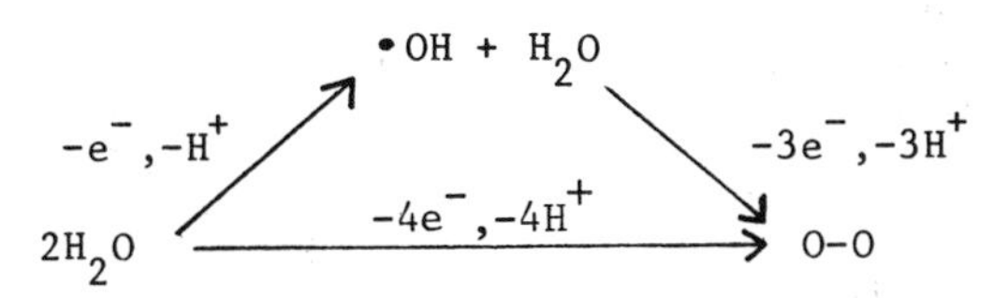

The direct pathway for hydrocarbon oxidation involving O-atom insertion into the C-H bond would avoid high energy intermediate radical formation but it places a significant demand on the catalyst and requires a two electron change coupled with the transfer of oxygen. For H_2O oxidation, the mechanistic cost of avoiding unstable intermediates is even higher with a requirement for the loss of $4H^+$ and $4e^-$ from two H_2O molecules with concomitant formation of an O-O bond.

The goal of this account is to describe the higher oxidation state chemistry of ruthenium and osmium based on oxo complexes and how their chemical properties encompass some of the mechanistic demands implied in Schemes 1-4.

HIGHER OXIDATION STATES. METAL OXO COMPLEXES

There is an extensive coordination chemistry of six-coordinate polypyridyl complexes of M(II) and M(III) (M=Os,Ru) based on ligands like 2,2'-bipyridine (bpy),1,10-phenanthroline (phen), or 2,2'(6,6'),2"-terpyridine (trpy).

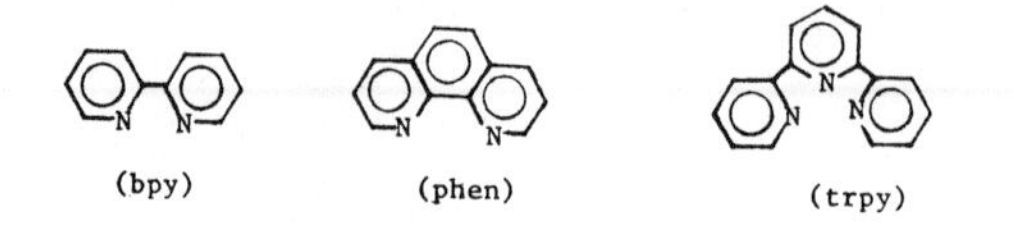

Polypyridyl complexes of ruthenium have proven to be remarkably versatile chemically and their properties have provided bases for reactions which extend from the catalytic reduction of carbon dioxide,[1] to an elaborate nitrosyl chemistry,[2-4] to the remarkably adaptable photochemical and photophysical properties of complexes like $[Ru(bpy)_3]^{2+}$.[5-7] An equally versatile chemistry exists for aqua containing polypyridyl complexes where the higher oxidation states M(IV), M(V) and M(VI) are accessible based on oxidatively induced proton loss and formation of metal oxo complexes,[8-15] e.g.,

$$(bpy)_2(py)Ru^{II}\text{-}OH_2^{\,2+} \xrightarrow[-H^+]{-e^-} (bpy)_2(py)Ru^{III}\text{-}OH^{2+} \xrightarrow[-H^+]{-e^-} (bpy)_2(py)Ru^{IV}{=}O^{2+} \qquad (5)^8$$

In many cases the higher oxidation state complexes have been isolated and characterized by spectroscopic techniques or by x-ray diffraction while in others they have only been characterized in solution.

For the Ru(IV/III) and Ru(III/II) couples based on $[(bpy)_2(py)Ru(OH_2)]^{2+}$, reduction potentials at pH=7 are shown in the Latimer diagram in Scheme 7.

SCHEME 7
(at pH=7 vs. SCE)

$$(bpy)_2(py)Ru^{IV}{=}O^{2+} \xrightarrow{0.53} (bpy)_2(py)Ru^{III}\text{-}OH^{2+} \xrightarrow{0.42} (bpy)_2(py)Ru^{II}\text{-}OH_2^{\,2+}$$

$$\underbrace{\qquad\qquad\qquad\qquad}_{0.48}$$

Oxidation of Ru(II) to Ru(III) at pH 7 results in proton loss because of the enhanced acidity of bound water in the higher oxidation state. The key to the accessibility of Ru(IV) at such a relatively low potential lies with the oxo group and its role as an electronic donor to the metal. Oxidation from $(d_\pi)^6$ Ru(II) to $(d_\pi)^4$ Ru(IV) leaves electronic vacancies in the d_π levels (t_{2g} levels in O_h symmetry). The loss of protons from bound H_2O and then from bound OH^- to give O^{2-}, frees p-based electron density for donation into the vacancies in the d_π orbitals. The same phenomeon occurs in the 5-electron oxidation of $[Mn(H_2O)_6]^{2+}$ to MnO_4^-.

An advantage to ruthenium in its multiple oxidation state chemistry is apparent in the existence of the same basic, stable coordination environment in oxidation states II,III, and IV. By contrast, for a multiple electron couple like $CrO_4^{2-} \rightarrow Cr(H_2O)_6^{3+}$, the changes in coordination number between oxidation states often lead to significant activation barriers to redox reactions and to great difficulties in regenerating the higher oxidation states in a catalytic cycle. The reversible interconversion between Ru(II) and Ru(IV) also imparts a catalytic capability since Ru(II) can be reoxidized to Ru(IV) either at an electrode or by using a chemical oxidant.

The microscopic makeup of the Ru(IV)-oxo complex imparts some impressive reactivity characteristics in an implied sense. There is the two-electron acceptor capability of the Ru(IV/II) couple which, when combined with the oxo group as a lead-in atom, offer several mechanistic possibilities including O-atom transfer, hydride transfer, H-atom transfer or O-atom insertion into C-H bonds all of which have been observed mechanistically.[16-22]

It is not the goal of this account to describe in detail the metal oxo based chemistry of osmium and ruthenium, but it is important to realize that it is extensive. An additional example is shown in

SCHEME 8

(at pH=7 vs. SCE)

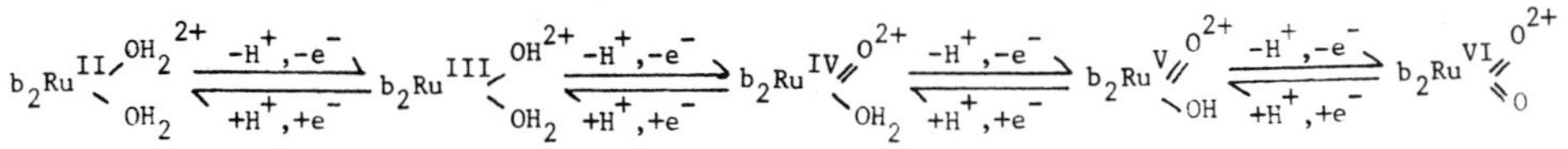

$$b_2Ru^{II}(OH_2)(OH_2)^{2+} \underset{+H^+,+e^-}{\overset{-H^+,-e^-}{\rightleftharpoons}} b_2Ru^{III}(OH)(OH_2)^{2+} \underset{+H^+,+e^-}{\overset{-H^+,-e^-}{\rightleftharpoons}} b_2Ru^{IV}(=O)(OH_2)^{2+} \underset{+H^+,+e^-}{\overset{-H^+,-e^-}{\rightleftharpoons}} b_2Ru^{V}(=O)(OH)^{2+} \underset{+H^+,+e^-}{\overset{-H^+,-e^-}{\rightleftharpoons}} b_2Ru^{VI}(=O)(=O)^{2+}$$

Scheme 8 where by replacing the pyridine ligand in the coordination sphere of $[(bpy)_2(py)Ru^{II}(H_2O)]^{2+}$ with a second water molecule, both Ru(V) and Ru(VI) appear because of the stabilization provided by the additional oxo group. In contrast to chemical oxidants like MnO_4^- or CrO_4^{2-}, a systematic basis exists in the underlying chemistry for preparing a family of closely related metal oxidants whose properties can be varied in a systematic way by changing the surrounding ligands. The implications for the control of chemical reactivity both in terms of rate and product specificity are profound.

THE EFFECTS OF PROTON COMPOSITION ON ENERGETICS AND MECHANISM

Because of the increased acidities in higher oxidation states, the redox potentials for the Ru(IV)/(III) and Ru(III)/(II) couples are pH dependent. An example is shown in Figure 1.

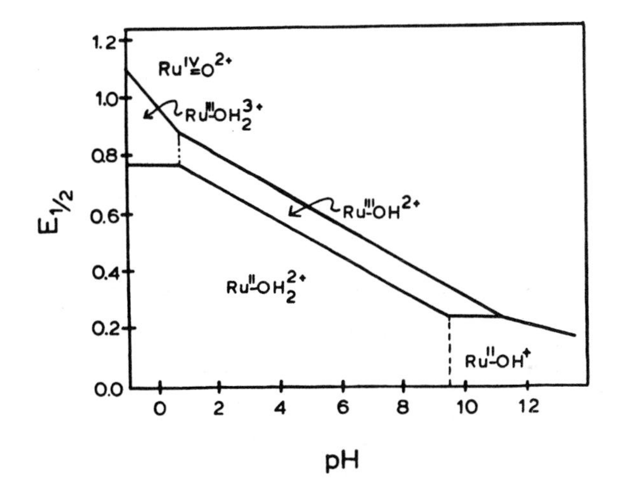

Figure 1. $E_{1/2}$ vs. pH diagram for the Ru(IV/III) and Ru(III/II) couples based on $[(bpy)_2(py)Ru(H_2O)]^{2+}$ at room temperature, I=0.1 M except below pH = 1. The reference electrode is the saturated sodium chloride calomel electrode (SSCE). The forms of the couples which are dominant in the various potential-pH regions are indicated with regard to oxidation state and proton composition. The vertical dashed lines are pK_a values for the oxidation state indicated.

The characteristic breaks that appear in such diagrams show where a pK_a occurs for one of the oxidation states of the couple. The general decrease in potentials as the pH is increased is a simple consequence of the fact that the higher oxidation states tend to be more acidic and the couples pH dependent, e.g.,

$$(bpy)_2(py)Ru^{III}-OH^{2+} + H^+ + e^- \rightarrow (bpy)_2(py)Ru^{II}-OH_2^{2+}$$

As predicted by the Nernst equation for a 1-H^+, 1-e^- change, $E_{1/2}$ for the

Ru(III/II) couple decreases by 59 mV/pH decade. For the Ru(IV/III) couple in acidic solution (pH<1), $E_{1/2}$ varies with a slope of 118 mV per pH decade because 2-H^+ are gained upon reduction to Ru(III),

$$(bpy)_2(py)Ru^{IV}{=}O^{2+} + 2H^+ + 1e^- \longrightarrow (bpy)_2(py)Ru^{III}{-}OH_2^{\ 3+}$$

Because of the differences in pH dependences above pH=12, where the Ru(III/II) couple is independent of pH, the potential-pH curve for the Ru(III/II) couple crosses the Ru(IV/III) couple, Ru(III) becomes a more powerful oxidizing agent than Ru(IV), and Ru(III) is unstable with respect to disproportionation into Ru(II) and Ru(IV),

$$OH^- + 2(bpy)_2(py)Ru^{III}(OH)^{2+} \dashrightarrow (bpy)_2(py)Ru^{IV}{=}O^{2+} + (bpy)_2(py)Ru^{II}(OH)^+ + H_2O$$

The role of pH and proton composition also plays a role in mechanism and reactivity as illustrated in Scheme 9 for the Ru(IV/III) couple. The

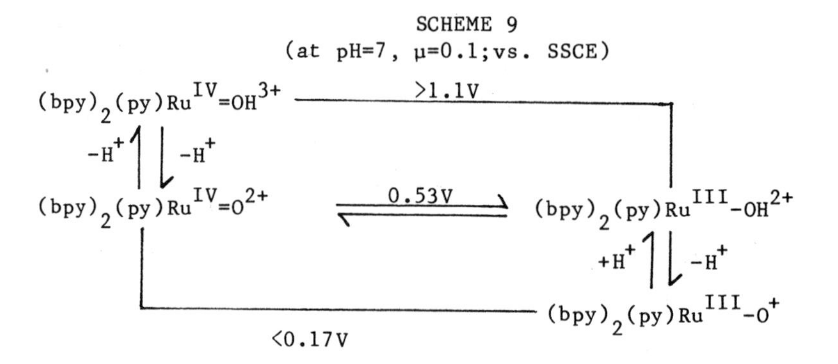

Ru(IV)/(III) couple has problems with simple electron transfer. If Ru(IV) is reduced by initial electron transfer to give $[(bpy)_2(py)Ru^{III}(O)]^+$, the absence of a proton in the lower oxidation state greatly decreases the available oxidative driving force. If initial protonation of the oxo group occurs followed by a 1-electron transfer, the low equilibrium concentration of the protonated form of the oxo complex also necessarily decreases the oxidative driving force. If the full oxidative capabilities of the Ru(IV)-oxo group are to be fully realized, the redox site must <u>at the same time</u> acquire both a proton and an electron in a single concerted step. As a consequence of the combined H^+/e^- demand of the couple, it should not be too suprising to discover that the mechanistic chemistry of the Ru(IV)-oxo group is both rich and diverse and that complex mechanistic pathways are often chosen over simple 1-electron steps.

A straightforward example appears in the comproportionation reaction between Ru(II) and Ru(IV),[17]

$$(bpy)_2(py)Ru^{II}{-}OH_2^{\ 2+} + (bpy)_2(py)Ru^{IV}{=}O^{2+} \longrightarrow 2\ (bpy)_2(py)Ru^{III}{-}OH^{2+} \quad (2)$$

Over a broad pH range, the rate law for the reaction is first order in both Ru(II) and Ru(IV) with $k(25^o;I=0.1\ \underline{M}) = 2.1 \times 10^5 \underline{M}^{-1}s^{-1}$ as measured by stopped-flow methods. The most striking feature about the reaction is the existence of a large solvent kinetic isotope effect, $[k(H_2O)/k(D_2O)] = 16.1$ at 25°, which, from a mole fraction study, arises from participation by a single proton. As shown in Scheme 10 the mechanism of the reaction appears to involve simultaneous proton/electron (H-atom) transfer from Ru(II) to Ru(IV) following initial preassociation between the reactants.

SCHEME 10

$$b_2(py)Ru^{IV}{=}O^{2+} + b_2(py)Ru^{II}{-}OH_2{}^{2+} \rightleftharpoons b_2(py)Ru^{IV}{=}O^{2+},H{-}\underset{H}{O}{-}\underset{py}{Ru^{II}}b_2{}^{2+}$$

$$b_2(py)Ru^{IV}{=}O^{2+},H{-}\underset{H}{O}{-}Ru^{II}(py)b_2{}^{2+} \longrightarrow b_2(py)Ru^{III}{-}OH^{2+},\underset{H}{O}{-}Ru^{III}(py)b_2{}^{2+}$$

(b is 2,2'-bipyridine)

OXIDATION MECHANISMS. ELECTROCHEMICAL CATALYSIS

The results of kinetic and mechanistic studies based on kinetics in aqueous and nonaqueous solvents, ^{18}O labelling, H-D kinetic isotope effects, and the observation of intermediates have revealed a rich and diverse oxidative mechanistic chemistry for Ru(IV)-oxo complexes. Several distinct classes of mechanisms have been identified. An example is oxygen atom transfer which occurs in the oxidation of some olefins,[18]

$$CH_3CN + (bpy)_2(py)Ru^{IV}{=}O^{2+} + PhCH{=}CH_2 \longrightarrow (bpy)_2(py)Ru^{II}{-}NCCH_3{}^{2+} + PhC\overset{O}{\frown}CH_2$$

and in the stepwise oxidation of dimethyl sulfide first to dimethyl sulfoxide[19]

$$b_2(py)Ru^{IV}{=}O^{2+} + Me_2S \longrightarrow b_2(py)Ru^{II}{-}O{=}SMe_2{}^{2+} \longrightarrow b_2(py)Ru^{II}{-}S(O)Me_2{}^{2+}$$

and subsequently to dimethyl sulfone.

$$CH_3CN + b_2(py)Ru^{IV}{=}O^{2+} + Me_2S{=}O \longrightarrow b_2(py)Ru^{II}{-}NCCH_3{}^{2+} + Me_2S(O)_2$$

In the oxidation of alcohols, a net two electron change occurs either by a direct $2e^-/2H^+$ (hydride) transfer to $Ru(IV){=}O^{2+}$ or via initial H-atom transfer followed by a second redox step before the one electron products can separate,[20]

SCHEME 11

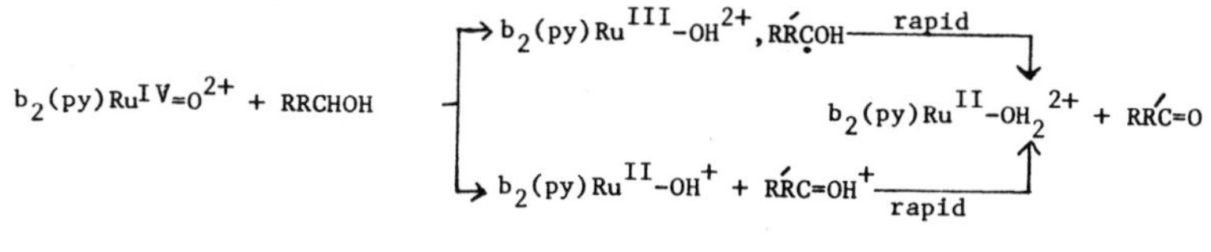

$$b_2(py)Ru^{IV}{=}O^{2+} + RR'CHOH \rightarrow \begin{cases} b_2(py)Ru^{III}{-}OH^{2+}, RR'\dot{C}OH \xrightarrow{\text{rapid}} b_2(py)Ru^{II}{-}OH_2{}^{2+} + RR'C{=}O \\ b_2(py)Ru^{II}{-}OH^{+} + RR'C{=}OH^{+} \xrightarrow{\text{rapid}} b_2(py)Ru^{II}{-}OH_2{}^{2+} + RR'C{=}O \end{cases}$$

A striking feature in the oxidation of alcohols is the appearance of extraordinarily high C-H/C-D kinetic isotope effects. For example, in the oxidation of benzyl alcohol to benzaldehyde by $[(bpy)_2(py)Ru^{IV}{=}O]^{2+}$ the rate of oxidation of $C_6H_5CH_2OH$ compared to $C_6H_5CD_2OH$ is greater by a factor of 50 at 25°.[20a]

Several organic oxidations are known to occur by <u>net</u> C-H insertion although the individual mechanistic details for different substrates may be quite different. As shown by ^{18}O isotopic labelling studies, C-H insertion occurs in the allylic oxidation of olefins, for example,[21]

$$2\,b_2(py)Ru^{IV}{=}{}^{18}O^{2+} + C_6H_{10} \rightarrow b_2(py)Ru^{III}{-}{}^{18}O(H){-}C_6H_9{}^{2+} + b_2(py)Ru^{III}{-}{}^{18}OH^{2+}$$

$$\xrightarrow{+2CH_3CN}\ 2\,b_2(py)Ru^{II}{-}NCCH_3{}^{2+} + H_2{}^{18}O + {}^{18}O{=}C_6H_8$$

in the oxidation of a series of phenols,[21]

$$2\,b_2(py)Ru^{IV}{=}{}^{18}O^{2+} + C_6H_5OH \rightarrow b_2(py)Ru^{II}{-}{}^{18}O{=}C_6H_4{=}O^{2+} + b_2(py)Ru^{II}{-}{}^{18}OH_2{}^{2+}$$

$$\xrightarrow{+\,2CH_3CN}\ 2\,b_2(py)Ru^{II}{-}NCCH_3{}^{2+} + H_2{}^{18}O + {}^{18}O{=}C_6H_4{=}O$$

which occur through detectable Ru(II) quinone intermediates, and in the oxidation of aldehydes,

$$CH_3CN + b_2(py)Ru^{IV}{=}{}^{18}O^{2+} + PhCHO \longrightarrow b_2(py)Ru^{II}{-}NCCH_3{}^{2+} + PhC^{18}OOH$$

Detailed mechanistic information is available about how $Ru(IV){=}O^{2+}$ carries out such extensive series of sequential oxidations as the conversion of aromatic hydrocarbons to the corresponding acids, $PhCH_3 \longrightarrow PhCH_2OH \longrightarrow PhCHO \longrightarrow PhCO_2H$, or the oxidations of anilines to aromatic nitro compounds, $ArNH_2 \longrightarrow ArNHOH \longrightarrow ArNO \longrightarrow ArNO_2$.[22] Competing pathways can exist for a given substrate, e.g., allylic oxidation vs. epoxidation for olefins, and in a multiple step sequence, $ArCH_3 \longrightarrow ArCH_2OH \longrightarrow \ldots$, the sequential steps may be competitive ratewise leading to more than one product. However, in general, oxidations based on $Ru(IV){=}O^{2+}$ are stoichiometrically clean and quantitative. The controlled nature of the oxidations appears to be a consequence of the mild nature of the oxidant and the fact that, in many cases, the key redox steps involve net two-electron changes such as C-H insertion or hydride abstraction and, therefore, avoid the one electron intermediates that often lead to multiple products.

The ruthenium complexes have provided a catalytic basis for the net electrochemical oxidation of a variety of organic functional groups.[23] As illustrated in Scheme 12

SCHEME 12

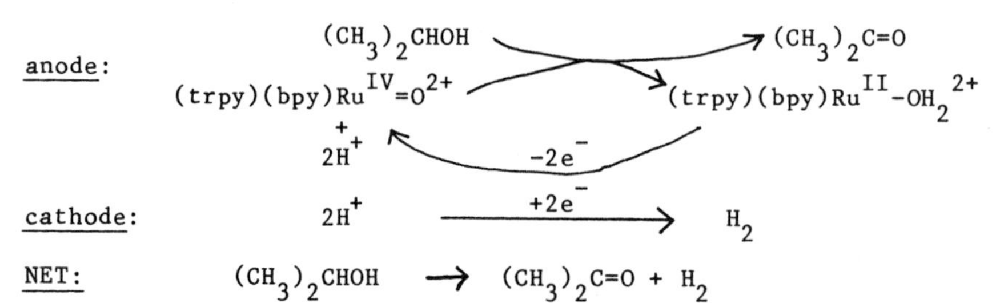

the catalysis is based on electrochemical regeneration via a "shuttle" mechanism where, following the redox step, the reduced form of the catalyst diffuses to the electrode surface and is reoxidized. For the case shown, the oxidation of isopropanol, the net reaction, which is nonspontaneous, involves dehydrogenation to give acetone and hydrogen with the required energy input coming from the applied potential difference across the electrodes. Any scheme like that in 12 could be operated spontaneously with the production of a current and a fuel cell like operation if the oxidative equivalents had their origin in a high current density oxygen electrode by combining reactions like 3 and 4.

In a more sophisticated version of Scheme 12 the catalytic sites would be directly attached to the electrode surface in order to: 1) enhance the local concentration of catalytic sites, 2) localize the catalyst in fixed sites, and 3) to allow for a flow through design. A number of different approaches have been developed for the incorporation of polypyridyl complexes of ruthenium within polymeric films on electrodes.[24] In one example, the Ru(III/II) aqua complex was incorporated into poly-4-vinylpyridine by binding to a fraction of the available pyridyl sites,

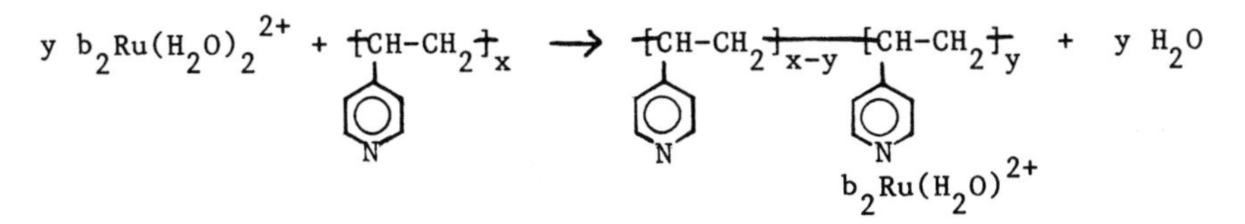

and the resulting soluble polymer subsequently evaporatively deposited onto carbon or platinum electrodes.[25] Electrochemical experiments on the resulting electrodes show that the redox chemistry of the Ru(III/II) and Ru(IV/III) couples is maintained in the films and that $Ru(IV)=O^{2+}$ maintains its oxidative catalytic capabilities as well although perhaps only in certain regions of the films.

CATALYTIC OXIDATION OF WATER TO DIOXYGEN

The mechanistic demands of a catalyst for the oxidation of water to dioxygen by a pathway which avoids high energy intermediates are severe. As shown in Scheme 6, the requirements of such a pathway include the loss of four electrons and four protons and O-O bond formation. Interestingly, in the stepwise oxidation of Ru(II) to Ru(IV) in reaction 5, two electrons are lost from orbitals largely metal d_{π} in character with the concomittant loss of two protons from bound water. If such a behavior could be combined with O-O coupling in a dimer, all of the ingredients implied by Scheme 6 would be present in the same molecule.

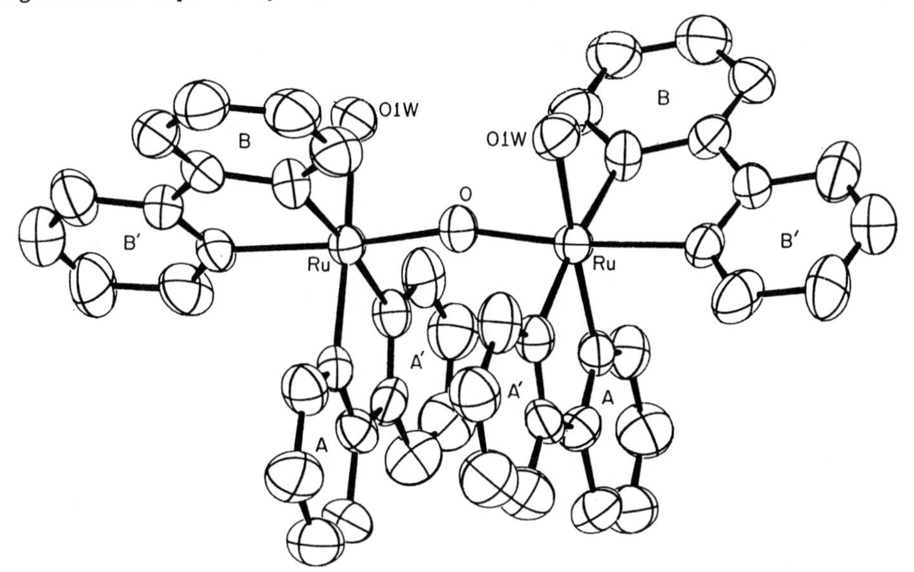

Figure 2. Structure of the μ-oxo cation in the salt $[(bpy)_2(H_2O)Ru^{III}ORu^{III}(H_2O)(bpy)_2](ClO_4)_4 \cdot 2H_2O$ with the O-atoms of the bound water molecules labelled as O1W, from ref. 10b

This strategy has proven to be successful. In Figure 2 is shown the structure of the deep blue, μ-oxo complex $[(bpy)_2(H_2O)Ru^{III}ORu^{III}(H_2O)(bpy)_2]^{4+}$ in which the formal oxidation state at Ru is III but where strong electronic coupling is known to exist between the metal ions across the oxide bridge.[10b] Electrochemical studies in aqueous solution as a function of pH have revealed an extensive multiple oxidation state chemistry for the μ-oxo ion as might have been expected given the properties of related monomers. In strongly acidic solution at pH=1, the pattern of redox events is the appearance first of a 1-electron 1-proton oxidation at $E_{1/2}$ = 0.79V,

$$b_2\underset{OH_2}{Ru^{III}}-O-\underset{OH_2}{Ru^{III}}b_2^{\ 4+} \xrightarrow[-H^+]{-e^-} b_2\underset{OH_2}{Ru^{III}}-O-\underset{OH}{Ru^{IV}}b_2^{\ 4+}$$

(b is bpy)

followed by a 3 electron-3 proton oxidative step at $E_{1/2}$ = 1.22V (vs. SCE),

$$b_2\underset{OH_2}{Ru^{III}}-O-\underset{OH}{Ru^{IV}}b_2^{\ 4+} \xrightarrow[-3H^+]{-3e^-} b_2\underset{O}{Ru^{V}}-O-\underset{O}{Ru^{V}}b_2^{\ 4+}$$

The product of the first oxidation is the isolable mixed-valence ion $[(bpy)_2(OH)Ru^{IV}ORu^{III}(H_2O)(bpy)_2]^{4+}$. The second oxidation gives the Ru(V)-Ru(V) ion $[(bpy)_2(O)RuORu(O)(bpy)_2]^{4+}$ which has only a transient existence in water and returns to the mixed-valence ion with a halftime of less than 1 min at room temperature. Electrochemical studies over a broad pH range show the existence of three pH dependent couples which interconnect μ-oxo ions in formal oxidation states III,III, III,IV, IV,V, and V,V.

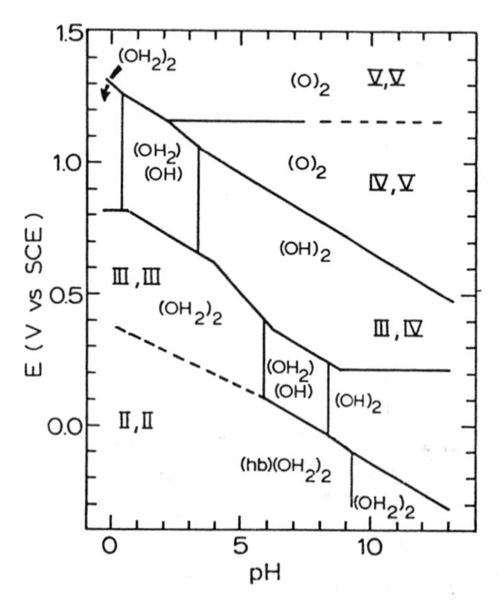

Figure 3. $E_{1/2}$-pH diagram for the V,V/IV,V, IV,V/III,IV, III,IV/III,III and III,III/II,II couples based on the μ-oxo ion $[(bpy)_2(H_2O)Ru^{III}ORu^{III}(H_2O)(bpy)_2]^{4+}$. The oxidation state and proton compositions of the various oxidation states in their potential-pH regions of dominant stability are indicated on the plot using abbreviations like III,IV $(OH_2)(OH)$ for $[(bpy)_2(OH)Ru^{IV}ORu^{III}(H_2O)(bpy)_2]^{4+}$. The vertical lines show pK_a values for the oxidation states.

The $E_{1/2}$-pH dependences of the various couples are shown in Figure 3. As for the monomeric couples in Figure 1, the complexities arise because of differences in proton content between different oxidation states at a given pH. Over the whole pH range, oxidation state IV,IV fails to appear showing that it is thermodynamically unstable with respect to disproportionation although it could be important mechanistically as a kinetic intermediate. Above pH~2 the most strongly oxidizing couple is the pH independent V,V/IV,V couple. Because of the differences in pH dependence between the V,V/IV,V and IV,V/III,IV couples, past pH = 2 the IV,V/III,IV couple becomes more strongly oxidizing than the V,V/IV,V couple, the IV,V ion becomes unstable with respect to disproportionation and only the $3e^-$ V,V/III,IV couple is observed electrochemically.

The V,V dimer is unstable in water because it oxidizes water to dioxygen and does so catalytically, Scheme 13

SCHEME 13
(at pH=1.0)

$$[b_2(H_2O)Ru^{III}ORu^{III}(H_2O)b_2]^{4+} \underset{+e^-,+H^+}{\overset{-e^-,-H^+}{\rightleftharpoons}} [b_2(H_2O)Ru^{III}ORu^{IV}(OH)b_2]^{4+} \underset{+3e^-,+3H^+}{\overset{-3e^-,-3H^+}{\rightleftharpoons}} [b_2(O)Ru^{V}ORu^{V}(O)b_2]^{4+}$$

$$2H_2O \rightarrow 4H^+ + O_2$$

The catalytic capability is somewhat restricted since after 10-25 turnovers of dioxygen production, the catalytic behavior ceases, in part because of oxidatively induced anation,[10b]

$$(bpy)_2(H_2O)Ru^{III}ORu^{IV}(OH)(bpy)_2^{4+} + X^- \rightarrow (bpy)_2(X)Ru^{III}ORu^{IV}(OH_2)(bpy)_2^{3+} + H_2O$$

Table 1. $\Delta G^{o'}$ Values for the Oxidation of Water at 22 $\pm$ 2°

Reaction	pH	$\Delta G^{o'}$,eV
Ru(V)-Ru(V)		
$[b_2(O)Ru^VORu^V(O)b_2]^{4+} + 2H_2O \rightarrow [b_2(H_2O)Ru^{III}ORu^{III}(H_2O)b_2]^{4+} + O_2$	1	-0.72
$+ 2H_2O \rightarrow [b_2(OH)Ru^{III}ORu^{III}(H_2O)b_2]^{3+} + O_2 + H^+$	7	-0.80
$+ 2H_2O \rightarrow [b_2(H_2O)Ru^{III}ORu^{IV}(OH)b_2]^{3+} + HO_2$	1	1.23
$+ H_2O \rightarrow [b_2(O)Ru^{IV}ORu^V(O)b_2]^{3+} + OH + H^+$	3	1.2
Ru(IV)-Ru(V)		
$2\ [b_2(O)Ru^{IV}ORu^V(O)b_2]^{3+} + 2\ H_2O \rightarrow 2\ [b_2(OH)Ru^{III}ORu^{IV}(OH)b_2]^{3+} + O_2$	7	-1.00
$[b_2(O)Ru^{IV}ORu^V(O)b_2]^{3+} + 2\ H_2O \rightarrow [b_2(OH)Ru^{III}ORu^{IV}(OH)b_2]^{3+} + H_2O_2$	7	0.58

(b is 2,2'-bipyridine)

From available reduction potential data, free energy changes for various possible net reactions in which water is oxidized to dioxygen are collected in Table 1. From the ΔG values: 1) It is only in oxidation state V,V that the full 4-electron requirement of the reaction can be met with a single μ-oxo ion. 2) Replacing Ru in the μ-oxo structure by less strongly oxidizing Os drops the oxidizing ability below the potential needed for water oxidation.[11a] 3) Both the one-electron oxidation of water to hydroxyl radicals by the Ru-based V,V ion and the 3-electron oxidation to perhydroxyl radical are considerably nonspontaneous thermodynamically. 4) The lower oxidation state IV,V ion is also thermodynamically capable of oxidizing water to oxygen, even up to pH 7, but the net reaction necessarily has a requirement of two equivalents of the IV,V ion.

The ability of both the V,V and IV,V ions as oxidants extends well beyond water oxidation. The V,V ion has been investigated as a catalyst for the oxidation of a variety of organics including a series of sugars and amino acids and it has been suggested that if properly attached to an electrode surface, it might serve a useful role as an electrochemical detector.[26] Oxidation of chloride ion to chlorine by the V,V ion in acidic solution occurs with k > 10^7 M^{-1} s^{-1} and the ability to oxidize chloride to chlorine extends to the dimer ion-exchanged into a polymeric film of polystyrene sulfonate on carbon electrodes.[27] The ability of the film coated electrodes to carry out the electrocatalyzed oxidation of chloride is impressive indeed. Even under conditions where electrochemical studies show that only a fraction of the μ-oxo ions are electrochemically active, current densities in excess of 100 mA/cm^2 are reached but only for short periods. The electrochemical activity ceases with 0.1 M added Cl^- after ~26,000 turnovers because, interesting enough, binding of the substrate, Cl^-, inhibits its oxidation to Cl_2.

The IV,V ion can be generated stoichiometrically in aqueous solutions by HOCl oxidation.[28] In water the ion is unstable with regard to water oxidation. It returns to the III,IV ion with the production of O_2 on a timescale of a few hours by parallel pathways zero order and first order in $[OH^-]$. The IV,V ion also typically undergoes rapid reactions with a variety of organic functional groups at considerably enhanced rates compared to $Ru(IV)=O^{2+}$, for example, with rate enhancements of >30, >42, and >6,000 for the oxidations of isopropanol, acetaldehyde and a water soluble olefin in water at 25°.[28]

Another interesting reaction occurs in the present of excess hypochlorite where the IV,V/III,IV couple acts as a very effective catalyst for the decomposition of hypochlorite

$$HOCl \longrightarrow 1/2\ O_2 + H^+ + Cl^-$$

The reaction proceeds very rapidly for a short period but, ultimately, the catalysis ceases as the μ-oxo-dichloro ion $[(bpy)_2ClRu^{III}ORu^{IV}Cl(bpy)_2]^{3+}$ appears via anation of the catalyst.

MECHANISMS OF WATER AND CHLORIDE OXIDATION

The mechanism or mechanisms by which water is oxidized by the μ-oxo ion are unknown but a number of observations have been made and it is of value to consider some of the mechanistic alternatives. One result based on an ^{18}O labelling study by D. Geselowitz at pH = 1, showed that oxidation of the $H_2{}^{18}O$ labelled III,IV dimer by three equivalents of Ce(IV) in normal water leads to the isotopic distribution ^{18}O-^{18}O(6%), ^{18}O-^{16}O(63%), and ^{16}O-^{16}O(33%) in the dioxygen product.[29] Although in small amount, the appearance of ^{18}O-^{18}O seemingly demands an intracoordination sphere coupling pathway. The amount of double labelled O_2 product becomes even more significant when it is realized that a number of exchange pathways exist by which the initial ^{18}O label could be lost from the complex.

Based on the labelling studies, there are a number of mechanistic possiblities by which the dimer could oxidize water to dioxygen and a tantalizing possibility is that more than one mechanism may be operating simultaneously, note Scheme 14.

SCHEME 14

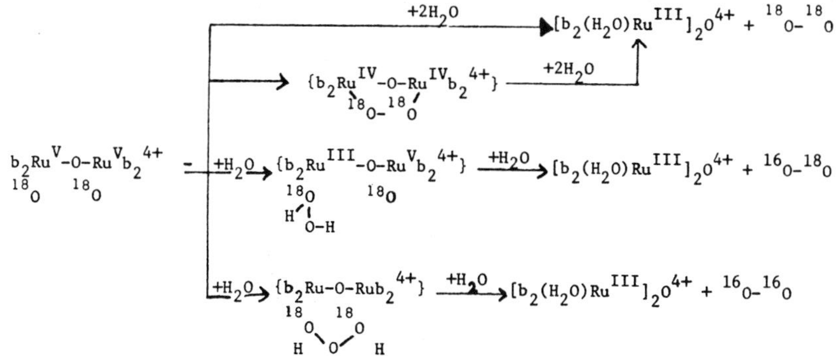

$$b_2Ru^V\text{-O-}Ru^Vb_2^{4+} \;(^{18}O,\ ^{18}O) \rightarrow \begin{cases} \xrightarrow{+2H_2O} [b_2(H_2O)Ru^{III}]_2O^{4+} + {}^{18}O\text{-}{}^{18}O \\ \rightarrow \{b_2Ru^{IV}\text{-O-}Ru^{IV}b_2^{4+}\}\ ({}^{18}O\text{-}{}^{18}O) \xrightarrow{+2H_2O} [b_2(H_2O)Ru^{III}]_2O^{4+} + {}^{18}O\text{-}{}^{18}O \\ \xrightarrow{+H_2O} \{b_2Ru^{III}\text{-O-}Ru^Vb_2^{4+}\}\ ({}^{18}O(H)\text{-}O\text{-}H,\ {}^{18}O) \xrightarrow{+H_2O} [b_2(H_2O)Ru^{III}]_2O^{4+} + {}^{16}O\text{-}{}^{18}O \\ \xrightarrow{+H_2O} \{b_2Ru\text{-O-}Rub_2^{4+}\}\ (H\text{-}{}^{18}O\text{-}O\text{-}{}^{18}O\text{-}H) \xrightarrow{+H_2O} [b_2(H_2O)Ru^{III}]_2O^{4+} + {}^{16}O\text{-}{}^{16}O \end{cases}$$

In Scheme 14 there are two possible routes shown to account for the appearance of ${}^{18}O\text{-}{}^{18}O$. One of them involves an intermediate peroxo structure and oxidation states IV,IV at the metals. In the second a synchronous 4-electron process occurs triggered by the attack of external water at the metals,O-O bond formation, and electron release to the two Ru(V) sites.

The pathway proposed to account for the appearance of ${}^{16}O\text{-}{}^{18}O$ is equally interesting. In that pathway, water attack occurs at the oxygen of an electron deficient Ru(V)-oxo site to give bound peroxide. Formation of the bound peroxide is followed by a second intramolecular 2-electron step to give the observed ${}^{16}O\text{-}{}^{18}O$ product.

Intramolecular mechanistic possiblities can also be written to account for the appearance of ${}^{16}O\text{-}{}^{16}O$ as a product. In the case shown the key is the attack of a water molecule on the oxygen of the electron deficient Ru(V)=O sites to give either a symmetrical or an unsymmetrical bridging structure in which the central oxygen atom becomes electron defficient. Following attack by a second molecule of water, dioxygen would be released. Pathways of this kind are especially interesting since they become "template" mechanisms where there is no necessity to make or break metal ligand bonds. Because of the absence of a requirement for a substitutional step, such mechanisms can be unusually facile.

The question of chloride oxidation to Cl_2 or HOCl is almost as interesting in a mechanistic sense as the oxidation of water. As mentioned above, the reaction is extraordinarily rapid. As shown in Scheme 15,

SCHEME 15

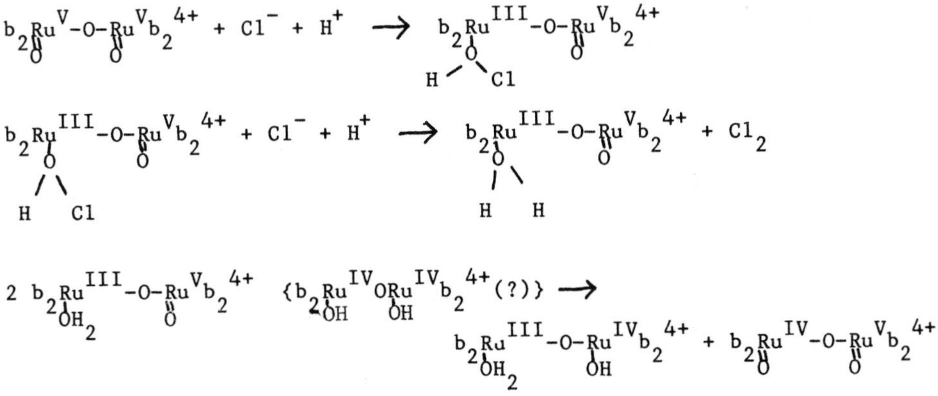

$$b_2Ru^V(O)\text{-O-}Ru^V(O)b_2^{4+} + Cl^- + H^+ \rightarrow b_2Ru^{III}(O(H)Cl)\text{-O-}Ru^V(O)b_2^{4+}$$

$$b_2Ru^{III}(O(H)Cl)\text{-O-}Ru^V(O)b_2^{4+} + Cl^- + H^+ \rightarrow b_2Ru^{III}(OH_2)\text{-O-}Ru^V(O)b_2^{4+} + Cl_2$$

$$2\ b_2Ru^{III}(OH_2)\text{-O-}Ru^V(O)b_2^{4+}\ \{b_2Ru^{IV}(OH)ORu^{IV}(OH)b_2^{4+}(?)\} \rightarrow b_2Ru^{III}(OH_2)\text{-O-}Ru^{IV}(OH)b_2^{4+} + b_2Ru^{IV}(O)\text{-O-}Ru^V(O)b_2^{4+}$$

an appealing possibility is that chloride attack occurs initially on the oxo group to give bound hypochlorite. Given the high rate at which the reaction proceeds, the second step may involve Cl^- attack on bound hypochlorite to give Cl_2 which above pH~4 is unstable with respect to disproportionation into Cl_2 and Cl^-. Although only a suggestion, the mechanism shown is especially appealing since it is another example of a template mechanism which might help to explain the rapid rate of the reaction.

Although some insight has been gained and will continue to be gained concerning the nature of the oxidation of H_2O and of Cl^-, much remains to be done in a mechanistic sense. However, the little that is known does allow a return to the question of the microscopic reverse of water oxidation, the activation of O_2 as in Scheme 3. Our hope is that by continued studies based on a family of related complexes, we may be able to understand in detail what mechanisms are involved. By making synthetic changes to lower the redox potentials of such systems, e.g., by replacing Ru by Os it may be possible to utilize the O_2-H_2O interconversion pathway or pathways for the activation of dioxygen.

Acknowledgments are made to the National Science Foundation under grant no. CHE-8601604 and to the National Institutes of Health under grant no. 5-RO1-GM32296-04 for the support of this research.

REFERENCES

1. Bolinger, C. M.; Sullivan, B. P.; Conrad, D.; Gilbert, J. A.; Story, N.; Meyer, T. J., J. Chem. Soc. Chem. Commun., 1985, 796.
2. Murphy, Jr., W. R.; Takeuchi, K.; Barley, M. H.; Meyer, T. J., Inorg. Chem., 1986, 25, 1041.
3. a) Keene, F. R.; Salmon, D. J.; Meyer, T. J., J. Amer. Chem. Soc., 1977, 99, 4821.
   b) Abruna, H. D.; Walsh, J. L.; Meyer, T. J.; Murray, R. W., J. Amer. Chem. Soc., 1980, 102, 3272.
4. a) Bowden, W. L.; Little, W. F.; Meyer, T. J., J. Amer. Chem. Soc., 1977, 99, 4340.
   b) Bowden, W. L.; Little, W. F.; Meyer, T. J., J. Amer. Chem. Soc., 1976, 98, 444.
5. a) Balzani, V.; Bolleta, F.; Gandolfi, M. T.; Maestri, M., Top. Curr. Chem., 1978, 75, 1.
   b) Kalyanasundaram, K., Coord. Chem. Rev., 1982, 46, 159.
   c) Crosby, G. A., Acc. Chem. Res., 1975, 8, 231.
6. Meyer, T. J., Pure and Applied Chemistry, 1986, 58, 1193.
7. a) Meyer, T. J., Acc. Chem. Res., 1978, 11, 94; Prog. Inorg. Chem., 1983, 30, 389.
   b) Sutin, N.; Creutz, C., Pure App. Chem., 1980, 52, 2717.
   c) Balzani, V.; Scandola, F., in "Energy Resources Through Photochemistry and Catalysis", Gratzel, M., Ed., Academic Press, 1983, Chapt. 1.
8. a) Moyer, B. A.; Meyer, T. J., J. Amer. Chem. Soc., 1978, 100, 3601.
   b) Moyer, B. A.; Meyer, T. J., Inorg. Chem., 1981, 20, 436.
   c) McHatton, R. C.; Anson, F. C., Inorg. Chem., 1984, 13, 3936.
   d) Roecker, L.; Kutner, W.; Gilbert, J. A.; Simmons, M.; Murray, R. W.; Meyer, T. J., Inorg. Chem., 1985, 24, 3784.
   e) Diamantis, A. A.; Murphy, Jr., W. R.; Meyer, T. J., Inorg. Chem., 1984, 23, 3230.
   f) Takeuchi, K. J.; Thompson, M. S.; Pipes, D. W.; Meyer, T. J, Inorg. Chem., 1984, 23, 1845.

9. a) Takeuchi, K. J.; Samuels, G. J.; Gersten, S. W.; Gilbert, J. A.; Meyer, T. J., Inorg. Chem., 1983, 22, 1407.
b) Che, W.-H.; Wong, K.-Y.; Leung, W.-H.; Poon, C.-K, Inorg. Chem., 1986, 25, 345.
c) Dobson, J.; Meyer, T. J., submitted.
10. a) Gersten, S. W.; Samuels, G. J.; Meyer, T. J., J. Amer. Chem. Soc., 1982, 104, 4029.
b) Gilbert, J. A.; Eggleston, D. S.; Murphy, Jr., W. R.; Geselowitz, D. A.; Gersten, S. W.; Hodgson, D. J.; Meyer, T. J., J. Amer. Chem. Soc., 1985, 107, 3855.
11. a) Gilbert, J. A.; Geselowitz, D. A.; Meyer, T. J., J. Amer. Chem. Soc., 1986, 108, 1493.
b) Geselowitz, D. A.; Kutner, W.; Meyer, T. J., Inorg. Chem., 1986, 25, 2015.
12. Dobson, J. C.; Takeuchi, K. J.; Pipes, D. W.; Geselowitz, D. A.; Meyer, T. J., Inorg. Chem., 1986, 25, 2357.
13. a) Che, C.-M.; Wong, K.-Y.; Mak, T. C. W., J. Chem. Soc. Chem. Commun., 1985, 546.
b) Yukawa, Y.; Aoyagi, K.; Kurihara, M.; Shirai, K.; Shimizer, K.; Mukaida, M.; Takeuchi, T.; Kakihara, H., Chem. Lett., 1985, 283.
14. Che, C.-M.; Cheng, W.-K., J. Chem. Soc. Chem. Commun., 1986, 1519.
15. a) Che, C.-M.; Yam, V. W.-W., J. Amer. Chem. Soc., 1987, 109, 1262.
b) Marmion, M. E.; Takeuchi, K. J., J. Amer. Chem. Soc., 1986, 108, 510.
16. Meyer, T. J., J. Electrochem. Soc., 1984, 131, 7, 221C.
17. a) Binstead, R. A.; Moyer, B. A.; Samuels, G. J.; Meyer, T. J., J. Amer. Chem. Soc., 1981, 103, 2897.
b) Binstead, R. A.; Meyer, T. J., J. Amer. Chem. Soc., 1987, 109, 3287.
18. Dobson, J. C.; Seok, W. K.; Meyer, T. J., Inorg. Chem., 1986, 25, 1514.
19. Roecker, L.; Dobson, J. C.; Vining, W. J.; Meyer, T. J., Inorg. Chem., 1987, 26, 779.
20. a) Roecker, L.; Meyer, T. J., J. Amer. Chem. Soc., 1987, 109, 746.
b) Thompson, M. S.; Meyer, T. J., J. Amer. Chem. Soc., 1982, 104, 4106.
21. Dobson, J. C.; Seok, W.; Meyer, T. J., submitted.
22. Seok, W., work in progress.
23. a) Thompson, M. S.; DeGiovani, W. F.; Moyer, B. A.; Meyer, T. J., J. Org. Chem., 1984, 25, 4972.
b) Moyer, B. A.; Thompson, M. S.; Meyer, T. J., J. Amer. Chem. Soc., 1980, 102, 2310.
24. a) Abruna, H. D.; Denisevich, P.; Umana, M.; Meyer, T. J.; Murray, R. W., J. Amer. Chem. Soc., 1981, 103, 1.
b) Murray, R. W., "Chemically Modified Electrodes", in Electroanalytical Chemistry, Vol. 13, Bard, A. J., ed., M. Dekker, New York, 1984.
25. Samuels, G. J.; Meyer, T. J., J. Amer. Chem. Soc., 1981, 103, 307.
26. Kutner, W.; Gilbert, J. A.; Tomaszewski, A.; Meyer, T. J., Murray, R. W., J. Electroanal. Chem., 1986, 205, 185.
27. a) Vining, W. J.; Meyer, T. J., Inorg. Chem., 1986, 25, 2023.
b) Vining, W. J.; Meyer, T. J., J. Electroanal. Chem., 1985, 195, 183.
28. Raven, S., work in progress.
29. Geselowitz, D., unpublished results.

KINETICS OF FORMATION OF

BIOLOGICAL OXYGEN CARRIERS

Ralph G. Wilkins

Department of Chemistry
New Mexico State University
Las Cruces, NM 88003

INTRODUCTION

At this point in time, there are only three known types of respiratory proteins involving only two transition elements (Chart I).[1-4] These function to transport and store oxygen and therefore play a key role in biochemical processes. They can increase by up to one hundred fold the amount of oxygen delivered to tissues. The types include, with reference to recent reviews (a) the hemoglobins and myoglobins[5-7] (b) the hemocyanins[8,9] and (c) the hemerythrins and myohemerythrins.[10,11] In spite of their name, (b) and (c) do not contain the heme center.

Chart I. Naturally-Occurring Carriers of Oxygen

Hemoglobin and Myoglobin
Iron porphyrin-containing proteins found extensively in the animal kingdom and in some plants. They are present in erythrocytes or free in solution. Hemoglobin is very familiar as the tetramer (M.W. $\sim$ 6.4×10^4) but the monomeric myoglobin and extracellular hemoglobins (M.W. $\sim$ 4×10^6) are also widespread.

Hemocyanin
Large copper-containing proteins (M.W. 5×10^5 - 10^7) which occur in hemolymphs of many invertebrate species. Hemocyanins from molluscs (octopus, snail) and arthropods (lobster, crab) have been especially studied. Molluscan hemocyanins are cylindrical oligomers with subunit M.W. $\sim$ 4×10^5. Arthropod hemocyanins consist of hexamers or multihexamers with subunit M.W. $\sim$ 7.5×10^4. There is no evidence for a function (oxygen storage) for monomeric protein.

Hemerythrin and Myohemerythrin
Iron-containing proteins found widely but only in brachiopods, sipunculids, priapulids and a few species of annelids. Hemerythrins from the sipunculids, *Phascolopsis gouldii* and *Themiste zostericola* have been mostly studied. Hemerythrin occurs in the coelomic fluid usually as octamer but in some species it occurs in lower polymeric form. It is present in the muscle of *T. zostericola* as the monomer (myohemerythrin). The subunit M.W. is $\sim1.35\times10^4$

Chart II. Schematic Representation of Active Site for Oxy-Forms of the Naturally-Occurring Carriers

Oxyhemoglobin (scarlet)

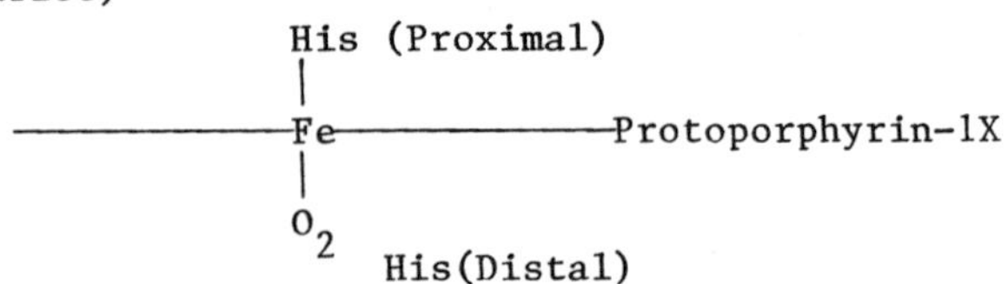

In deoxyhemoglobin the iron atom is ∿0.6Å out of the heme plane. On oxygenation, (possibly producing a $Fe(III)-O_2^-$ moiety) the iron moves into the plane, this pulling the proximal His with it.[12,13] This, in turn, transmits an effect to the chain resulting in rupturing the chain salt links and transforming T into R state.

Oxyhemocyanin (blue)

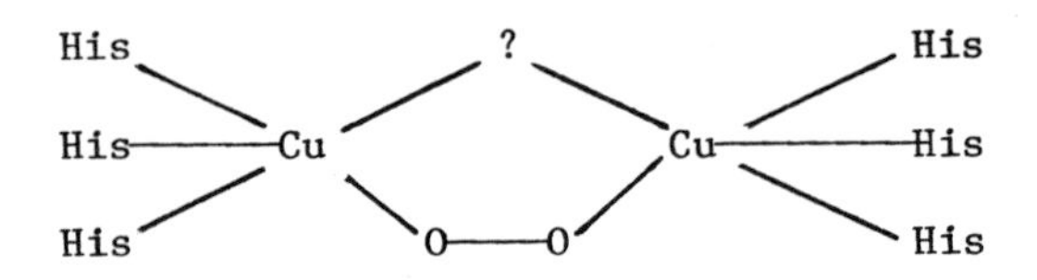

X-ray crystallography, EXAFS and vibrational spectroscopy are consistent with a $Cu(II)-O_2^{2-}-Cu(II)$ site. Another bridging group (OH?) is probably present.[14,15]

Oyhemerythrin (wine red)

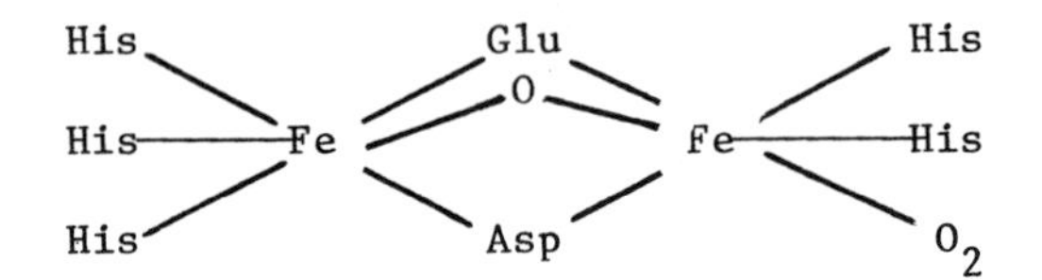

X-ray crystallography and multi-spectroscopic investigations are consistent with binuclear iron, and $Fe(III)-Fe(III)-O_2^{2-}$ probably H^+ bonded to bridged-O-moiety.[16]

The molecular structure and the active site of all three types are known with varying degrees of exactness.[1-4,12-16] The elucidation of the three dimensional structures of myoglobin and hemoglobin is a landmark in molecular biology. The active site in hemoglobin and hemerythrin has been well characterized and the structural changes on oxygenation of the proteins are well understood.[16,17] One knows, at present, much less about hemocyanin. The structural aspects of the oxygen carriers are dealt with elsewhere. We show the active site characteristics in Chart II.

It is not possible in the limited space to give a detailed account of the kinetics and mechanism of oxygen interaction with the respiratory proteins. Full surveys are available in References 2-7, 11 and 18. However, a short appraisal of the present state of this topic will be attempted emphasizing comparative behavior and the techniques and approaches used. With the respiratory proteins, the equilibria and structural aspects tend to be emphasized in standard texts and monographs. Certainly the equilibria data are vital in setting up the appropriate conditions for the kinetics experiments. In general however, kinetics provide at least twice the information obtained from equilibria studies (i.e. forward and reverse reaction rates and sometimes transient conformational changes). If the kinetics are not well understood, then it is likely that a true understanding of the equilibria is missing also, whether this is realized or not.

Techniques

High O_2-binding rates are invariably encountered for respiratory proteins of all types and therefore specialized techniques have been required and developed for their measurements. The invention of the flow method for the study of rapid reactions in solution stems, in fact, from its need to study the hemoglobin interactions with O_2 and CO.[19] The method has been continually developed since 1923 and nowadays stopped-flow equipment, interfaced with computers, are available for the accurate determination of ligand binding rate constants. Fascinating accounts of the early history of these mixing methods and their application to hemoglobin kinetics are available.[20,21]

Since however some of the reaction steps in the oxygenation process are in the submillisecond time frame, it has been necessary to circumvent the mixing limitation of the flow techniques. The development in the 40-50's of the relaxation techniques by Eigen and of the flash photolysis method pioneered by Norrish and Porter was therefore timely and fortunate. It was known in the last century that the O_2 and CO adducts of hemoglobin can be photodissociated.[22] The first application of flash photolysis to the CO adduct (because its quantum efficiency of photolysis is much higher than for the O_2 adduct) showed, dramatically, different conformations of the unliganded and liganded hemoglobin and detected interconversion processes.[23,24] More recent developments using nano-, pico- and subpico-second imposed laser pulses[25-27] as well as Raman and infrared spectroscopy for monitoring events[28,29] have allowed a glimpse of happenings at or near the heme site when O_2 adds to, or is removed from, the iron and long before it reaches the solvent outside the protein. Obviously very sophisticated (and expensive!) equipment is needed for this type of study. The data obtained, particularly for the very short times, tend to be equivocal and give information on structural, rather than detailed kinetic aspects. The results have not been correlated yet with those from nanosecond laser photolysis nor with the "overall" parameters for ligand binding. In contrast, photolytic perturbation methods have hardly been applied to hemocyanin and hemerythrin systems, in part perhaps because of the severe problem in monitoring events with these weaker absorbing proteins, but also because of the inherently greater interest in mammalian respiratory proteins.

With the monomeric proteins (only one binding site per molecule) or with any hemerythrin, the binding of O_2 to the protein P can be simply represented as (1)

$$P + O_2 \rightleftharpoons PO_2 \qquad k_{on},\ k_{off},\ K \qquad (1)$$

Stopped-flow or temperature-jump studies of the equilibrium (1) yields accurate values of k_{on} and usually approximate values of k_{off}. Accurate values of k_{off} are best obtained by a scavenging method, i.e. by adding to PO_2 either $S_2O_4^{2-}$ which reacts with O_2 or CO which adds to P. When successive binding steps are involved, as with the multisited proteins, the solution is extremely tedious, and possibly unsolvable, and approximate approaches are used (next section). A full discussion of the methods and the analysis of results for the reaction of hemoglobin with ligands is contained in References 5, 18 and 30.

Table I. Kinetic Data for Oxygen-Binding to Myoglobin and Hemoglobin. Most Conditions: 20°, pH ∿ 7.0

Species	k_{on} $\mu M^{-1} s^{-1}$	k_{off} s^{-1}	Ref.
MONOMERIC			
Myoglobin			
sperm whale	15	10	5, 31
horse heart	14[a]	11[b]	5
Asian elephant	18	18	32
Leghemoglobin a	118	5.5	33
Glycera dibranchiata Hb-I	190[c]	$2.8x10^{3}$[d]	34
POLYMERIC			
Human HbA			35, 36
T-state (α)	2.9	180	
R-state (α)	59	12	
T-state (β)	11.8	$2.5x10^{3}$	
R-state (β)	59	21	
Isolated α-chains	50	28	5
Isolated β-chains	60	16	5

[a] $\Delta H^{\neq}=4.9$ $\Delta S^{\neq}=-9.3$ [b] $\Delta H^{\neq}=18.4$ $\Delta S^{\neq}=9$ [c] $\Delta H^{\neq}=5.8$ $\Delta S^{\neq}=4$ [d] $\Delta H^{\neq}=18.4$ $\Delta S^{\neq}=19$

HEMOGLOBIN AND MYOGLOBIN

Kinetic data for the binding of O_2 to the globins are shown in Table I.[5,31-37] Binding of ligands to the monomeric proteins is, as expected, a uniphasic process and interpreted in terms of a single second-order, first order reversible reaction (1). The forward and reverse rate constants for O_2 binding to the different myoglobins shown in Table I do not vary much even although Gln replaced distal His in elephant myoglobin.[32] The much higher stability ($K=k_{on}/k_{off}$) of the (monomeric) leghemoglobin-O_2 adduct arises mainly from a large value for k_{on}. This in turn is believed to result from the distal histidine (Chart II) being much further away from the active site than this same residue in sperm whale myoglobin (and hemoglobin).[33] Replacement of distal histidine by leucine in *Glycera dibranchiata* hemoglobin may also lead to the easier access by O_2 and the large k_{on} value (Table I).[34] Distal effects on rates have been recently thoroughly studied.[37]

Many hemoglobins consist of four polypeptide chains, each containing the iron center, held together by noncovalent interactions. Hemoglobin A (the principal human hemoglobin and much studied) consists of two α- and two β-chains. The three dimensional structures of myoglobin and the subunits of hemoglobin are very similar. In contrast to the monomeric proteins however, hemoglobin shows cooperativity, in which the binding of O_2 to one site affects the binding of subsequent O_2 to another binding site on the tetramer. This phenomenon has been thoroughly studied. About the best model (and not a bad one) which we have to describe the binding of O_2 to hemoglobin is that of Monod, Wyman and Changeux (MWC model),[38] supported by the equilibria studies of Edelstein[39] and the kinetic application of Hopfield, Shulman and Ogawa.[40] It may be regarded as two non-cooperative Adair schemes linked by conformational equilibria (Chart III).[18] For normal human hemoglobin in the absence of ligands, the protein is present as a special state

Chart III. MWC Model for Ligand Binding to Four Iron Centers in Hemoglobin

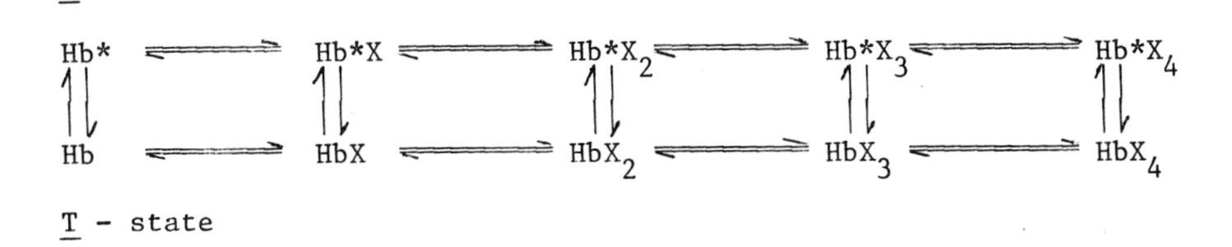

termed T (tense) state and symbolised Hb in Chart III. This is the low affinity form. This is in equilibrium with a high affinity form R (relaxed), Hb*, which is the predominant form at high ligand saturation (i.e. at Hb^*X_4). The binding of O_2 to each subunit is both thermodynamically and kinetically assumed to be independent of the extent of binding of other subunits, within the R or T states. Thus "intrinsic" rate constants for the "on" and "off" processes in the R and T states can be assessed. The kinetics of the T and R states may best be determined separately by working at very low O_2 fractional saturation (for T state) or at very low fractional photolysis with high O_2 saturations (for R state). In this way study of Hb and $Hb^*(O_2)_4$ is assured.[18,30] The model has been modified to include possible differences in the α- and β-chains and the latest results on this basis are shown in Table I.[36] The two-state model shown in Chart III does not appear completely adequate to explain all kinetic data,[18] particularly some irritating slow relaxations.[41] Nevertheless some conclusions appear warranted. Binding constants are $\sim 10^7 - 10^8\ M^{-1}s^{-1}$ and independent of the allosteric (T or R) states and oxygen-dissociation rates represent the major source of difference between the two states. The conformational transitions in human hemoglobin are usually assumed to be rapid compared with the ligational rates.

Apart from the intrinsic problems outlined above, the study of the binding of ligands including oxygen has been beset with periodic setbacks.[41] These include the discovery of 2,3-diphosphoglycerate (which reduces the O_2 affinity of cellular hemoglobin)[42] and α- and β-chains[43] (and their possible kinetic inequivalences) to mention two.

By using the simpler monomeric proteins, it has been possible to understand in a more intimate manner the oxygen binding (and dissociation) process. Application to oxymyoglobin or oxyleghemoglobin of either short, very intense, laser pulses or alternatively a combination of laser flash photolysis and lowered temperatures (down to 10°K in water/glycerol to slow the processes) is believed to produce transients such as B and C shown in Chart IV.[44-47] Rapid absorbance changes following the pulse are ascribed to the decay of B and C and rate parameters can be estimated for these changes. It should be emphasized that the mechanism as shown is almost certainly a simplified one with further states existing between A and B and between B and D distinct possibilities.[25,26,48] A very recent study yields the rate constants shown in Chart IV.[47] If these data hold up to the test of time and other investigators efforts (!) they show:

a) the rate limiting step for the overall association of O_2 with myoglobin is the formation of B; the overall rate of O_2 dissociation is limited by the A →B step as well as competition between geminate recombination and migration from the distal pocket, i.e. $k_3/(k_3 + k_2)$.

b) the values of k_3 are almost invariant for the Mb - O_2, NO and CO

Chart IV. Kinetic Parameters for O_2 Binding to Sperm Whale Myoglobin (20°C, pH 7.0)[2]

$$\underset{A}{PFeO_2} \underset{k_2}{\overset{k_1+k_{hv}}{\rightleftarrows}} \underset{B}{PFe...O_2} \underset{k_4}{\overset{k_3}{\rightleftarrows}} \underset{C}{PFe.....O_2} \underset{k_6}{\overset{k_5}{\rightleftarrows}} \underset{D}{PFe + O_{2(solv)}}$$

A	B	C	D
O_2 bound to heme iron	"Geminate" state O_2 embedded in protein near heme and not Fe bound	O_2 still embedded in protein but further from Fe	Separated reactants in solution
Distinctive Spectrum	Spectrum like PFe	Spectrum like PFe	Spectrum of PFe
$k_1 = 92s^{-1}$	$k_3 = 120\mu s^{-1}$	$k_5 = 14\mu s^{-1}$	$k\ (D \rightarrow B)$
$k_2 = 490\mu s^{-1}$	$k_4 = 8.5\mu s^{-1}$	$k_6 = 43\mu M^{-1}s^{-1}$	$16.6\mu M^{-1}s^{-1}$

systems whereas k_2 varies over three orders of magnitude for the same ligands. Since the overall quantum yield Q is determined primarily by k_2/k_3, then Q also varies widely with the Mb-O_2, Mb-NO and Mb-CO adducts.

It is hoped that by experiments such as these one may also understand the causes of the variable rate constants shown in Table I. For instance, the larger values of k_{on} (Table I) for leghemoglobin and Hb-I from the blood worm *Glycera dibranchiata* may arise, at least partly, from faster migration of O_2 from solvent to the heme pocket.[34,46]

HEMOCYANIN

There have been relatively few equilibria or kinetic studies of the oxygenation of hemocyanin and these mainly by Antonini and Brunori and their colleagues.[49,50] The most extensive data comes from hemocyanins from the (mollusc) Roman snail *Helix pomatia* and the (arthropod) spiny lobster *Panulirus interruptus* and are collected in Table II.[49-51] The ligand-binding equilibria and kinetics data have been rationalized (fairly well) in terms of the two-state MWC model. The O_2-binding behavior of the monomer or of the oligomer at low O_2 saturation gives information on the low O_2-affinity (T) state of the protein. Experiments on the oligomer at high (>90%) fractional saturation of O_2 give data for binding to the R-state. Binding to the R-state appears to be simpler than to the T-state.[2] In addition, with the high M.W. hemocyanin from *Helix pomatia*, slower relaxations may be interpretable in terms of a slow R ↔ T interconversion in the millisecond time frame.[2,4]

It can be seen from Table II that hemocyanin from *P.interruptus* differs from hemoglobin in that isolated subunits behave similarly to the T- and not the R-state. Like hemoglobin however, the oxygen association rate constants of the T- and R-states are similar, whereas the dissociation constants differ substantially. This means that with both hemoglobin and hemocyanin the lower O_2-affinity of the T states (k_{on}/k_{off}) resides in a higher dissociation rate constant. This appears

Table II. Kinetic Data for Oxygen-Binding to Hemocyanin
Most Conditions: 25°, pH∿8.5-9.5

Species	k_{on} $\mu M^{-1} s^{-1}$	k_{off} s^{-1}	Ref.
Helix pomatia			
T-state	5.0	700	49
R-state	5.0	5.0	49
Panulirus interruptus			
T-state (monomer)	37[a]	∿10^3[b]	50
	57	10^2	51
R-state (hexamer)	31[c]	60[d]	50

[a] $\Delta H^{\neq}=7.4 \Delta S^{\neq}=1$ [b] $\Delta H^{\neq}=18.1$ $\Delta S^{\neq}=17$ [c] $\Delta H^{\neq}=3\pm2 \Delta S^{\neq}=-14\pm6$ [d] $\Delta H^{\neq}=14$ $\Delta S^{\neq}=-3$

to be true generally for a number of hemocyanins.[2,4]

The question of the (potentially) biphasic binding of O_2 at the dinuclear copper site (Chart II) arising simply from its attachment to two coppers appears not to have been addressed.

HEMERYTHRIN

The folding of the polypeptide chain in a subunit of the octamer or of the trimer is virtually identical with that of the monomer. This hemerythrin fold has been observed in other proteins with different amino acid sequences and functions.[11] There is rarely evidence for cooperativity or for the operation of a Bohr effect (decrease in O_2 affinity with pH decrease) when deoxyhemerythrin reacts with oxygen. Hemerythrin from some brachiopods binds oxygen cooperatively.[52] Thus, so far, the kinetics of reaction even of the octameric forms lack the complexity shown by the hemoglobins and hemocyanins. Stopped-flow and temperature-jump methods have been used to obtain the data of Table III.[53,54] The large rate constant for binding of O_2 to deoxymyohemerythrin[54] suggested addition of O_2 directly to the iron, rather than substitution of coordinated water. This is a similar behavior to that shown by the other respiratory proteins, and it was

Table III. Kinetic Data for Oxygen-Binding to Hemerythrin
Most Conditions: 25°, pH 7-8

Species	k_{on} $\mu M^{-1} s^{-1}$	k_{off} s^{-1}	Ref.
Themiste zostericola			
monomer	78[a]	315[b]	54
octamer	7.5	82	54
Phascolopsis gouldii			
octamer	7.4[c]	51[d]	53

[a] $\Delta H^{\neq}=4$ $\Delta S^{\neq}=-11$ [b] $\Delta H^{\neq}=16.8$ $\Delta S^{\neq}=9$ [c] $\Delta H^{\neq}=8.2$ $\Delta S^{\neq}=1$ [d] $\Delta H^{\neq}=20.6$ $\Delta S^{\neq}=19$

Table IV. Kinetic Data for Binding of Deoxyhemerythrin (*P. gouldii*) with Protonated Ligands
Conditions: 25°C and I=0.5M

Ligand	HN_3	HCNO	HF
$k_1 (M^{-1} s^{-1})$	3.0×10^4	5.8×10^4	5×10^3
$\Delta H_1^{\neq}$ (kcal. mol^{-1})	6.1	7.9	4.9
$\Delta S_1^{\neq}$ (cal.mol^{-1}deg^{-1})	-17	-6.9	-24
$k_{-1} (s^{-1})$	0.10	0.012	0.010
$\Delta H_{-1}^{\neq}$ (kcal.mol^{-1})	12.2	13.9	12.5
$\Delta S_{-1}^{\neq}$ (cal.mol^{-1}deg^{-1})	-22.3	-20.3	-25

gratifying when subsequent structural studies strongly suggested one five-coordinated iron(II) in deoxyhemerythrin.[16] A few photodissociation experiments on oxyhemerythrin have been performed.[55] The second-order rate constants for binding O_2 from such experiments agree well with those obtained by temperature-jump. There appears to be no "unusual" form of deoxyhenerythrin produced by photodissociation but further experiments using short, very intense, laser pulses would be worthwhile.

In addition to binding O_2, deoxyhemerythrin has the remarkable ability to bind, quite strongly, N_3^-, CNO^- and F^-. This binding is assisted by one proton per anion and on the basis that the undissociated acids HN_3, HCNO and HF attack (Eqn (2)), the equilibria and kinetic data shown in Table IV are obtained.[58]

$$Hr^{O} + HX \rightleftharpoons Hr^{O}X^{-}(H^{+}) \qquad k_1, k_{-1} \qquad (2)$$

Protonation of the anion allows entry of a neutral entity into the hydrophobic core and the undissociated acids HX are thus simulating the behavior of O_2. The proton remains associated with the protein even at a pH as high as 9.0 and possibly associates with, or even breaks, the μ-OH bridge believed present in deoxyhemerythrin.[16]

SUMMARY

The rate constants for binding oxygen to all respiratory proteins are large and near the diffusion-controlled limit. The rate constants for oxygen dissociation are by contrast more variable. Apparently only three radically different reaction sites are needed in nature to fulfill the oxygen binding function (Chart II). All appear to have one unoccupied position in the deoxygenated form of the protein and at this site oxygen can add. In the product the oxygen moiety is transformed-to peroxide ion in hemerythrin and hemocyanin and possibly (still unsettled) superoxide ion in the globins.

Oxygen reacts in a straightforward fashion with monomeric forms of the proteins but with the oligomeric forms of hemoglobin and hemocyanin, oxygen-binding is a complicated process. This behavior can be fairly well understood in terms of the MWC two-state model, in which the protein can exist in a low affinity (T) or high affinity (R) state. The binding rate constant is almost independent of the allosteric state but the dissociation rate constant depends strongly on the quaternary structure and therefore the process of oxygen removal controls differences in the T and R states. The transformation of these states is very fast compared with the O_2 binding for hemoglobin and low molecular weight hemocyanins. It may however be rate limiting for some hemocyanins from arthropods.

For all respiratory proteins, the enthalpies of activation for the dissociative process (14-20 kcal. mol^{-1}) are always larger than for the formation (3-8 kcal. mol^{-1}) meaning that oxygenation is always exothermic. The entropies of activation are always more positive for oxygen dissociation than oxygenation by 12-18 entropy units, which can be understood from simple considerations.

These three respiratory proteins have been known since the early 1800's. It is clear however that we still do not understand by any means the mechanisms of oxygen binding to them and the multivarious nuances encountered. Aid in this will undoubtedly come from the study of model complexes. Cleverly constructed models simulate both the structural and the R and T state oxygen-binding behavior of hemoglobin.[57] Models which resemble the hemocyanin and hemerythrin sites have also been synthesized.[58,59] These have structural but not functional similarities.

REFERENCES

1. J. Lamy and J. Lamy, Eds. "Invertebrate Oxygen-Binding Proteins: Structure, Active Site and Function" Marcel Dekker, New York, (1981).
2. M. Brunori, B. Giardina and H. A. Kuiper, Oxygen-Transport Proteins, in "Inorganic Biochemistry, Vol III" H. A. O. Hill, ed., Royal Society of Chemistry, London (1982), pp. 126-182 (excludes hemerythrin).
3. A. G. Sykes, Functional Properties of the Biological Oxygen Carriers, in Adv. Inorg. Bioinorg. Mech. 1:119-176 (1982).
4. M. Brunori, M. Coletta and B. Giardina, Oxygen Carrier Proteins, in "Metalloproteins-Part 2: Metal Proteins with Non-redox Roles", P. M. Harrison, ed., Verlag Chemie, Weinheim (1985) Chapter 6.
5. E. Antonini and M. Brunori, "Hemoglobin and Myoglobin in Their Reactions with Ligands", North-Holland, Amsterdam (1971).
6. Q. H. Gibson, The Oxygenation of Hemoglobin, in "The Porphyrins, Vol. V, Part C", D. Dolphin, ed., Academic Press, New York (1978) Chapter 5.
7. C. Ho, ed. "Hemoglobin and Oxygen Binding" Elsevier Biomedical, New York (1982).
8. K. E. van Holde and K. I. Miller, Hemocyanins, Quart. Rev. Biophys., 15:1-129 (1982).
9. H. D. Ellerton, N. F. Ellerton and H. A. Robinson, Hemocyanin-A Current Perspective, Prog. Biophys. Molec. Biol., 41:143-248 (1983).
10. I. M. Klotz and D. M. Kurtz, Binuclear Oxygen Carriers:Hemerythrin, Accts. Chem. Res., 17:16-22 (1984).
11. P. C. Wilkins and R. G. Wilkins, The Coordination Chemistry of the Binuclear Iron Site in Hemerythrin, Coordn. Chem. Revs., in press (1987).
12. S. E. V. Phillips, Structure and Refinement of Oxymyoglobin at 1.6 Å Resolution, J. Mol. Biol. 142:531-554 (1980).

13. B. Shaanan, The Iron-Oxygen Bond in Human Haemoglobin, Nature (London) 296:683-684 (1982).
14. W. P. J. Gaykema, W. G. J. Hol, J. M. Vereijken, N. M. Soeter, H. J. Bak and J. J. Beintema, 3.2Å Structure of the Copper-Containing Oxygen-Carrying Protein Panulirus interruptus haemocyanin, Nature (London), 309:23-29 (1984).
15. G. L. Woolery, L. Powers, M. Winkler, E. I. Solomon and T. G. Spiro, EXAFS Studies of Binuclear Copper Site of Oxy-, Deoxy-, Metaquo-, Metfluoro-, and Metazidohemocyanin from Arthropods and Molluscs, J. Am. Chem. Soc., 106:86-92 (1984).
16. J. Sander-Loehr, Oxygen-Binding to the Binuclear Iron Center of Hemerythrin, in "Frontiers in Bioinorganic Chemistry", A. V. Xavier, ed., VCH Publishers, Weinheim (1986) pp. 574-583.
17. M. F. Perutz, Stereochemical Mechanism of Oxygen Transport by Haemoglobin, Proc. Roy. Soc. (London) B208:135-162 (1980).
18. L. J. Parkhurst, Hemoglobin and Myoglobin Ligand Kinetics, Ann. Rev. Phys. Chem. 30:503-546 (1979).
19. H. Hartridge and F. J. W. Roughton, Method of Measuring the Velocity of Very Rapid Chemical Reactions, Proc. Roy. Soc. (London), 104A:376-394 (1923); Kinetics of Hemoglobin, II. Velocity with which Oxygen Dissociates from its Combination with Hemoglobin, Proc. Roy. Soc. (London), 104A:395-430 (1923); Kinetics of Hemoglobin. III. Velocity with which Oxygen Combines with Reduced Hemoglobin, Proc. Roy. Soc. (London), 107A:654-683 (1925).
20. F. J. W. Roughton, The Origin of the Hartridge-Roughton Rapid Reaction Velocity Method, in "Rapid Mixing and Sampling Techniques in Biochemistry" B. Chance, R. H. Eisenhardt, Q. H. Gibson and K. K. Lonberg-Holm, eds., Academic Press, New York (1964) pp. 5-13.
21. Q. H. Gibson, Application of Rapid Reaction Techniques to the Study of Biological Oxidations, Ann. Rev. Biochem. 35:435-456 (1966).
22. S. Claesson, ed. "Fast Reactions and Primary Processes in Chemical Kinetics (Nobel Symposium V), Interscience Publishers, New York, 1967; lectures by R. G. W. Norrish (pp. 33-67), G. Porter (pps. 141-161) and M. Eigen (pps. 333-367).
23. Q. H. Gibson, The Photochemical Formation of a Quickly Reacting Form of Haemoglobin, Biochem. J., 71:293-303 (1959).
24. C. A. Sawicki and Q. H. Gibson, Quaternary Conformational Changes in Human Hemoglobin Studied by Laser Photolysis of Carboxyhemoglobin, J. Biol. Chem. 251:1533-1542 (1976).
25. L. Eisenstein and H. Frauenfelder, Introduction to Hemoproteins, and L. J. Noe, The Study of the Primary Events in the Photolysis of Hemoglobin and Myoglobin, in "Biological Events Probed by Ultrafast Laser Spectroscopy", R. R. Alfano, ed., Academic Press, New York (1982), Chapters 14 and 15.
26. J. M. Friedman, T. W. Scott, G. J. Fisanick, S. R. Simon, E. W. Findsen, M. R. Ondrias and V. W. Macdonald, Localized Control of Ligand Binding in Hemoglobin: Effect of Tertiary Structure on Picosecond Geminate Recombination, Science, 229:187-190 (1985).
27. M. A. West, Flash and Laser Photolysis in "Investigation of Rates and Mechanisms of Reactions, Part II: Investigation of Elementary Reaction Steps in Solution and Fast Reaction Techniques", C. F. Bernasconi, ed., Wiley-Interscience, New York (1986).
28. T. G. Spiro and J. Terner, Heme Protein Structure and Dynamics, Studied by Resonance Raman Spectroscopy Pure Appl. Chem. 55:145-149 (1983).
29. J. Terner and M. A. El-Sayed, Time-Resolved Resonance Raman Spectroscopy of Photobiological and Photochemical Transients, Acc. Chem. Res. 18:331-338 (1985).
30. E. Antonini, L. Rossi-Bernardi and E. Chiancone, eds. "Hemoglobin" Methods in Enzymology, volume 76, Academic Press (1981) Section VII.

31. J. LaGow and L. J. Parkhurst, Kinetics of Carbon Monoxide and Oxygen Binding for Eight Electrophoretic Components of Sperm-Whale Myoglobin, Biochemistry, 11:4520-4525 (1972).
32. A. E. Romero-Herrara, M. Goodman, H. Dene, D. E. Bartnicki and H. Mizukami, An Exceptional Amino Acid Replacement on the Distal Side of the Iron Atom in Probescidean Myoglobin, J. Molec. Evolution, 17:140-147 (1981).
33. C. A. Appleby, J. H. Bradbury, R. J. Morris, B. A. Wittenberg, J. B. Wittenberg and P. E. Wright, Leghemoglobin, J. Biol. Chem. 258:2254-2259 (1983).
34. L. J. Parkhurst, P. Sima and D. J. Goss, Kinetics of Oxygen and Carbon Monoxide Binding to the Hemoglobins of Glycera dibranchiata, Biochemistry, 19:2688-2692 (1980).
35. J. S. Olson, M. E. Andersen and Q. H. Gibson, The Dissociation of the First Oxygen Molecule from Some Mammalian Oxyhemoglobins, J. Biol. Chem., 246, 5919-5923 (1971).
36. C. A. Sawicki and Q. H. Gibson, Properties of the T State of Human Oxyhemoglobin Studied by Laser Photolysis, J. Biol. Chem. 252:7538-7547 (1977).
37. M. P. Mims, A. G. Porras, J. S. Olson, R. W. Noble and J. A. Peterson, Ligand Binding to Heme Proteins, J. Biol. Chem. 258:14219-14232 (1983).
38. J. Monod, J. Wyman and J. Changeux, On the Nature of Allosteric Transitions: A Plausible Model, J. Mol. Biol. 12:88-118 (1965).
39. S. J. Edelstein, Extensions of the Allosteric Model for Haemoglobin Nature (London) 230:224-227 (1971).
40. J. J. Hopfield, R. G. Shulman and S. Ogawa, An Allosteric Model of Hemoglobin: I, Kinetics, J. Mol. Biol., 61, 425-443 (1971).
41. Q. H. Gibson, Dynamics of Ligand Binding in Reference 7, pp. 321-327.
42. R. Benesch and R. E. Benesch, The Effect of Organic Phosphates from the Human Erythocytes on the Allosteric Properties of Hemoglobin Biochem. Biophys. Res. Commun. 26:162-167 (1967); A. Chanutin and R. R. Curnish, Effect of Organic and Inorganic Phosphates on the Oxygen Equilibrium of Human Erythrocytes, Arch. Biochem. Biophys. 121:96-102 (1967).
43. E. Bucci and C. Fronticelli, A New Method for the Preparation of α and β Subunits of Human Hemoglobin, J. Biol. Chem. 240:551-552 (1965).
44. R. H. Austin, K. W. Beeson, L. Eisenstein, H. Frauenfelder and I. C. Gunsalus, Dynamics of Ligand Binding to Myoglobin, Biochemistry, 14:5355-5373 (1975).
45. W. Doster, D. Beece, S. F. Bowne, E. E. DiIorio, L. Eisenstein, H. Frauenfelder, L. Reinisch, E. Shyamsunder, K. H. Winterhalter and K. T. Yue, Control of pH Dependence of Ligand Binding to Heme Proteins, Biochemistry, 21:4831-4839 (1982).
46. F. Stetzkowski, R. Banerjee, M. C. Marden, D. K. Beece, S. F. Bowne, W. Doster, L. Eisenstein, H. Frauenfelder, L. Reinisch, E. Shyamsunder and C. Jung, Dynamics of Dioxygen and Carbon Monoxide Binding to Soybean Leghemoglobin, J. Biol. Chem. 260:8803-8809 (1985).
47. Q. H. Gibson, J. S. Olson, R. E. McKinnie and R. J. Rohlfs, A Kinetic Description of Ligand Binding to Sperm Whale Myoglobin, J. Biol. Chem. 261:10228-10239 (1986).
48. H. Frauenfelder and P. G. Wolynes, Rate Theories and Puzzles of Hemeprotein Kinetics, Science, 229:337-345 (1985).
49. M. Brunori, H. A. Kuiper, E. Antonini, C. Bonaventura and J. Bonaventura, in Reference 1, p. 693; R. vanDriel, H. A. Kuiper, E. Antonini and M. Brunori, Kinetics of the Co-operative Reaction of Helix pomatia Hemocyanin with Oxygen, J. Mol. Biol. 121:431-439 (1978).

50. E. Antonini, M. Brunori, A. Colosimo, H. A. Kuiper and L. Zello, Kinetics and Thermodynamic Parameters for Oxygen Binding to the Allosteric States of Panulirus Interruptus Hemocyanin, Biophys. Chem. 18:117-124 (1983).
51. G. D. Armstrong and A. G. Sykes, Reactions of O_2 with Hemerythrin, Myoglobin and Hemocyanin: Effects of D_2O on Equilibration Rate Constants and Evidence for H-Bonding, Inorg. Chem. 25:3135-3139 (1986).
52. C. Manwell, Oxygen Equilibrium of Brachiopod Lingula Hemerythrins, Science, 132:550-551 (1960); D. E. Richardson, R. C. Reem and E. I. Solomon, Cooperativity in Oxygen Binding to Lingula reevii Hemerythrin: Spectroscopic Comparison to the Sipunculid Hemerythrin Coupled Binuclear Iron Active Site, J. Am. Chem. Soc., 105:7780-7781 (1983).
53. D. J. A. deWaal and R. G. Wilkins, Kinetics of the Hemerythrin-Oxygen Interaction, J. Biol. Chem. 251:2339-2343 (1976).
54. A. L. Petrou, F. A. Armstrong, A. G. Sykes, P. C. Harrington and R. G. Wilkins, Kinetics of the Equilibration of Oxygen with Monomeric and Octameric Hemerythrin from Themiste zostericola, Biochim. Biophys. Acta, 670:377-384 (1981).
55. P. Wilkins and R. G. Wilkins, unpublished experiments.
56. P. C. Wilkins and R. G. Wilkins, Acid-Assisted Anion Interaction with Deoxyhemerythrin, Biochim. Biophys. Acta, in press (1987).
57. K. S. Suslick and T. J. Reinert, The Synthetic Analogs of O_2-Binding Heme Proteins, J. Chem. Educ. 62:974-983 (1985).
58. K. D. Karlin, B. I. Cohen, J. C. Hayes, A. Farooq and J. Zubieta, Models for Methemocyanin Derivatives: Structural and Spectroscopic Comparisons of Related Azido-Coordinated (N_3^-) Mono- and Dinuclear Copper(II) Complexes, Inorg. Chem., 26:147-153 (1987) and refs. therein.
59. S. J. Lippard, The Bioinorganic Chemistry of Rust, Chem. in Britain 22:222-229 (1986) and refs. therein.

SYNTHETIC DIOXYGEN CARRIERS FOR DIOXYGEN TRANSPORT

Daryle H. Busch

Professor of Chemistry
The Ohio State University
120 West 18th Avenue
Columbus, Ohio 43210

INTRODUCTION

Synthetic Dioxygen Carriers, a Key Area for the 1990s

The promising application areas for synthetic dioxygen carriers range from internal medicine and small devices to the commodity gas market and basic fuel production and there seems little doubt this area will impact the lives of most people in the developed nations during the coming decades. Government and societal leaders, both Nationally and internationally,[1,2,3] continue to look to synthetics as possible eventual sources of dioxygen transport materials for temporary whole blood substitutes, envisioning such scenarios as those associated with major disasters and military engagements. Existing research has been focused on portable devices[4] to provide dioxygen enriched atmospheres for those suffering such maladies as emphysema and for the very different area of underwater dioxygen supply.[5,6] Dioxygen electrode systems for batteries are attractive targets on the full range of scales from tiny hearing aid cells through electric automobiles to fuel cells for the storage of off-peak energy by electric utility companies. For many large scale uses, for example foundry operation, a moderate enrichment of the dioxygen level is adequate and this is an especially attractive target area for separation techniques based on the use of transition metal dioxygen carriers.[7,8] The cleansing of contaminated atmospheres is a less than obvious but related area for application. Using the same basic science and technology, control of very low levels of O_2 is possible with such materials since the variability of O_2 affinities of carriers spans many orders of magnitude (at least 6 and possibly 10 or 12). Commodity level applications are most dramatically shown by the potential demands of the synfuel industry as revealed by industrial response to the synfuel goals set by the Carter administration.[9,10,11] It was concluded by American dioxygen-supplying industry that the existing cryogenic technology could not be expanded fast enough to meet the needs of the then projected synfuel industry and that at least one new major technology would have to be exploited. The first attempts to exploit transition metal dioxygen carriers were military.[12,13]

The most promising undeveloped resource for the separation of dioxygen from the air, or other fluids, is the chemistry of synthetic transition metal dioxygen carriers. The eventual success of such technologies depends on the design, synthesis and availability of molecular species with the critical chemical and physical properties that are dictated by the practical and economic demands of the specific applications. The requirements vary with the applications and some of them relate to such fundamental performance parameters of dioxygen carriers as their equilibrium constants for dioxygen binding and their rates of association with and dissociation from O_2. Others depend more critically on such practical matters as, specific operating conditions, the cost of the carrier and its lifetime in the operating environment. In view of the enormous range of dioxygen affinities and, presumably, kinetic parameters exhibited by known dioxygen carriers, performance parameters should not limit the use of the science.

Autoxidation--the Bane of All Dioxygen Carriers

The bane of all known dioxygen carriers is autoxidation by the very molecular entity they are designed to manage, O_2. Even the most marvelous O_2 carrier of nature, hemoglobin, autoxidizes and some of the most exciting demonstrations of O_2 binding in the laboratory have involved species that autoxidize very quickly under extremely moderate conditions.[14,15] While expensive specialty uses may emerge even when limited to costly short-lived O_2 carriers, the widespread application of this chemistry can be expected to blossom only when the problem of autoxidation is solved. From the standpoint of scientific research, this problem is very attractive since there is little prospect of controlling the autoxidation of dioxygen carriers unless the mechanisms of the deleterious processes are thoroughly understood as the result of the inevitable fundamental research. Combined with the need for fundamental understanding of autoxidation mechanisms is the attendant requirement for the design of molecules that will resist those harmful mechanisms. This implies basic studies in molecular design, an equally exciting area of fundamental chemistry.

Three very general mechanistic areas operative in the autoxidation of cobalt(II) dioxygen carriers are (1) irreversible formation of 2:1 peroxo-bridged dimers by carriers that function reversibly as 1:1 carriers; (2) irreversible ligand oxidation and (3) central atom oxidation. These three general processes may occur in concert as simultaneous and/or consecutive reactions and other entirely independent autoxidation mechanisms may well take place in some systems.[a] For many cases, ligand oxidation is the most harmful since additional chemistry may be applied more easily to reverse the effects of the other two general mechanisms; i.e., reduce the oxidized central atom.

Instructive examples of ligand oxidation have been reported.[16-21] When solutions of diaquo(5,7-dimethyl-1,4,8,11-tetraazacyclotetradeca-4,7-diene)cobalt(II) (structure I) are exposed to dioxygen, the ligand is oxidized forming the conjugated ketone shown in structure II.[16] A similar reaction was reported by Weiss and Goedken with complex III being oxidized to IV.[17] Ligand autoxidations have also been observed for iron(II) complexes.[18-20]

The most notable ligand that has been reported to undergo autoxidation, from the standpoint of its well known dioxygen carrier chemistry, is the Schiff base formed between ethylenediamine and 2,4,-pentanedione. Even the nickel(II) complex is autoxidized and two especially interesting products have been reported (structures V and VI).[21] It is also significant that the tetradentate ligand in the nickel complex is cleaved by hydrogen peroxide at the saturated dimethylene linkage.[21] However, in the presence of suitable axial

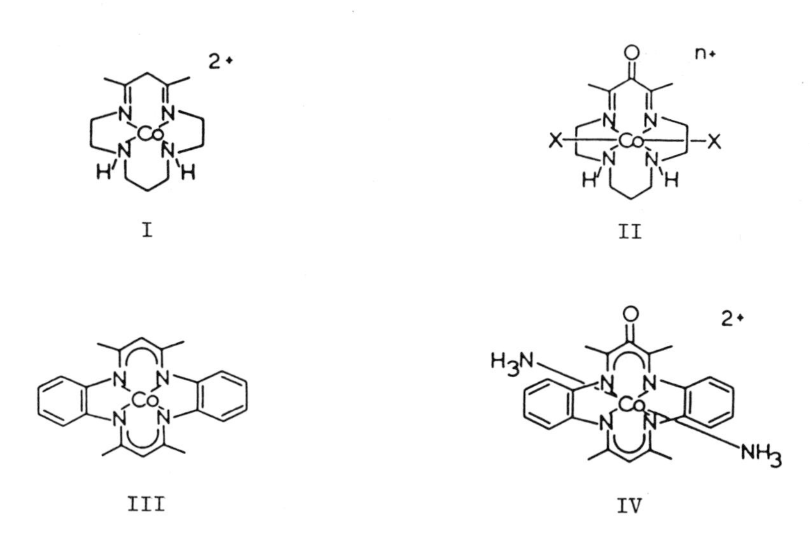

ligands, the cobalt complex can be oxidized to stable cobalt(III) derivatives. While the cobalt derivatives appear to be oxidized without ligand alteration, there is no guarantee such reactions would not be observed if recurrent oxidations and reductions of the system were carried out. In a molecular design aimed at producing outstanding O_2 carriers, this suspected weakness would need to be confronted.

The Design of Optimal Dioxygen Carriers

The studies to be summarized here have emphasized ligand design to produce optimal dioxygen carriers and, in addition to the usual concern about electronic relationships,[22-27] three broad concepts have been heavily used.[28] The first and most general is inclusion chemistry which entails the principles relating to the chemistry of molecules containing permanent voids.[b] The molecules in question are either macrocycles or macrobicycles and the cyclic components of the structures are incorporated to (a) provide the advantages attributable to macrocyclic ligands in the binding of the metal ion and/or (b) produce a distinct molecular chamber within which the binding of molecular dioxygen will be controlled.

The second principle guiding the design and synthesis of the cobalt(II) dioxygen carriers discussed here is that of <u>ligand superstructure.</u>[29] The traditional role of the ligand is to bind to the metal ion and determine its chemical and physical properties by electronic and topological relationships <u>Ligand superstructure</u> composes additional structural components appended to the parent ligand for the purpose of performing additional functions. The fusion of a ring that is perpendicular to the coordination plane of a tetradentate ligand bound to a planar metal ion produces an enclosure of variable volume and surface area in the vicinity of the coordination site (structure VII). Such a ring might be designed to produce complete inaccessability to the coordination site within it or it might be designed to admit small ligand molecules and exclude larger ones, or it might be designed to harbor an uncoordinated nucleophile, or a proton source, in the vicinity of the coordination site, etc. The fused perpendicular ring is superstructure on the parent macrocycle and the chemistry that this structural feature controls is inclusion chemistry.

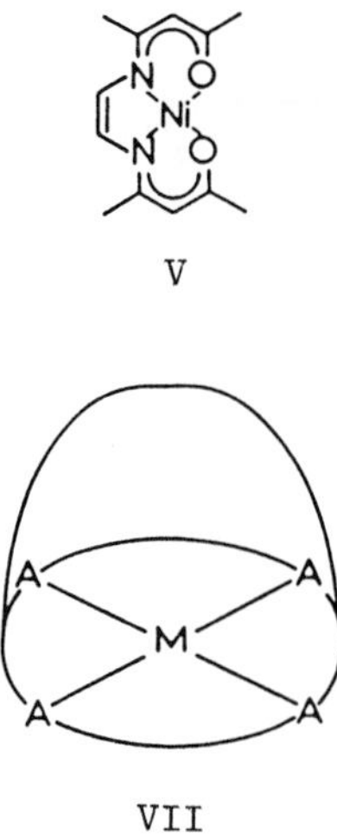

V

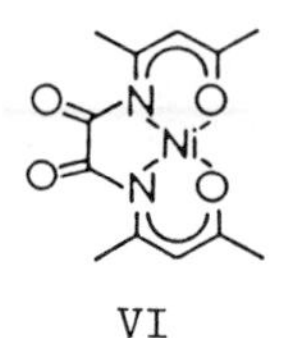

VI

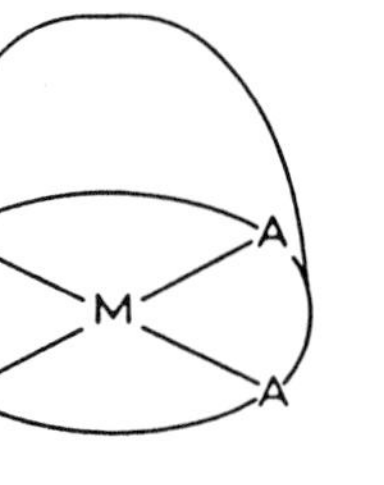

VII

VIII

The addition of superstructures to ligands of some complexity may require special synthetic strategies and these may profit from the presence of the metal ion during reactions that involve the building of the superstructure. This is especially evident in cases where the prior coordination of the metal ion produces conformations of the ligand that are favorable for formation of the superstructure. These reactions are correctly identified as examples of coordination template reactions, processes that were first produced in these laboratories many years ago.[28,30,31]

In the discussion that follows, two general families of dioxygen carriers are described. The synthetic strategies leading to their preparations and significant features displayed during their characterization will be presented. The efficacies of these compounds as dioxygen carriers is then reviewed in detail. The first are the cyclidene complexes a class of dioxygen carrier developed at Ohio State which show the power of inclusion chemistry in the design of good O_2 carriers. Remarkable control over the performance parameters of these carriers is afforded by the lacuna, or permanent void, within which the O_2 is bound. In addition, the lacuna inhibits autoxidation of the O_2 adducts under a variety of conditions. The principles applied or learned during the study of the cyclidene complexes are then applied to the lacunization of the familar Schiff base complexes of the acacen type[c,32] (structure VIII) and an optimal dioxygen carrier is designed and characterized. Finally, promising new directions for dioxygen carriers are considered.

During the earliest stages of the evolution of the idea of superstructure for expansion of ligand function, explorations were underway for families of tetradentate and/or pentadentate ligands that would facilitate the appending of such groups in advantageous ways. A first such development occurred with the copper(II) and nickel(II) complexes of the tetradentate tetraaza macrocycle called TAAB (Structure IX).[33] The electrophilic nature of the carbon atoms of the C=N groups facilitated the addition of nucleophiles at this point and alkoxides, amines, and enolate carbons were added to give bis(nucleophile) adducts.[34] Further, these reactions were used to produce bridges between trans methine groups (structure X). As promising as this work was, it was abandoned when the cyclidenes were discovered.

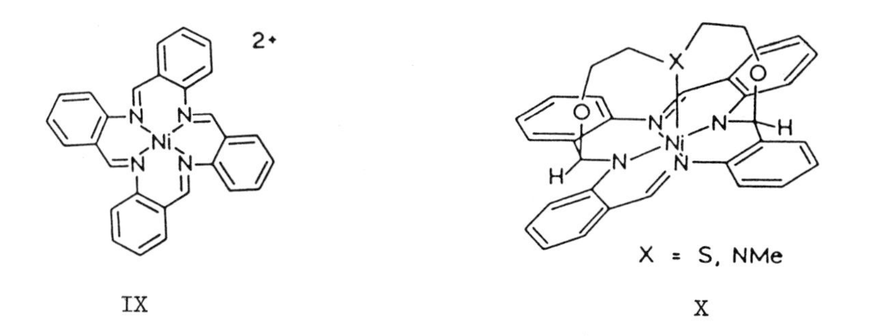

THE LACUNAR COBALT(II) CYCLIDENE DIOXYGEN CARRIERS

Hipp and Mokren[35] discovered that very strong methylating agents react with Jager's macrocyclic complex of structure XI by adding at the oxygen atom of the acyl substituent (structure XII). Even more interesting, the resulting methoxo group is readily replaced by other nucleophiles according to what appeared to be an addition-elimination reaction. These new reactions converged with x-ray structural results to produce the essential information making possible the facile synthesis of an extremely broad range of lacunar complexes that we call cyclidenes.[29,36,37] The x-ray structural determination on the methoxo compound (XII) showed that the new functional groups enjoy relative orientations that are extremely propitious for the closing of a ring between them.[29] This is evident in the deeply clefted saddle shape shown in the 3-dimensional view (structure XIII). The subsequent synthesis of the lacunar cyclidenes followed Scheme 1.[36,37] The complete synthesis of the lacunar cyclidenes from simple diamines, diketones, and triethylorthoformate makes use of two template reactions. The formation of the Jager macrocycle from the linear tetradentate precursor (see Scheme 2) provides the first template reaction. The second template reaction takes advantage of the fact that the nickel(II) produces a deeply clefted conformation of the methylated precursor complex that facilitates the second ring closure.

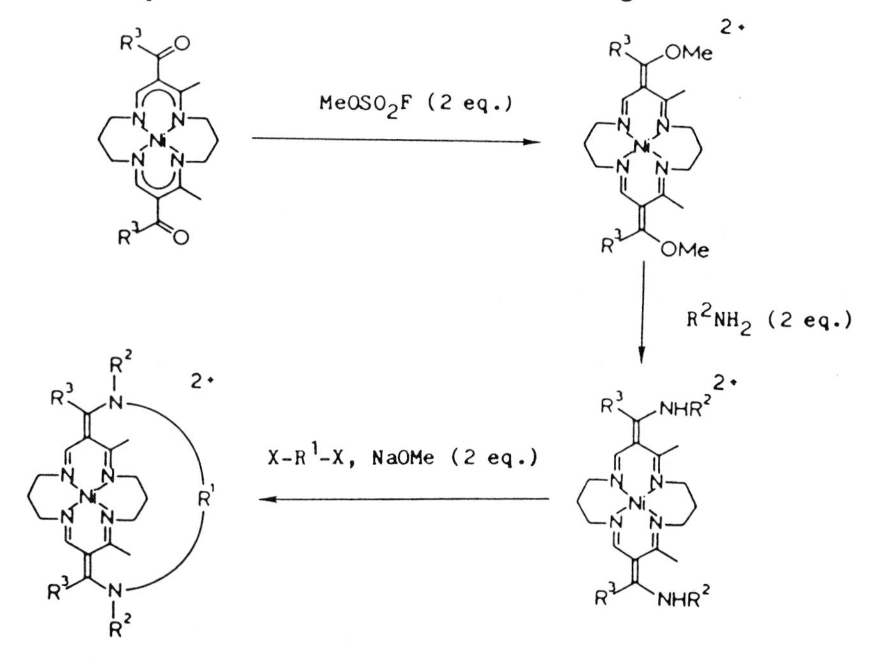

Scheme 1. Preparation of nickel(II) cyclidene complexes.

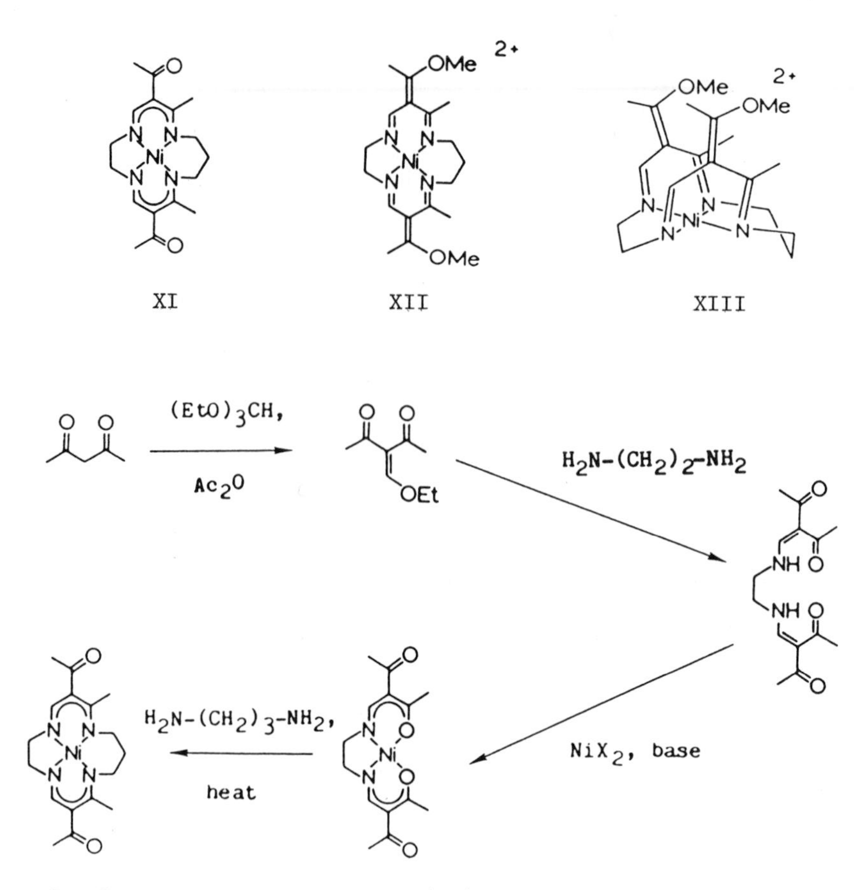

Scheme 2. Preparation of the nickel(II) 15-membered Jager macrocycle.

Flat and 3-dimensional representations of the general lacunar cyclidene structure are shown in structures XIV and XV. Many variations of the structure[d,36,37,41,42] have been produced; e.g., R^1 and/or R^2 = Me, Et, Pr, Bu, t-Bu, Ph, substituted Ph, $PhCH_2$, $PhCH_2CH_2$, $-N(CH_2CH_2)_2NH$, $-N(CH_2CH_2)_2CH_2$ (and similar rings), and the bridging groups have included polymethylene groups continuously from tri- through octa- plus dodeca-, m- and p-xylylene, 5-substituted-m-xylylenes, $-CH_2CH_2OCH_2CH_2-$, $-CH_2CH_2CH_2SCH_2CH_2CH_2-$, $-CH_2$-m-Ph-CH_2-m-Ph-CH_2, and others. Further, whereas the initial discoveries were made using the 15-membered ring Jager complex,[35] almost all of the studies on O_2 carriers have been performed on derivatives having a 16-membered parent macrocycle.

Structural analysis shows that the cleft is most pronounced with the larger parent macrocycle. In contrast, the 14-membered macrocycle is expected to exhibit a Z-shape with one unsaturated chelate ring on each side of the coordination plane rather than the deep cleft produced when the two unsaturated rings rise from the same side of the coordination plane.[43]

While most of the available physical methods have been applied to the characterization of the lacunar cyclidene complexes of nickel(II)[29,36,37,41], NMR, electrochemistry and x-ray studies have been most revealing. Eight x-ray crystal structure determinations have been

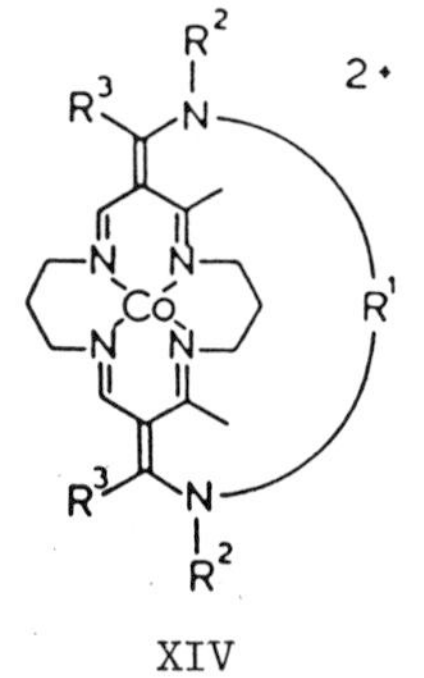

XIV

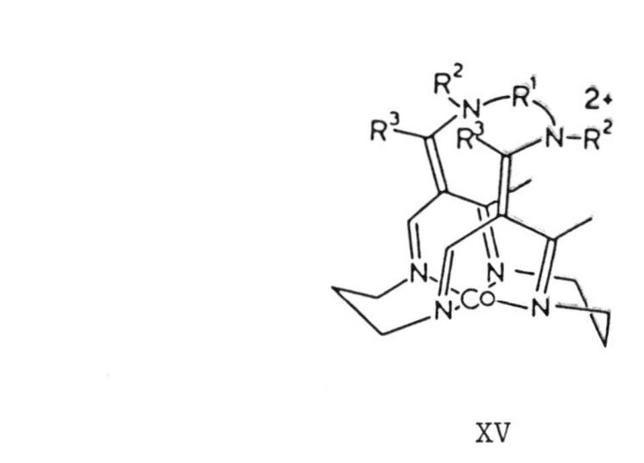

XV

reported on lacunar cyclidene complexes.[29,41,44-47] Further, a recent publication has analyzed the detailed stereochemical relationships associated with Jager's complexes and the precursors and unbridged cyclidenes on the basis of the results of 7 crystal structure determinations.[43] Because the nickel(II) complexes are invariably diamagnetic and, we believe, square planar, NMR techniques have provided both structure proof and a convenient control technique for purity and for the progress of syntheses.[36,37,41] Carbon-13 enrichment was used to confirm assignments[48] and, most recently, the powerful new 2D techniques have been used to assign all proton and carbon resonances for a number of the compounds.[49] Electrochemical studies revealed quasi-reversible formation of nickel(III) species at potentials independent of the nature and length of the bridging group R^1, a most significant fact that will arise in later discussions (see Table 1).[37,41] One may conclude at this point that the lacunae of the nickel(II) complexes are unoccuppied in solvents such as acetonitrile and water. The potential of the $Ni^{III/II}$ couple is sensitive to electronic effects, however (Table 1).

The cobalt(II) complexes were prepared from the previously isolated ligand salts.[42,44,50-53] Scheme 3 shows the removal of the ligand from the templating nickel(II) ion and its coordination to cobalt(II). The procedure is essentially the same for all cobalt(II) complexes. Under the usual conditions, the cobalt(II) complexes crystallize as the 4-coordinate square planar species (Fig. 1); however, the 5-coordinated complexes containing a mole of axial ligand are readily crystallized in the presence of an excess of the appropriate Lewis base.[50-53]

In the presence of axial bases, the cobalt(II) cyclidene complexes are invariably low spin and 5-coordinate. ESR spectra show the typical

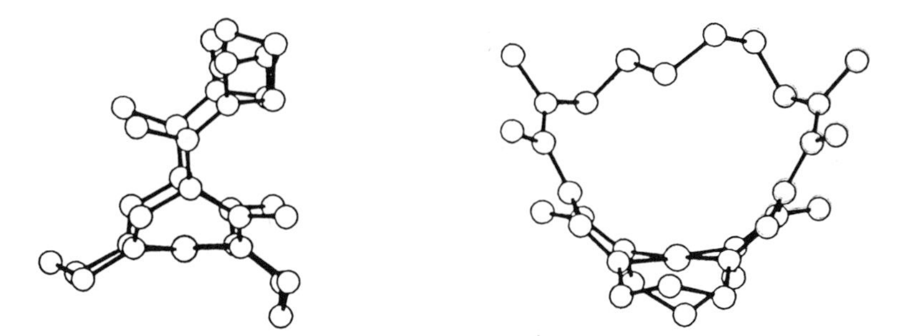

Figure 1. Stereo representations of $[Co(MeMeC_6cyclidene)]^{2+}$

Table 1. Potentials for the First Oxidation Process for the Lecunar Cyclidene Complexes of Nickel(II)[a]

R^1	R^2	R^3	$E_{1/2}$,V	E_p(ox),V	E_p(red),V	$E_{3/4-1/4}$,mV	ref.
m-xylylene	CH_3	H	0.94	1.00	0.87	80	b
m-xylylene	CH_3	C_6H_5	0.94	0.99	0.89	70	b
m-xylylene	CH_3	CH_3	0.78	0.83	0.74	70	c
m-xylylene	H	CH_3	0.93	0.97	0.89	67	c
m-xylylene	$CH_2C_6H_5$	CH_3	0.82	0.85	0.77	70	c
m-xylylene	$CH_2C_6H_5$	C_6H_5	0.99	1.05	0.94	80	b
$(CH_2)_4$	CH_3	C_6H_5	0.91	0.94	0.87	60	b
$(CH_2)_5$	CH_3	C_6H_5	0.91	0.94	0.86	70	b
$(CH_2)_6$	CH_3	H	0.97	1.00	0.93	60	b
$(CH_2)_6$	CH_3	C_6H_5	0.92	0.95	0.88	70	b
$(CH_2)_6$	H	n-C_7H_{15}	0.92	0.96	0.87	85	b
$(CH_2)_6$	H	C_6H_5	0.94	0.98	0.91	75	b
$(CH_2)_6$	H	t-C_4H_9	0.78	0.81	0.74	60	b
$(CH_2)_7$	CH_3	C_6H_5	0.90	0.94	0.87	70	b
$(CH_2)_8$	CH_3	C_6H_5	~0.90	broad	-	-	b
CH_3	H	$(CH_2)_6$	0.89	0.92	-	90	d
CH_3	H	$(CH_2)_7$	0.89	0.92	-	70	d
CH_3	H	$(CH_2)_8$	0.92	0.95	-	60	d
CH_3	CH_3	$(CH_2)_7$	0.68	0.72	-	-	d
CH_3	CH_3	$(CH_2)_6$	0.60	0.63	-	70	d
CH_3	CH_3	$(CH_2)_5$	0.67	0.70	-	70	d

[a] Conditions: in CH_3CN solution, 0.1 M $(n\text{-}Bu)_4NBF_4$ supporting electrolyte, $Ag^0/AgNO_3$ (0.1 M) reference electrode.

[b] B. Korybut-Daszkiewcz, M. Kojima, J. H. Cameron, N. Herron, M. Chavan, A. J. Jircitano, B. K. Coltrain, G. L. Neer, N. W. Alcock, and D. H. Busch, Inorg. Chem., 23, 903 (1984).

[c] D. H. Busch, S. C. Jackels, R. C. Callahan, J. J. Grzybowski, L. L. Zimmer, M. Kojima, D. J. Olszanski, W. P. Schammel, J. C. Stevens, K. A. Holter, and J. Mocak, Inorg. Chem., 20, 2834 (1981).

[d] J. H. Cameron, M. Kojima, B. Korybut-Daszkiewicz, B. K. Coltrain, T. J. Meade, N. W. Alcock, and D. H. Busch, Inorg. Chem., 26, 427 (1987).

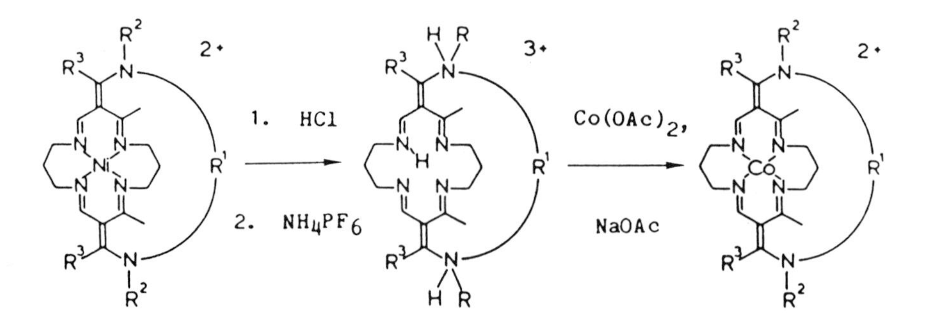

Scheme 3. Preparation of cobalt(II) cyclidene complexes.

axial patterns of tetragonal pyramidal, low spin d^7 (Fig. 2, Table 2).[50-52] While the crystal structure of the square planar cobalt(II) complex shows the presence of the vacant cavity, the ability of the cavity to accept small ligands has been proven by x-ray crystal structure determinations on a variety of lacunar cyclidene complexes, including two bis(thiocyanato)cyclidenecobalt(III) complexes,[44,46] a carbon monoxide adduct of iron(II)[45] and a dioxygen adduct of cobalt.[53] Figure 3a shows the presence of NCS in a pentamethylene bridged cavity as viewed by looking directly into the opening while Fig. 3b displays the bending enforced on the Co-N-C group by interaction between the small ligand and the atoms of the bridge. Thus the cavity both provides a sheltered chamber for the small ligand and interacts with that ligand in a manner that may modify the thermodynamics and kinetics of its binding.

The 1:1 dioxygen adducts of cobalt(II) are very simply identified by their highly characteristic ESR spectra.[22,24] These are usually recorded on frozen solutions at the boiling point of liquid nitrogen. Fig. 2b shows a typical spectrum for the cobalt(II) cyclidene complexes with $g_1 \cong g_2 \cong 2$ and $g_3 \cong 2.1$ and with hyperfine splitting due ^{59}Co appearing in all branches (Table 3). The x-ray crystal structure has

Table 2. Electron Spin Resonance Spectroscopic Parameters for the Lacunar Cobalt(II) Dioxygen Carrires[a]

Cyclidene Compound R^1	R^2	R^3	$g_\perp$	$g_{\parallel}$	$A_{\parallel}$ (G)	$A_\perp$ (G)
$(CH_2)_4$	CH_3	CH_3	2.292	2.003	99	15
$(CH_2)_5$	CH_3	CH_3	2.303	2.007	100	13
$(CH_2)_6$	CH_3	CH_3	2.303	2.003	99	13
$(CH_2)_7$	CH_3	CH_3	2.312	2.011	-	-
$(CH_2)_{12}$	CH_3	CH_3	2.310	2.014	100	13
$(CH_3)_2$	CH_3	CH_3	2.300	2.009	102	15
$(CH_2)_6$	H	$t\text{-}C_4H_9$	2.313	2.016	-	-
$(CH_3)_2$	CH_3	$t\text{-}C_4H_9$	2.290	2.009	97	13
$(CH_2)_6$	CH_3	H	2.313	2.016	-	-
$(CH_2)_7$	H	CH_3	2.309	2.010	-	-
$(CH_2)_6$	H	C_6H_5	2.319	2.018	-	-
$(CH_2)_6$	CH_3	C_6H_5	2.305	2.007	-	-
$(CH_3)_2$	CH_3	$(CH_2)_5$	2.311	2.015	-	-
$(CH_3)_2$	CH_3	$(CH_2)_6$	2.316	2.014	-	-
$(CH_3)_2$	CH_3	$(CH_2)_7$	2.319	2.009	104	15
$(CH_3)_2$	CH_3	$(CH_2)_8$	2.286	2.007	100	13
$(CH_3)_2$	H	$(CH_2)_7$	2.286	2.006	102	13
m-xylyl	H	C_6H_5	2.312	2.015	-	-
m-xylyl	CH_3	C_6H_5	2.306	2.009	100	15

[a]Spectra recorded on frozen CH_3CN glass at -196°C; 1-methylimidazole axial ligand; estimated standard deviations: $g_{\parallel}$, $g_\perp$ ±0.002; $A_{\parallel}$, $A_\perp$ ±0.5.

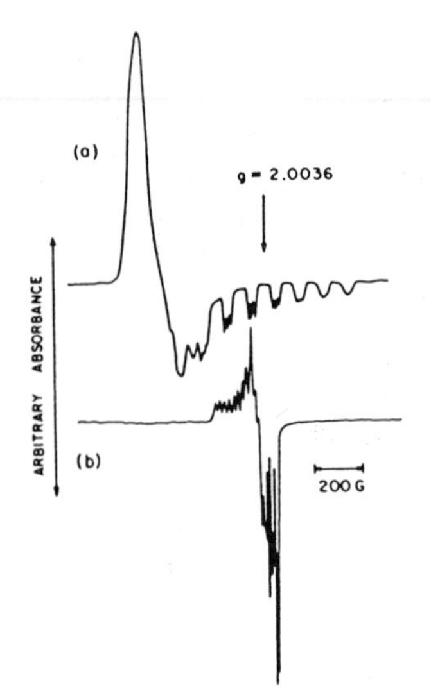

Figure 2. ESR spectra of a cobalt(II) cyclidene complex in frozen water containing excess N-methylimidazole (2.5M) at -196°C: a) [Co(MeMeC_6cyclidene)] b) The dioxygen adduct of a). (Reproduced with permission).

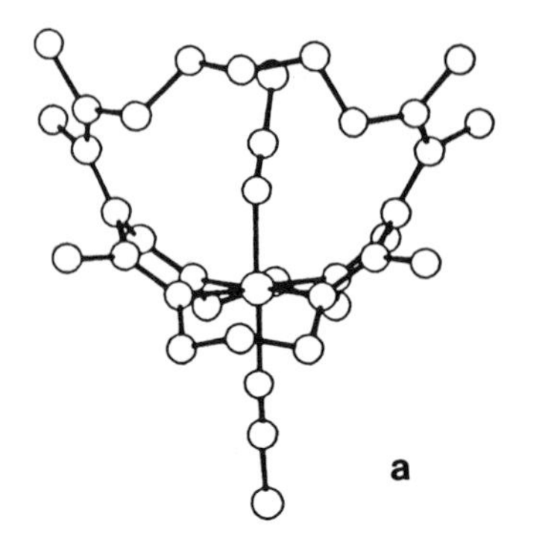

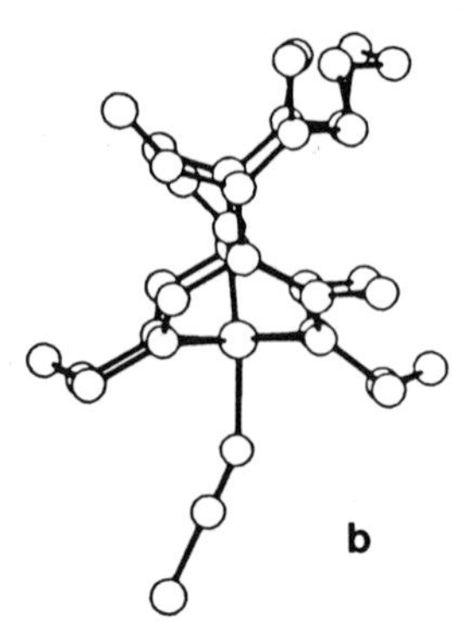

Figure 3. Stereo representations of $[Co(MeMeC_5cyclidene)(NCS)_2]^+$

been determined for the dioxygen adduct of $[Co\{Me,Me,(CH_2)_6cyclidene\}N\text{-}MeIm](PF_6)_2$. Figure 4 shows details of the structure.[53]. The O-O distance, 1.32(2)Å, is typical of superoxide, as is usually the case for such dioxygen adducts.[54] The bond angle of 121(1)° is also unexceptional.[54] The drawings show the disordering of the N-methyl imidazole axial ligands.

The novel structure of the lacunar cyclidene ligands has produced dioxygen carriers of unusually favorable capabilities.[50-53] They invariable show evidence only for the formation of 1:1 complexes with no indication of the complications that would be associated with an accompany 2:1 equilibrium. This is a direct consequence of the ligand design. Because of the character of the lacuna, it successfully selects between the intra-lacunar and extra-lacunar axial sites on the basis of ligand size. The usual axial ligands, such aromatic bases as pyridine and substituted imidazoles, are too large to enter the lacunae. Consequently, the propensity of the cobalt(II) to accept a fifth ligand is satisfied by axial ligation at the extra-lacunar site.[e] This

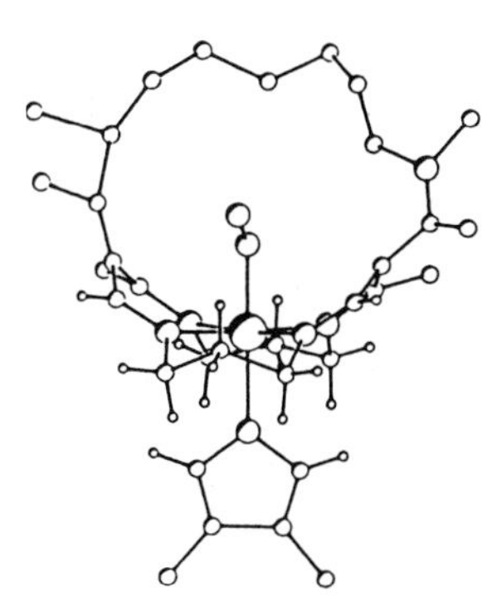
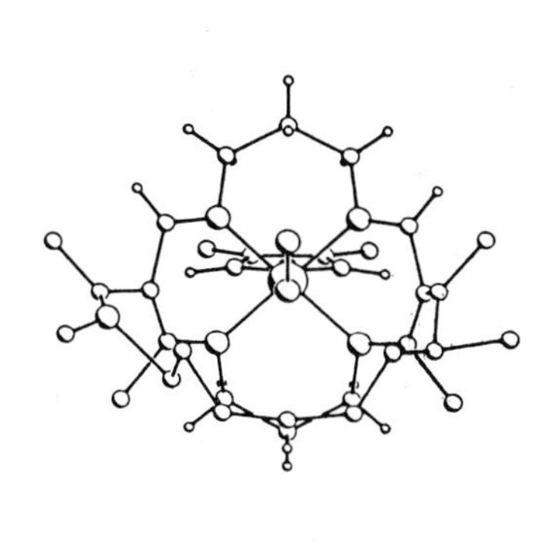
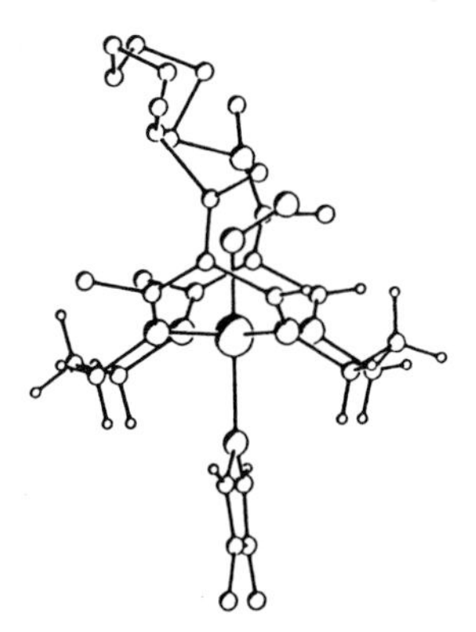

Figure 4. Stereo representations of the dioxygen adduct of $[Co(MeMeC_6cyclidene)N\text{-}MeIm]^{2+}$

Table 3. Electron Spin Resonance Spectroscopic Parameters for the Dioxygen Adducts of the Lacunar Cobalt(II) Complexes[a]

Cyclidene Compound R^1	R^2	R^3	$g_\perp$	$g_{\|\|}$	$A_\perp$ (gauss)	$A_{\|\|}$ (gauss)
$(CH_2)_4$	CH_3	CH_3[a]	1.999	2.098	15	20
$(CH_2)_5$	CH_3	CH_3[a]	1.999	2.091	15	20
$(CH_2)_6$	CH_3	CH_3[a]	1.999	2.088	14	20
$(CH_2)_7$	CH_3	CH_3	2.019	2.091	11	26
$(CH_2)_{12}$	CH_3	CH_3	2.016	20.91	10	20
$(CH_3)_2$	CH_3	CH_3	2.018	2.076	10	18
$(CH_2)_6$	H	$t\text{-}C_4H_9$	2.020	2.094	-	-
$(CH_3)_2$	CH_3	$t\text{-}C_4H_9$	2.016	2.093	9	20
$(CH_2)_6$	CH_3	H	2.008	2.081	10	18
$(CH_2)_6$	H	C_6H_5	2.016	2.086	9	16
$(CH_2)_6$	CH_3	C_6H_5	2.008	2.088	10	17
$(CH_3)_2$	CH_3	$(CH_2)_7$	2.014	2.081	8	18
$(CH_3)_2$	CH_3	$(CH_2)_8$	2.010	2.086	10	18
m-xylyl	CH_3	C_6H_5	2.012	2.088	8	16
Schiff Base Compound						
$[Co\{Me_2(anisoyl)Me_2malMeDPT\}]\cdot O_2$			2.100	2.004	20.5	15.0
$[Co\{Me_2(\text{"}C_6\text{"})Me_2malMeDPT\}]\cdot O_2$			2.096	1.997	20.5	15.0
$[Co\{Me_2\text{"p-xylylene"}]Me_2malMeDPT\}]\cdot O_2$			2.092	1.999	20.7	15.4

[a]Spectra recorded on frozen glass at -196°C; 1-methylimidazole axial ligand; estimated standard deviations: $g_{||}$, $g_\perp$ ±0.002; $A_{||}$, $A_\perp$ ±0.5

restricts O_2 binding to the interior position which, in turn, prevents a single O_2 moiety from simultaneously binding to two cobalt(II) atoms. In the case of such small axial ligands as the solvent acetonitrile and thiocyanate, doubt remains as to which site (extra-lacunar or intra-lacunar) dominates as the metal ion achieves its axial ligation. It follows that one might anticipate considerably more sensitivity to peroxo-bridged dimer formation in the absence of large axial base molecules. Axial ligand binding studies have shown that the equilibrium can be saturated (>99%) in ~1.5M N-methylimidazole in acetonitrile solution and ~2.5M N-MIm in aqueous solution.[50]

The dioxygen affinities of many of the lacunar cobalt(II) cyclidene complexes have been determined and selected data are summarized in Table 4.[50-53] It should be noted that a large number of these measurements have been made in the vicinity of room temperature and even above. This attests to both unusually large dioxygen affinities for these 1:1 adducts and to the dilatory nature of their autoxidation processes. The magnitudes of the larger of the dioxygen affinities can be realized by comparing them with values for other dioxygen carriers. In aqueous solution, the dioxygen affinity of $[Co\{Me,Me,(CH_2)_6cyclidene\}N\text{-}MIm]^{2+}$ is essentially the same as that of iron-containing myoglobin, despite the fact that the values for iron-O_2 adducts usually exceed those of the corresponding cobalt-O_2 adducts by a factor approaching 100.[27,55] Thus, the cyclidene ligand confers on cobalt(II) an exceptionally great affinity for the O_2 molecule.

Equally striking[50] is the fact that the O_2 affinity depends very greatly on the detailed nature of the bridging group R^1. Consider the variation of K_{O_2} with length of the polymethylene chain for $[Co\{Me,Me,(CH_2)_ncyclidene\}N\text{-}MIm]^{2+}$ (Table 4). At a given temperature, the values are about constant for n = 7 and above; however, K_{O_2} decreases regularly as the chain becomes shorter, until at $(CH_2)_4$ the number is so small it could be measured with the available technique only at relatively low temperatures. There is no evidence that a small ligand can enter the cavity for the still more restricted trimethylene bridged case. The xylylene bridges also produce very low dioxygen affinities and, in fact, seem to approximate those of the tetramethylene case. Both the magnitude and the origin of this effect are surprising. The overall change in dioxygen affinity that can be applied by this structural variation easily exceeds 5 orders of magnitude in the equilibrium constant and may be much larger since it is difficult to place a numerical value on the vanishingly small O_2 affinity of the limiting trimethylene bridged case. A principal contribution to this effect almost certainly must come from steric interactions relating to the diminishing of the cavity size to the point where, eventually, no small ligand can enter. Studies on capped and strapped porphyrins have also revealed large decreases in ligand affinity because of the restraint of crowded binding sites.[56-58] The contrast in the sizes of the cavities that are produced by different bridges is dramatized by Fig. 5 which represents structures that will be reported elsewhere.

Substituent effects are also reasonably clear (Table 4). Replacing the remaining protons on the bridge nitrogen atoms with methyl groups increases the O_2 affinity by a small factor (2 - 4) while replacing the methyl group on the vinyl carbon atom by a phenyl group decreases the dioxygen affinity by a factor of about 5 to 10. The effects of the substituents and of the bridging group on the equilibrium constant for O_2 binding appear to be approximately independent of each other. The overall result is a rather great ability to control the dioxygen

Table 4. Equlibrium Constants for Dioxygen Binding by Lacunar Cobalt(II) Complexes

Cyclidene Complex[a]				
R^1	R^2	R^3	T(°C)	K_{O_2} ($torr^{-1}$)
$(CH_2)_4$	CH_3	CH_3	-40	.0020
$(CH_2)_5$	CH_3	CH_3	20	.0094
$(CH_2)_6$	CH_3	CH_3	20	.155
$(CH_2)_7$	CH_3	CH_3	19.4	.62
$(CH_2)_8$	CH_3	CH_3	20	.65
$(CH_2)_6$	CH_3	H	0	.010
$(CH_2)_6$	H	CH_3	0	.138
$(CH_2)_6$	H	C_6H_5	1.0	.41
$(CH_2)_6$	CH_3	C_6H_5	0	.085
$(CH_2)_6$	H	C_4H_9	-0.4	~200
H	CH_3	$(CH_2)_7$	0	.0020
CH_3	CH_3	$(CH_2)_8$	-37.6	.020
$(CH_2)_6$	CH_3	CH_3	-10	4.6
$(CH_2)_6$	CH_3	CH_3	1.0	1.3
$(CH_2)_6$	CH_3	CH_3	2.1	.98
$(CH_2)_6$	CH_3	CH_3	15	.25

Schiff Base Complex	T(°C)	K_{O_2} ($torr^{-1}$)
[Co{Me_2(anisoyl)Me_2malMeDPT}][b]	25	0.15 ± 0.02
[Co{Me_2("C_6")Me_2malMeDPT[b]	25	0.15 ± 0.02
[Co{Me_2("p-xylylene")Me_2malMeDPT}][b]	-20	$5.9 \pm 0.3 \times 10^{-2}$
	-10	$2.71 \pm 0.04 \times 10^{-2}$
	0	$9.3 \pm 0.2 \times 10^{-3}$
	5	$6.7 \pm 0.3 \times 10^{-3}$
	10	$4.7 \pm 0.3 \times 10^{-3}$
	25	$1.1 \pm 0.1 \times 10^{-3}$
[Co{Me_2(anisoyl)Me_2malen}][c]	-15	2.9 ± 0.4
[Co{Me_2("C_8")Me_2malen}][d]	21	0.37 ± 0.05

[a]In 1.5M N-methylimidozole in CH_3CN. [b]In toluene. [c]In 2% pyridine in toluene. [d]In 2% 4-(t-butyl)pyridine in toluene.

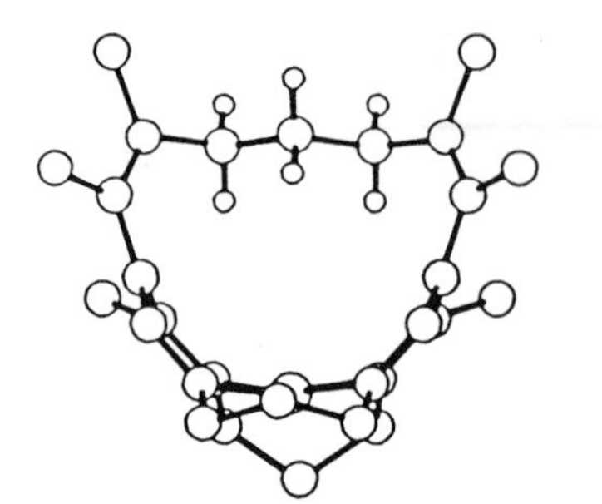

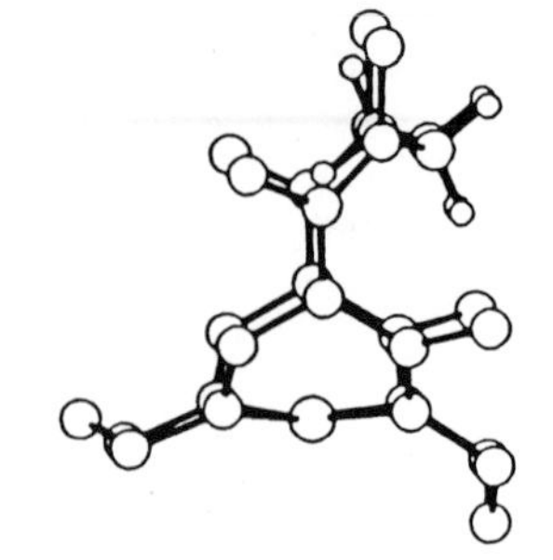

a

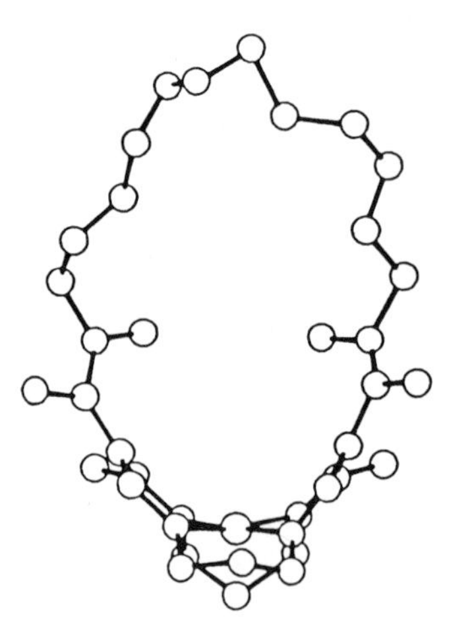

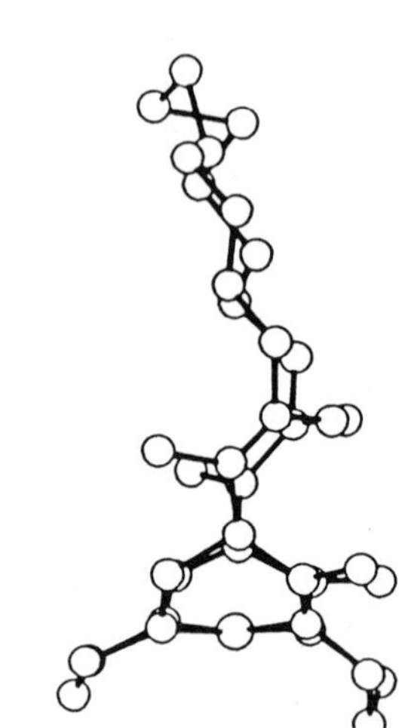

b

Figure 5. Stereo representations of cyclidene complexes:
a) $[Cu(MeMeC_3cyclidene)]^{2+}$
b) $[Ni(MeMeC_{12}cyclidene)]^{2+}$

affinity by varying these structural parameters. In effect, it is possible to produce a structure that will exhibit any desired value of K_{O_2}, within a wide span of possible values.

The values of K_{O_2} also show a strong dependence of the binding constant on solvent and on axial base.[42,50] Values measured in water exceed those from acetonitrile by a factor of from 5 to 10, as expected on the basis of the increased polarity of the solvent.[59,60] The axial ligand, N-methylimidazole, produces dioxygen affinities some 3-10 greater than those due to pyridine while those obtained with acetonitrile as the axial ligand are much smaller. Among the more noteable properties of the cobalt(II) cyclidene complexes is their ability to function smoothly as 1:1 dioxygen carriers in aqueous solvent.[50]

The thermal parameters for O_2 binding, summarized in Table 5, display entropy changes of 64±2, for 14 cases, which is within experimental error of the value for gaseous molecular dioxygen (62eu, standard state of 1 torr[61]). This is consistent with the cessation of rotations as well as translation upon binding and with the shielding of the bound dioxygen from strong interaction with solvent. The values of ΔS of O_2 binding observed by other investigators for non-lacunar Schiff base cobalt(II) adducts are substantially larger, possibly reflecting stronger solvational effects.[22,23]

Table 5. Thermodynamic Parameters for Dioxygen Adducts Formation by Lacunar Cobalt(II) Complexes

Cyclidene Compound R^3	R^2	R^1	ΔH (kcal/mole)	ΔS (eu/mole)
H	Me	$(CH_2)_6$[a]	-14.8(8)	-63.3
Me	H	$(CH_2)_6$[a]	-16.6(3)	-64(1)
Me	Me	$(CH_2)_5$[a]	-16.2(6)	-65(2)
Me	Me	$(CH_2)_6$[a]	-17.2(4)	-62(1)
Me	Me	$(CH_2)_7$[b]	-18.6(9)	-65(3)
Me	Me	$(CH_2)_8$[b]	-17.3(3)	-60(1)
C_6H_5	H	$(CH_2)_6$[b]	-17.2(6)	-65(2)
C_6H_5	Me	$(CH_2)_6$[b]	-17.5(4)	-64(2)
$(CH_2)_7$	H	CH_3[a]	-13.8(6)	-62(2)
Schiff Base Compounds				
[Co{Me_2("C_6")Me_2malMeDPT}]			-16(1)	-58(2)
[Co{Me_2("p-xylylene")Me_2malMeDPT}]			-13(1)	-57(3)

[a]J. C. Stevens, thesis, The Ohio State University, 1979.

[b]M. Kojima, unpublished results.

[c]J. H. Cameron, M. Kojima, B. Korybut-Daszkiewicz, B. K. Coltrain, T. J. Meade, N. W. Alcock, and D. H. Busch, Inorg. Chem., 26, 427 (1987).

A family of closely related dioxygen carriers formed by attaching the bridging group to the vinyl carbons rather than the nitrogen atoms has been labeled the retro-bridged cyclidenes (structure XVI).[62] While generally showing similar dependences and parameters, the dioxygen affinities of these complexes are lower than those of corresponding members of the previously discussed bridged cyclidene complexes. For example (Table 4), the octamethylene retro-bridged species has as many atoms in its bridge as the hexamethylene bridged cyclidene complex and the value of K_{O_2} for the former in 1.5M Im/CH_3CN at -25°C is 0.0037 $torr^{-1}$ while that for the latter in the same medium at the somewhat higher temperature of -10°C is orders of magnitude greater at 4.6 $torr^{-1}$. A complicated steric effect has been observed with the retro-bridged species which causes the shorter chain derivative with a $NHCH_3$

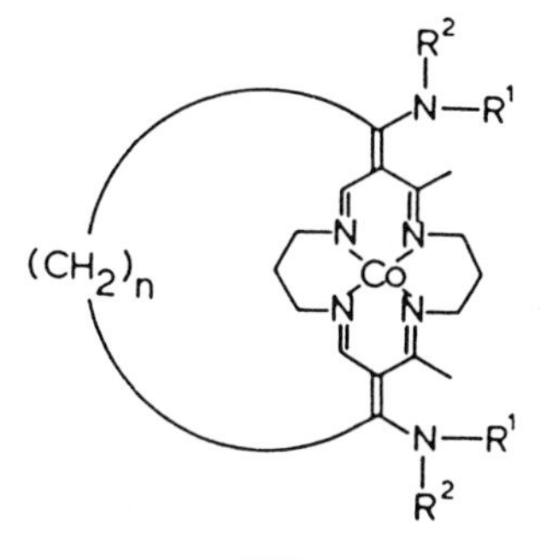

XVI

group to have a much greater dioxygen affinity than the $N(CH_3)_2$ derivative having a longer chain. In the case of the usual cyclidene complexes, both replacement of the NH by NCH_3 and the increase in bridge length would contribute to increases in dioxygen affinity. For the retro-bridged complexes, the second amino methyl group displaces the bridging group from the position in which it can best accommodate the O_2 molecule.

The mode of autoxidation of these complexes is not well understood.[63] At 640 torr of O_2 and 30°C in 1.5M N-MIm/CH_3CN, the half-lives of the different lacunar cyclidene complexes of cobalt(II) vary from a few hours to a few days. There is evidence that the ligand may be susceptible to destructive autoxidation since pure cobalt(III) compounds could be isolated only when certain axial ligands were added, such as thiocyanate or cyanate. This suggests that it may be necessary to accommodate a ligand in the lacuna when the metal ion is oxidized in order to stabilize its higher state and prevent an accompanying or subsequent ligand oxidation. Electrochemical studies support this point of view.[64]

As the polymethylene chain forming the bridge in lacunar cyclidene cobalt(II) complexes increases from trimethylene, regularly to octamethylene, and then jumps first to dodecamethylene and then, effectively, to an infinite length (open structure), the potential of the first electrochemical oxidation process shifts regularly to more negative potentials, spanning a range of approximately 500 mV (Table 6).[64] In contrast, the same change in structure for the corresponding nickel(II) complexes is accompanied by a change in potential of only 50 mV. Studies on the nature of the oxidation products ($[NiL]^{3+}$ and $[CoL]^{3+}$) revealed that (1) small lacunae inhibit the formation of 6-coordinate oxidized complexes and favor the formation of species in which the ligand has been oxidized, $[Ni^{II}(L^{\ddagger})]^{3+}$ and $[Co^{II}(L^{\ddagger})]^{3+}$, and large lacunae permit the binding of two axial ligands, which stabilizes complexes containing the trivalent metal ions, $[Ni^{III}(L)An_2]^{3+}$ and

Table 6. Variation of Potential with Chain Length for the Lacunar Complexes of Cobalt(II) and Nickel(II)[a]

R^1	M	$E_{1/2}$,V	M	$E_{1/2}$,V
$(CH_3)_2$	Co	-0.15	Ni	0.74
$(CH_2)_{12}$		--		0.74
$(CH_2)_8$		0.05		0.77
$(CH_2)_7$		0.08		0.78
$(CH_2)_6$		0.20		0.76
$(CH_2)_4$		0.34		0.79
$(CH_2)_3$		0.34		0.79

[a] $R^3 = R^2 = CH_3$ and R^1 is varied. Data from M. Y. Chavan, T. J. Meade, D. H. Busch and T. Kuwana, Inorg. Chem., 25, 314 (1986). Determined from rotating platinum-disk electrode (vs. $Ag/AgNO_3$, 0.1 M) results in acetonitrile with 0.1 M tetra-n-butylammonium fluoroborate as supporting electrolyte.

$[Co^{III}(L)An_2]^{3+}$ (An is acetonitrile). (2) Strong axial ligands favor the trivalent derivative; i.e., in the case of the hexamethylene bridge, $[Co^{III}(L)AnAn']^{3+}$ exists in acetonitrile but $[Co^{II}(L^{\dot{+}})]^{3+}$ exists in acetone. (3) Higher temperatures favor ligand oxidized species, possibly by labilizing axial ligands. The difference in the potentials for the first oxidation processes of the tetramethylene bridged complex and for the unbridged cobalt cyclidene complex is clearly due to the stabilization of the cobalt(III) state in the latter case.

It is a fascinating and somewhat ironic fact that the lacuna is at once a principal advantage of these dioxygen carriers and also the source of a major limitation. When the cavity is fairly restricted but large enough to permit O_2 binding, autoxidation is favored by a pathway involving oxidation of the cyclidene ligand. In contrast, when the cavity is removed, or so large as to be effectively removed, the cyclidene ligand may be safe from oxidation but this is only because the metal ion is stabilized in its oxidized state. In this case, the O_2 binding capability is impaired just as it was in the first case; however, the impairment may not be permanent since the metal ion might be reduced again and the ability to bind O_2 restored. Clearly improved design would involve reducing the susceptibility of the cyclidene ligand toward oxidation.

In summary, the cobalt(II) cyclidene dioxygen carriers exhibit exceptional dioxygen affinities, having values approaching those of some of the stronger natural iron(II) dioxygen binders. The complexes enjoy exceptional stability toward the usual pathways of autoxidation, especially that involving the formation of peroxo bridged dinuclear species. The structure is particularly well suited to the control of such performance parameters as dioxygen affinity since extensive variations in structural components are easily achieved. Further, dioxygen affinity can be controlled over many orders of magnitude by varying the length and/or nature of the bridge and the substituents on the parent ligand. The complexes have been a useful resource for the study of the fundamental relationships that control dioxygen affinity and autoxidation of dioxygen carriers. Despite the vastly improved behavior of the cyclidene O_2 carriers, they still suffer from the principal limiting factor of all dioxygen carriers -- autoxidation.

THE NEW LACUNAR COBALT(II) SCHIFF BASE COMPLEXES

Especially efficacious new cobalt(II) dioxygen carriers have been designed and prepared by incorporating several molecular design features into a familar class of Schiff base complexes. The design profits from the relationships disclosed during the extensive studies on the lacunar cyclidene complexes but it profits from the incorporation of additional features that favor reversible dioxygen binding. The dioxygen complexes of the Schiff base 4,4'-ethanediyldinitrilo)bis(2-pentanonato)cobalt(II) (see structure XVII), which is commonly called ethylenediaminebis-(acetylacetonato)cobalt(II) and abbreviated [Co(acacen)],[c] and the family of O_2 carriers it represents were the second major group of

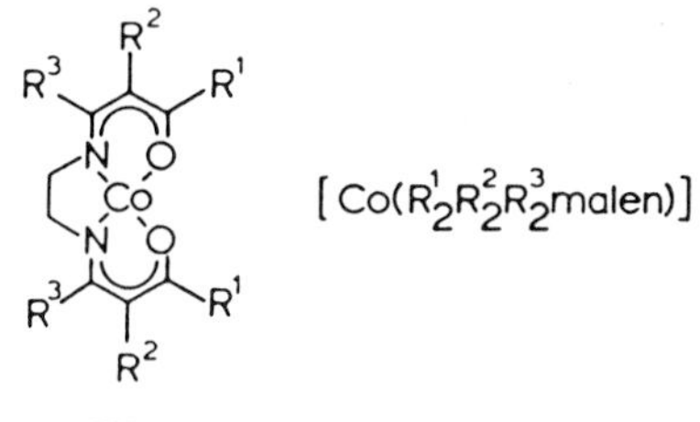

XVII

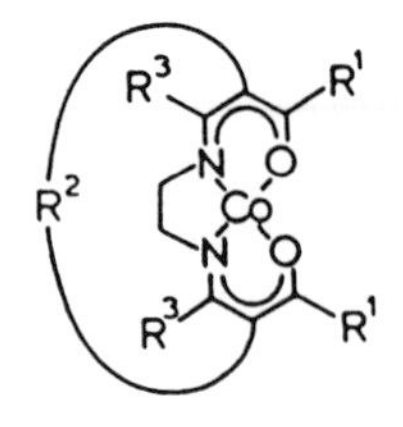

XVIII

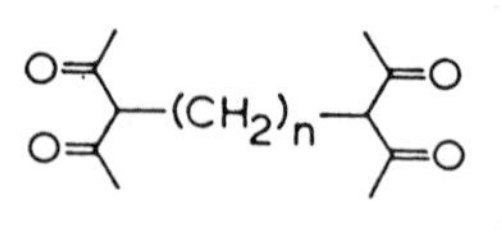

XIX

cobalt(II) species to display this property. The first was bis(salicylidene)ethylenediimine (salen) whose cobalt(II) complex was identified as a dioxygen carrier in 1938.[65] Early studies with the substituted malen complexes (i.e., acacen and its congenors)[c] covered a wide range of structural variations and helped reveal many of the relationships between structure and dioxygen affinity, including the correlation with electrode potentials and with pK values.[59,60] Reversible binding was generally limited to relatively low temperatures with autoxidation occurring as the temperature was raised. Further, recent studies suggest that the peroxo-bridged dinuclear complex forms in addition to the expected 1:1 complex.[66,67]

Our first experiments in this area simply involved the appending of a bridge to the familar complex [Co(acacen)] (structure XVIII). The preparation of this ligand made use of an interesting polymethylenebis (β-diketone) that was first developed for the study of polymeric coordination compounds (structure XIX).[68] Unfortunately this new lacunar Schiff base complex proved to be very sensitive to autoxidation even at $-10^{o}C$. The rapid oxidation was attributed to donation of electron density by the alkyl substituents. This was confirmed by synthesis and study of the non-lacunar bis(n-butyl) derivative. The cobalt(II) complex of that species autoxidized still more rapidly. Clearly the presence of lacunae alone is no panacea to the failings of cobalt(II) dioxygen carriers.

The report by Kida et al.[69] that the presence of acetyl groups in the R^2 position leads to dioxygen complexes that are stable in solution at room temperature offered great promise. That study also revealed

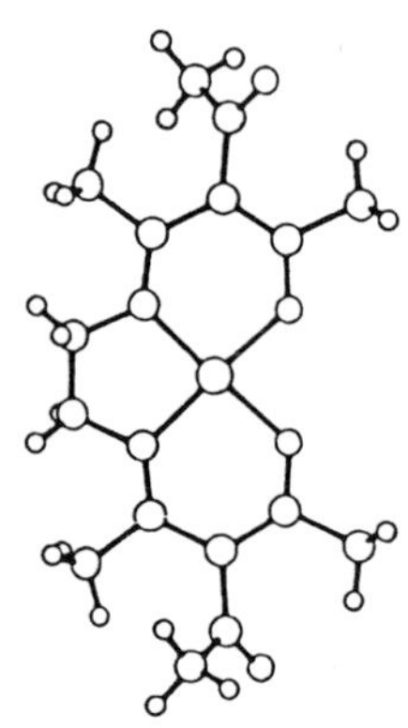
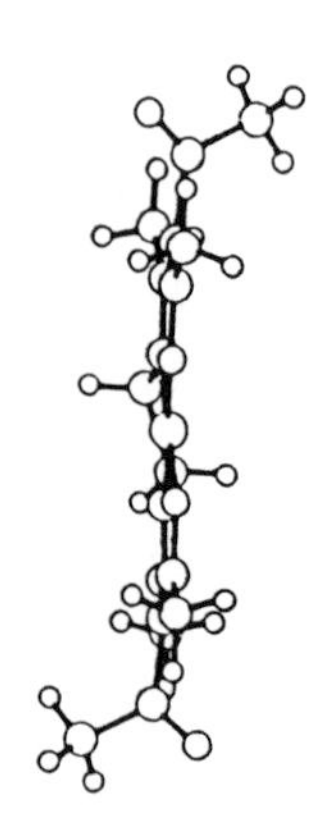
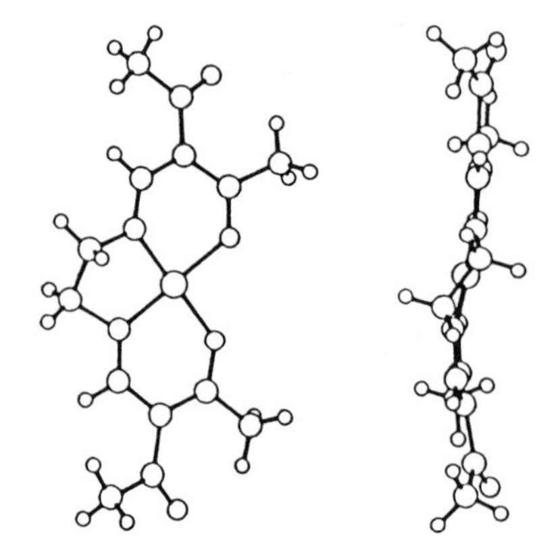

Figure 6. Stereo representation of Schiff base complexes:
a) $[Ni(Me_2Ac_2Me_2malen)]$
b) $[Ni(Me_2Ac_2H_2malen)]$

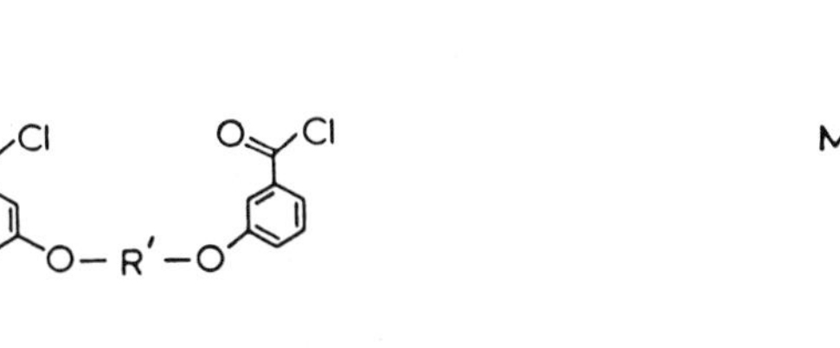

XX XXI

that the 2:1 equilibrium accompanies the desired 1:1 process suggesting that the lacunization of such species might produce good 1:1 dioxygen carriers. Our subsequent studies[32] showed a useful relationship between substituents at the R^1 and R^3 positions and an acyl group at the R^2 position (Fig. 6). When methyl groups are present at both positions, the acyl group is forced to orient itself essentially perpendicular to the coordination plane. In contrast, if a methyl group is replaced by a hydrogen atom, the acyl group lies almost coplanar with the coordination plane. This provides guidance with respect to the subsequent design of lacunar species. In addition, Goldsby's electrochemical studies[32] showed that the bis(acetylated) complexes autoxidize at potentials substantially more positive than those for the species that are unsubstituted at the R^2 position, but still more important, in the

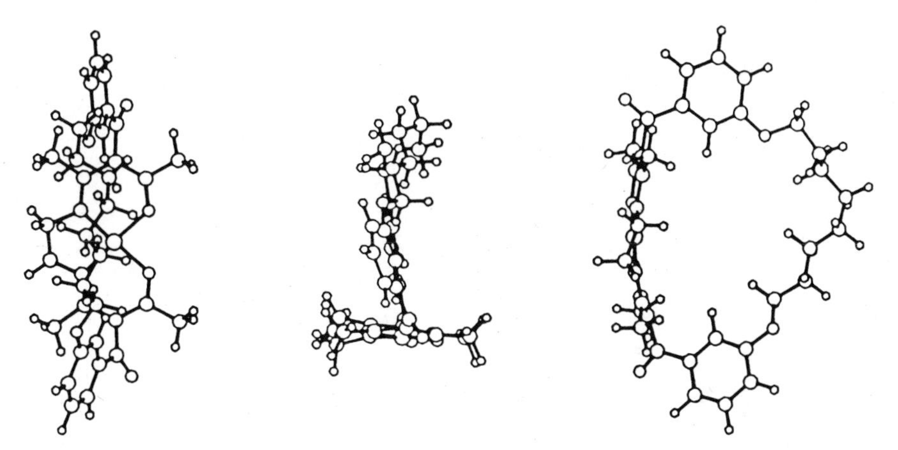

Figure 7. Stereo representations of [Ni(Me$_2$"C$_8$"Me$_2$malen)]

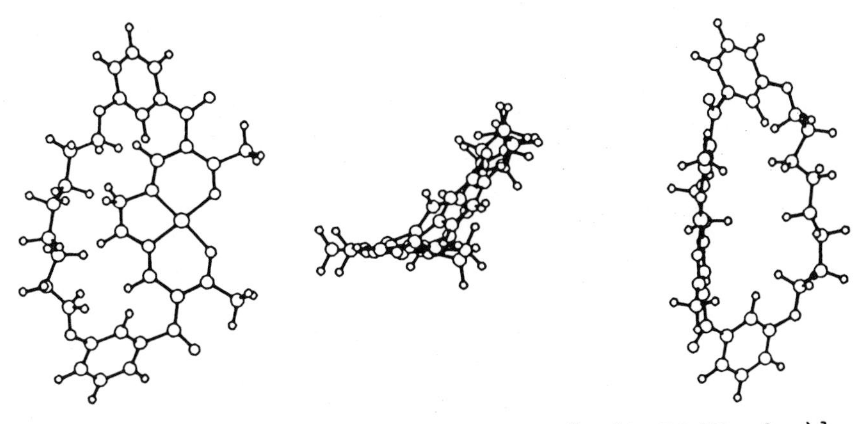

Figure 8. Stereo representations of [Ni(Me$_2$"C$_8$"H$_2$malen)]

former cases the oxidations are reversible while those for the latter are completely irreversible. This is consistent with the expectation that the limitations of R^2 unsubstituted and alkyl substituted cobalt malens are due to ligand oxidation.

Initial attempts to produce lacunization with long chain polymethylenebis(acylchloride) reagents were frustrated by competing reactions leading to ketenes and their dimers. This inspired the design[70] of the bridging group shown in structure XX. Replacement of the alkyl group with the benzene ring removed the competing elimination reaction and enhanced both the reactivity of the reagent (XX) and the oxidation stability of the ligand. The ether oxygen at the meta position provides a desirable flexibility to the bridge while minimizing electron donation to the benzene ring. Various groups can be placed at the R' position, including polymethylene and xylylene.[70,71]

The results of x-ray crystal structure determinations[70] on two of the new lacunar tetradentate complexes of this class are shown in Figs. 7 and 8. The species in Fig. 7 shows the upright orientation of the bridge expected[32] for a complex having bulky groups at the R^1 and R^2 positions, while that shown in Fig. 8 indicates how the bridge can be displaced from the vertical in the absence of the steric restraint. The ESR spectra of these complexes are typical of low spin tetragonal cobalt(II), showing both hyperfine splittings due to ^{59}Co and superhyperfine couplings from the single axial nitrogen bases.[71] Further, exposure of these complexes to dioxygen in the presence of axial ligands produces ESR spectra typical of the 1:1 O_2 adducts. However, at O_2 partial pressures <0.2 torr the intensities of the ESR spectra are much diminished as would be expected if the dinuclear peroxo complex were formed. Similar observations were made with both lacunar complexes and with related nonlacunar complexes having two anisoyl substituents at the R^2 positions. As would be expected, this complicates the calculation of equilibrium constants for dioxygen binding. It should be pointed out that these new lacunar complexes exhibited greater resistance to autoxidation at temperatures as high as room temperature than the bis(anisoyl) derivative.

The complexities associated with saturating the axial ligand equilibrium, the possibility of coordinating the axial ligand on the alternate lacunar side of the coordination plane, and the possible formation of the peroxo-bridged dinuclear dioxygen complex are all intertwined. Deconvolution accompanies elimination of the need for a separate axial ligand by incorporation of a fifth donor atom into the structure of the principal ligand in the dioxygen carrier.[71] In early studies, Cummings et al, showed that this could be done with the parent complexes (structure XXI) and the resulting cobalt(II) complexes include some of the best O_2 carriers known prior to the development of lacunar complexes,[72] as revealed, for example, by interests in their exploitation.

Incorporation of a fifth donor in the evolving design of lacunar Schiff bases of the malen class has produced optimal cobalt(II) dioxygen carriers.[71] The complex shown in structures XXII (flat) and XXIII (projection) was prepared as shown in Scheme 4, using the reactions developed in conjunction with the synthesis of the corresponding tetradentate complexes. However, it was necessary to use the copper(II) complex in this synthesis in order to obtain good yields. The complex has been characterized by elemental analyses, infrared and mass spectroscopy and, as this is written, single crystals have just been mounted for x-ray study. The free ligand has been identified by NMR spectroscopy as well.

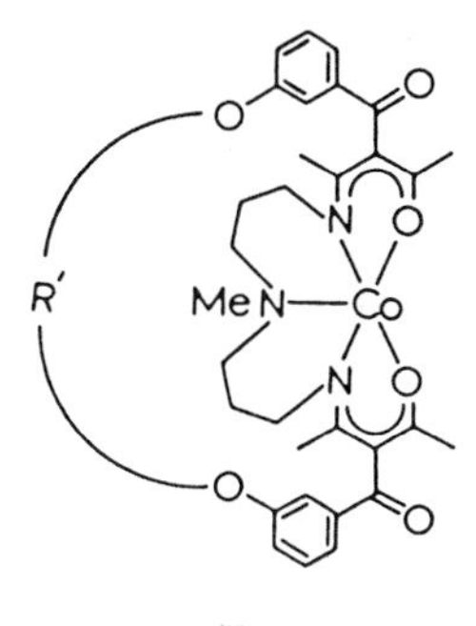

XXII

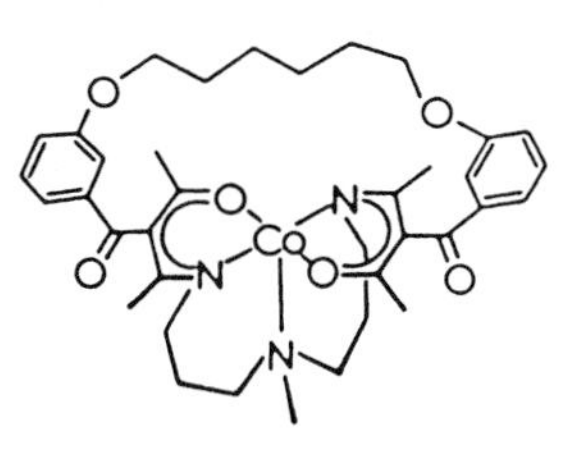

XXIII

Stereochemical considerations led to the choice of the hexamethylene group as R', the link in the bridge, and a more restrictive para-xylylene bridge has also been used to provide examples having different cavity sizes for comparative purposes. The 1:1 dioxygen adducts were identified by ESR spectroscopy and the equilibrium constants were measured by the usual electronic spectroscopy.[50,51] The 1:1 complexes are clearly the only dioxygen adducts formed by these species and they are stable enough to study even at temperatures as high as 50°C.[71] The ESR hyperfine splitting constants suggest that the dioxygen affinity of the hexamethylene derivative is large and this is confirmed by the values of K_{O_2} given in Table 4. Remarkably, the binding constants are still quite large even at room temperature. From the thermal parameters, (Table 5) it is evident that a favorable entropy change upon O_2 adduct formation is responsible for the high dioxygen affinity. This may arise from decreased solvation of the oxygen complex in the lacunar case. Also,

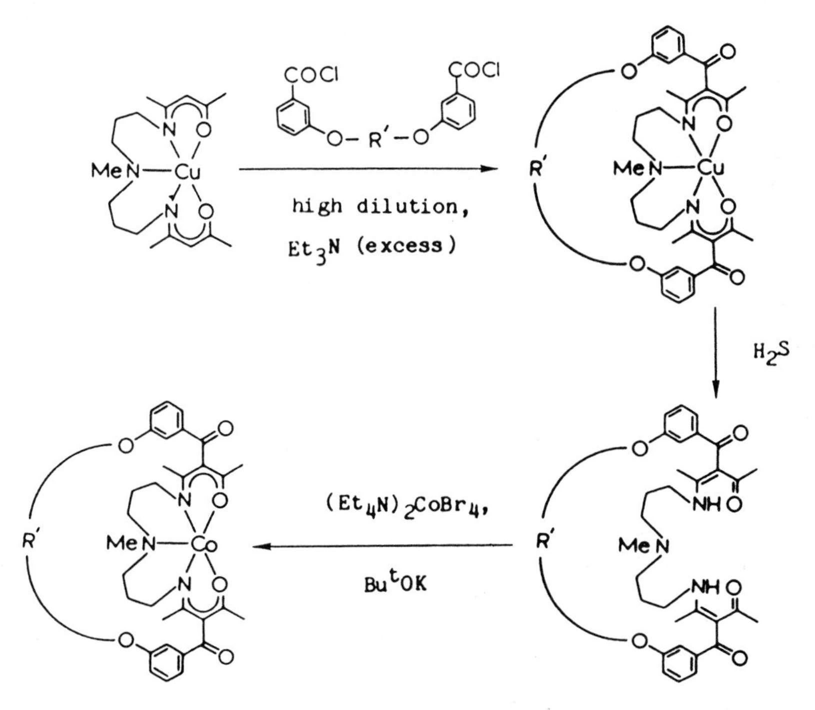

Scheme 4. Preparation of bridged pentadentate cobalt(II) complexes.

complete reversibility is found when the O_2 is removed from the adduct, even at 25°C. In contrast, the corresponding bis(anisoyl) derivative shows both incomplete reversibility and evidence for peroxo bridged 2:1 complex formation.

Replacing the hexamethylene group by the p-xylylene bridge leads to a decrease in dioxygen affinity by 2 to 3 orders of magnitude. This establishes the opportunity for control of dioxygen affinity in this optimal family of dioxygen carriers. Both the p-xylylene and hexamethylene derivatives show slow rates of autoxidation at high temperatures and O_2 pressures.

The conclusion is offered that the application of the various lessons learned in the study of lacunar cobalt(II) dioxygen carriers has led to the design of an optimal family of dioxygen carriers. The complexes of the general structure exemplified by XXII and XXIII enjoy the advantages of having all of the donor groups already present in the molecule while also having O_2 coordination site protected by its presence within a special molecular chamber whose volume and extent of wall area can be controlled. This facilitates the control of dioxygen affinity and, possibly, other performance parameters for the O_2 adduct. The species is further favored by incorporation of the electron withdrawing acyl substituents that provide some electronic inhibition of ligand oxidation. While autoxidation does occur it is quite slow at room temperature and is presently under study at 50°C using ordinary spectrophotometric techniques.

The successful application of the principles learned during study of lacunar cyclidene dioxygen carriers to the design of optimal Schiff base carriers of the lacunar malX Schiff base class confirms their generality. Further, the enhanced good properties of the new class of O_2 carriers suggests that still more autoxidation resistant cobalt(II) dioxygen carriers should be possible. The nature of the autoxidation process acting on the new Schiff base compounds must be learned. Other families of ligands are under investigation to continue the optimization of cobalt(II) dioxygen carriers. It is expected that the bugaboo of autoxidation will eventually be overcome as the destructive mechanisms become well understood and then inhibited by appropriate molecular designs.

ACKNOWLEDGEMENT

The research summarized here is the work of my fine collaborators, including graduate and undergraduate students, visiting scholars from about the world and senior colleagues who have chosen to share in this endeavor. These efforts have been made possible by the following sponsors: The National Institutes of Health, National Science Foundation, and NATO. We are truly grateful.

FOOTNOTES

a. Precedent exists for unexpected mechanisms in O_2 carrier autoxidation.[73,75] In the parallel area of autoxidation of iron(II) dioxygen carriers, a peroxo-bridged pathway has long been known and is strongly documented. This mechanism cannot be involved in autoxidation of such natural species as hemoglobin or myoglobin because the globular protein isolates each metal ion. Whereas it has commonly been assumed that the natural carriers autoxidize by dissociation of O_2^- from the O_2 adducts, extensive studies in these laboratories provide compelling support for electron transfer between O_2 and Fe^{II} without adduct formation as the competing mechanism.

b. The definition given is restricted to the inclusion chemistry of discrete molecular entities. There is a complementary field that deals with continuous solid structures containing voids of molecular dimensions. For those unfamilar with the subject reference might be made to the Journal of Inclusion Chemistry or the biennial international meetings on this exciting young research subject (4th Internl. Symp. on Inclusion Phenomena, Univ. Lancaster, 20-15 July, 1986).

c. A more useful set of abbreviations than the older acacen sort has been devised to provide easy recognition of the particular substitution pattern that applies to any given example of a Schiff base ligand of this class. The root is mal from malonaldehyde, the unsubstituted parent dialdehyde, and the form is $R^1{}_2R^2{}_2R^3{}_2$malX, where R^1, R^2, and R^3 are the successive substituents on the β-diketone moieties, starting with the position nearest the remaining ketone moiety. X represents the di- or triamine residue and is en for an ethylenediamine derivative. Thus acacen becomes $Me_2H_2Me_2$malen and the ethylenediamine derivative of benzoylacetone is $Ph_2H_2Me_2$malen. The derivative from acetylacetone and 3,3'-diaminodipropylamine is $Me_2H_2Me_2$MalDPT since DPT is the usual abbreviation for the triamine.

d. In addition to the lacunar dioxygen carriers produced from these cyclidene ligands three other noteable families of superstructured ligands have been characterized: (1) The so-called vaulted cylcidenes which contain much enlarged cavities that are adequate to accommodate various organic host molecules along with such small ligand molecules as O_2 (see reference 38 and literature cited therein); novel dimeric derivatives having two face-to-face cyclidene complexes separated by a persistent void (see reference 39 and literature cited therein); and a family of clathrochelate complexes formed by rearrangement reactions (see reference 40).

e. This propensity of low spin cobalt(II) to exist in 5-coordinate structures has recently been dramatically illustrated in the rearrangements of certain of the cyclidene derivatives. Whereas the iron(II) complex which tends to form 6-coordinate low spin structures rearranges a lacunar cyclidene to a sexadentate clathrochelate, the cobalt only rearranges one side of the complex and produces a pentadentate half-clathro complex.[76]

REFERENCES

1. Advances in Blood Substitute Research, Ed. by R. B. Bolin, R. P. Geyer, and G. J. Nemo, Alan. R. Liss, Inc., New York, 1983.
2. International Symposium on Blood Substitutes, June 19-20, 1987, Bari, Italy.
3. D. H. Busch, Critical Care Medicine, 10, No.4, 246 (1982).
4. R. R. Gagne, private communication.
5. J. Bonaventura and C. Bonaventura, U.S. Patent No. 4,343,715, 10 August, 1982.
6. "Hemosponge: Unlimited Oxygen Under the Sea," High Technology, 75, May 1984.
7. I. Roman, U.S. Patent Application 393711, 30 June 1982.
8. "A Membrane Path to Oxygen-Enriched Air," Science News, 151, March 6, 1982.
9. "Synfuels Market Makes Oxygen Producers Giddy," Chem. Eng. News, 82, April 21, 1980.
10. D. P. Burke, Chem. Week., 18, July 9, 1980.
11. "The World's Biggest Oxygen Producers Battle for Synfuels Business," Chem. Week, 23, Sept. 3, 1980.

12. R. F. Steward, P. A. Estep, and J. J. S. Sebastian, U.S. Bur. Mines Inform., Circ. No. 7906 (1959).
13. A. J. Adduci, Chemtech, 575 Sept. 1976.
14. J. E. Baldwin and J. Huff, J. Am. Chem. Soc., 95, 5757 (1973).
15. N. Herron and D. H. Busch, J. Am. Chem. Soc, 103, 1236 (1981).
16. B. Durham, T. J. Anderson, J. A. Switzer, J. F. Endicott, and M. D. Glick, Inorg. Chem., 16, 271 (1977).
17. M. C. Weiss and V. L. Goedken, J. Am. Chem. Soc., 98, 3389 (1976).
18. D. P. Riley and D. H. Busch, Inorg. Chem., 22, 4141 (1983).
19. V. L. Goedken and D. H. Busch, J. Am. Chem. Soc., 94, 7355 (1972).
20. J. C. Dabrowiak and D. H. Busch, Inorg. Chem., 14, 1881 (1975).
21. S. Dilli, A. M. Maitra, and E. Patsalides, Inorg. Chem., 21, 2832 (1982).
22. R. D. Jones, D. A. Summerville, and Fred Basolo, Chem. Rev., 79, 139 (1979)
23. E. C. Niederhoffer, J. H. Timmons, and A. E. Martell, Chem. Rev., 84, 137 (1984).
24. T. D. Smith and J. R. Pilbrow, Coord. Chem. Revs., 39, 295 (1981).
25. B. S. Tovrog, D. J. Kitko, and R. S. Drago, J. Am. Chem. Soc., 98, 5144 (1976).
26. J. E. Newton and M. B. Hall, Inorg. Chem., 23, 4627 (1984).
27. T. G. Traylor and P. S. Traylor, Ann. Rev. Biophys. Bioeng., 11, 105 (1982).
28. D. H. Busch and C. J. Cairns, in "Synthesis of Macrocycles -- The Design of Selective Complexing Agents," Ed. by R. M. Izatt and J. J. Christensen, pp. 1-52, Wiley-Interscience, New york, 1987.
29. W. P. Schammel, K. S. B. Mertes, G. G. Christoph, and D. H. Busch, J. Am. Chem. Soc., 101, 1622 (1979).
30. M. C. Thompson and D. H. Busch, J. Am. Chem. Soc., 86, 3651 (1964).
31. E. L. Blinn and D. H. Busch, Inorg. Chem., 7, 2426 (1968).
32. K. A. Goldsby, A. J. Jircitano, D. M. Minahan, D. Ramprasad, and D. H. Busch, Inorg. Chem., in press.
33. G. A. Melson and D. H. Busch, J. Am. Chem. Soc., 86, 4834 (1964).
34. V. Katovic, L. T. Taylor, and D. H. Busch, J. Am. Chem. Soc., 91, 2122 (1969).
35. P. W. R. Corfield, J. D. Mokren, C. J. Hipp, and D. H. Busch, J. Am. Chem. Soc., 95, 4465 (1973).
36. D. H. Busch, D. J. Olszanski, J. C. Stevens, W. P. Schammel, M. Kojima, N. Herron, L. L. Zimmer, K. A. Holter, and J. Mocak, J. Am. Chem. Soc., 103, 1472 (1981).
37. D. H. Busch, S. C. Jackels, R. C. Callahan, J. J. Grzybowski, L. L. Zimmer, M. Kojima, D. J. Olszanski, W. P. Schammel, J. C. Stevens, K. A. Holter, and J. Mocak, Inorg. Chem., 20, 2834 (1981).
38. T. J. Meade, Whei-Lu Kwik, N. Herron, N. W. Alcock, and D. H. Busch, J. Am. Chem. Soc., 108, 1954 (1986).
39. N. Hoshino, K. A. Goldsby, and D. H. Busch, Inorg. Chem., 25, 3000 (1986).
40. N. Herron, J. J. Grzybowski, N. Matsumoto, L. L. Zimmer, G. G. Christoph, and D. H. Busch, J. Am. Chem. Soc., 104, 1999 (1982).
41. B. Korybut-Daszkiewicz, M. Kojima, J. H. Cameron, N. Herron, M. Y. Chavan, A. J. Jircitano, B. K. Coltrain, G. L. Neer, N. W. Alcock, and D. H. Busch, Inorg. Chem., 23, 903 (1984).
42. M. Kojima, D. Nosco, C. J. Cairns, A. J. Jircitano, D. Ramprasad, and D. H. Busch, unpublished results.
43. N. W. Alcock, W.-K Lin, A. Jircitano, J. D. Mokren, P. W. R. Corfield, G. Johnson, G. Novotnak, C. Cairns, and D. H. Busch, Inorg. Chem., 26, 440 (1987).
44. J. C. Stevens, P. J. Jackson, W. P. Schammel, G. G. Christoph, and D. H. Busch, J. Am. Chem. Soc., 102, 3283 (1980).

45. D. H. Busch, L. L. Zimmer, J. J. Grzybowski, D. J. Olszanski, S. S. C. Jackels, R. C. Callahan, and G. G. Christoph, Proc. Natl. Acad. Sci., USA, 78, 5919 (1981).
46. P. J. Jackson, C. Cairns, W.-K. Lin, N. W. Alcock, and D. H. Busch, Inorg. Chem., 25, 4015 (1986).
47. N. Herron, L. L. Zimmer, J. J. Grzybowski, D. J. Olszanski, S. C. Jackels, R. W. Callahan, J. H. Cameron, G. G. Christoph, and D. H. Busch, J. Am. Chem. Soc., 105, 6585 (1983).
48. K. J. Takeuchi, D. H. Busch, and N. W. Alcock, J. Am. Chem. Soc., 105, 4261 (1983).
49. T. J. Meade, C. M. Fendrick, and D. H. Busch, submitted for publication.
50. J. C. Stevens and D. H. Busch, J. Am. Chem. Soc., 102, 3284 (1980).
51. J. C. Stevens, thesis, The Ohio State University, 1979.
52. P. J. Jackson, thesis, The Ohio State University, 1981.
53. D. H. Busch, J. C. Stevens, P. J. Jackson, D. Nosco, N. Matsumoto, M. Kojima, and N. W. Alcock, submitted for publication.
54. W. P. Schaefer, B. T. Huie, M. G. Kurilla, and S. E. Ealick, Inorg. Chem., 19, 340 (1980).
55. J. P. Collman, T. R. Halbert, and K. S. Suslick, Metal Ion Activation of Dioxygen, Ed. by T. G.Spiro, John Wiley & Sons, New York, 1980.
56. J. R. Budge, P. E. Ellis Jr., R. D. Jones, J. E. Linard, F. Basolo, J. E. Baldwin and R. L. Dyer, J. Am. Chem. Soc., 101 4760 (1979).
57. M. Sabat and J. A. Ibers, J. Am. Chem. Soc., 104, 3715 (1982).
58. T. G. Traylor and D. V. Stynes, J. Am. Chem. Soc., 102, 5938 (1980).
59. A. L. Crumbliss and F. Basolo, J. Am. Chem. Soc., 921, 55 (1970).
60. M.J. Carter, D.P. Rillema, and F. Basolo, J. Am. Chem. Soc., 96, 392 (1974).
61. Handbook of Chemistry and Physics, 53rd Ed. by R. C. Weast, p. D-55, Chemical Rubber Co., Cleveland OH, 1973.
62. J. H. Cameron, M. Kojima, B. Korybut-Daszkiewicz, B. K. Coltrain, T. J. Meade, N. W. Alcock, and D. H. Busch, Inorg. Chem., 26, 427 1987).
63. W. Evans, D. Nosco, N. Stephenson, and D. H. Busch, unpublished results.
64. M. Y. Chavan, T. J. Meade, D. H. Busch and T. Kuwana, Inorg. Chem., 25, 314 (1986).
65. T. Tsumaki Bull. Chem.Soc., Jpn., 13, 252 (1938).
66. M. W. Urban, Y. Nonaka, and K. Nakamoto, Inorg. Chem., 21, 1046 (1982).
67. K. Kakamoto, Y. Nonaka, T. Ishiguro, M. Suzuki, M. Kozuka, Y. Nishida, and S. Kida, J. Am. Chem. Soc., 104, 3386 (1982).
68. D. F. Martin, W. C. Fernelius and M. Shamma, J. Am. Chem. Soc., 81, 130 (1959).
69. K. Kurokura, H. Okawa and S. Kida, Bull.Chem.Soc., Jpn., 54, 2036 (1978).
70. D. Ramprasad, W.-K. Lin, K. A. Goldsby and D. H. Busch, submitted for publication.
71. R. Delgado, D. Ramprasad, M. W. Glogowski, N. A. Stephenson, D. M. A. Minahan, and D. H. Busch, submitted for publication.
72. M. D. Braydich, J. J. Fortman and S. C. Cummings, Inorg. Chem., 22, 484 (1983).
73. L. Dickerson, thesis, The Ohio State University, 1986.
74. N. Herron, L. Dickerson and D. H. Busch, J. Chem. Soc., Chem. Commun., 884 (1983).
75. N. Herron, L. Dickerson, C. M. Fendrick and D. H. Busch, unpublished results.
76. M. L. Caste, C. J. Cairns, J. Church, W.-K. Lin, J. C. Gallucci, and D. H. Busch, Inorg. Chem., 26, 78 (1987).

FORMATION AND DEGRADATION OF COBALT DIOXYGEN COMPLEXES

Arthur E. Martell

Department of Chemistry
Texas A&M University
College Station, Texas 77843

INTRODUCTION

The extensive literature dealing with cobalt dioxygen complexes has been summarized in several reviews[1-6]. The earliest known synthetic dioxygen complex is probably the decamminedicobalt dioxygen adduct described by Werner and Myelius[7] in 1889. Forty years later Tsumaki[8] discovered the oxygen complexing properties of cobalt Schiff bases. This work was followed up through extensive investigations by Calvin and coworkers[9-14] and Diehl et al.[15] These studies established the fact that oxygen may be reversibly bound by the cobalt(II) complexes of tetradentate and pentadentate Schiff bases formed by diamines and triamines with salicylaldehyde and its derivatives. Later work by Basolo and coworkers, and others, extended the Schiff bases to include those of acetylacetone and analogous compounds.[3] These complexes form in solution, sometimes with an appropriate axial base as an auxiliary ligand, or in the solid state. A typical binuclear cobalt complex is illustrated by formula **1** for the Schiff base ligand "fluomine". For this complex, and for all the dioxygen complexes described in this paper, the Co^{3+}-O_2^- or Co^{3+}-$\bar{O}$-$\bar{O}$-Co^{3+} formalism is employed for convenience, although it realized that dioxygen complexes involve more or less stable intermediate states between Co^{2+}-dioxygen, and Co^{3+}-superoxide or $2Co^{3+}$-peroxide systems.

At about the same time as that of the investigation of Calvin and Diehl and coworkers, Burk, Hearon, et al.[16-19] reported that the bishistidinatocobalt complex is capable of reversibly binding oxygen, in aqueous solution, to form an adduct illustrated by formula **2**. The coordinate bonding indicated in **2** is based on the current formalism, described above. With the observation that these two fundamentally different types of dioxygen complexes (cobalt(II)-Schiff base and amino acid complexes) combine reversibly with

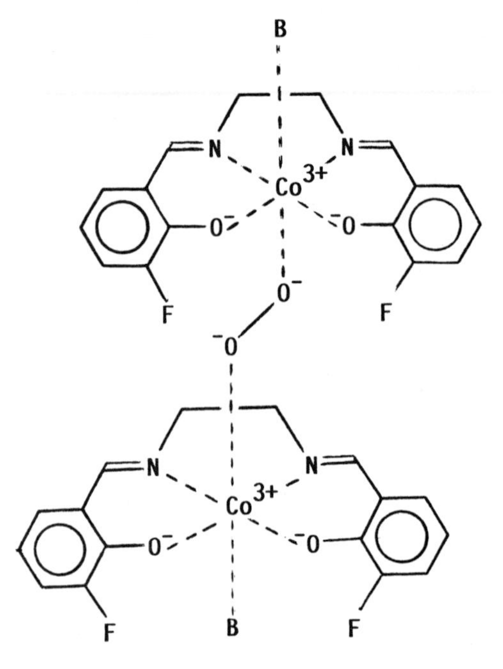

1 Binuclear cobalt dioxygen complex with fluomine and axial base B as ligands

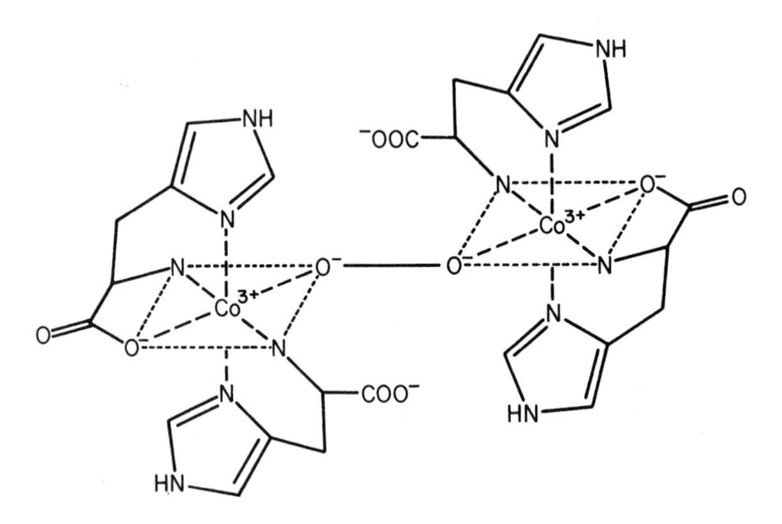

2 Bishistidinatocobalt(II)-dioxygen complex

dioxygen should have been sufficient to indicate that oxygenation of cobalt complexes is a general reaction which takes place with a wide variety of complexes varying considerably in chemical properties. At that time, however, dioxygen complex formation with synthetic complexes was considered an unusual rather than a general phenomenon and interest in such compounds slackened for about fifteen years. A renaissance of interest and activity in the field occurred in the mid to late 60's, and since then research on synthetic dioxygen complexes has continued to grow rapidly, mainly for the purpose of modeling biological oxygen carriers, and oxidase and oxygenase enzymes. Recently, interest has developed in the possible applications of dioxygen carriers as intermediates in catalysis, and for the purpose of oxygen separation and transport.

STABILITIES OF COBALT DIOXYGEN COMPLEXES IN AQUEOUS SOLUTION

Investigation of cobalt dioxygen complexes by the author's research group began with the work of Nakon[20-22] when it became apparent that the potentiometric techniques developed in this laboratory provides one of the most accurate and powerful methods for determining the dioxygen affinities of cobalt(II) complexes of polyamines, amino acids, peptides, etc., in aqueous solution. Since that time a large number of such complexes have been reported, along with the corresponding oxygenation equilibrium constants. The equilibria involved in the formation of such dioxygen complexes in aqueous solution are represented by equations (1)-(7).

$$HL^{(n-1)-} \rightleftharpoons L^{n-} + H^+$$

$$K_L^H = [HL^{(n-1)-}]/[H^+][L^{n-}] \qquad (1)$$

$$Co^{2+} + L^{n-} \rightleftharpoons CoL^{(2-n)+}$$

$$K_f = [CoL^{(2-n)+}]/[Co^{2+}][L^{n-}] \qquad (2)$$

for n = 0

$$CoL^{2+} + O_2 \underset{k_{-1}}{\overset{k_1}{\rightleftharpoons}} LCo^{3+}\text{---}O^-\text{—}O\bullet$$

$$K_{11} = [LCoO_2^{2+}]/[CoL^{2+}][O_2] \qquad (3)$$

$$LCo^{3+}\text{---}O^-\text{-}O\bullet + CoL^{2+} \underset{k_{-2}}{\overset{k_2}{\rightleftharpoons}} LCo^{3+}\text{---}O^-\text{—}O^-\text{---}Co^{3+}L$$

$$K_{12} = [L_2Co_2O_2^{4+}]/[LCoO_2^{2+}][CoL^{2+}] \qquad (4)$$

$$LCo^{3+}\text{---}O^-\text{—}O^-\text{---}Co^{3+}L \rightleftharpoons LCo^{3+}(\mu\text{-}O^-\text{—}O^-)(\mu\text{-}\bar{O}H)Co^{3+}L + H^+$$

$$K_a = [L_2Co_2O_2OH^{3+}][H^+]/[L_2Co_2O_2^{4+}] \qquad (5)$$

$$K_{O_2} = [Co_2L_2O_2^{4+}]/[CoL^{2+}]^2[O_2] \qquad (6)$$

$$K'_{O_2} = [Co_2L_2O_2OH^{3+}][H^+]/[CoL^{2+}]^2[O_2] \qquad (7)$$

Equation (1) represents the last of what is usually many acid dissociation steps of the multidentate ligand H_nL, with n protonation sites. Reaction (2) is usually strongly pH-dependent because it depends on the concentration of the ligand free base L, which is generally strongly protonated in aqueous solution. Similarly the oxygenation reactions (3) and (4), which do not involve hydrogen ions, are also usually very pH dependent because they depend on the pH-dependent concentration of the cobalt complex, CoL. Nearly all of the cobalt dioxygen complexes formed in aqueous solution are of the peroxo-bridged binuclear type, so that the mononuclear form indicated by equation (3) is generally an intermediate present at sufficiently low concentration as to be not detectable by potentiometric or spectrophotometric techniques. Exceptions are the cobalt(II) complexes of N-alkylated polyamines, which partially form mononuclear (superoxo) complexes in aquoeus systems, and produce higher yields of mononuclear dioxygen complexes in non aqueous solvents of moderate to low dielectric constant.

In cases were the peroxo-bridged dioxygen complex illustrated in equation (4) has one (or more) aquo donors on each metal ion, hydrolysis and olation occur (equation 5) to give a dibridged (μ-peroxo-μ-hydroxo) dioxygen complex. In most cases reactions (4) and (5) occur simultaneously with little build-up of the intermediate monobridged complexes, but exceptions have been noted.[4] Thus the two types of oxygenation constants generally reported for cobalt complexes in aqueous solution are represented by equations (6) and (7).

The overlap of the chemical reactions corresponding to equations (1)-(5) makes possible the determination of the oxygenation constants (6) and (7) through measurement of hydrogen ion concentrations, because of the fact that oxygen complex formation shifts all equilibria to the right and increases the competition between metal ion and hydrogen ion for the ligand in favor of the metal ion. An example of the type of experimental data obtained for such systems is illustrated by the p[H] profiles[5] in Figure 1. The depression of the buffer region of the TREN-Co^{2+} solution under nitrogen provides a qualitative measure of the affinity of the metal ion for the ligand (through hydrogen ion displacement). It is seen that the buffer region is further depressed in the presence of oxygen, and that the depression is a function of the oxygen concentration, thus demonstrating the sensitivity of hydrogen ion concentration to the concentration of dioxygen. The increased affinity of the metal ion for the ligand in the presence of dioxygen is good evidence for the increase in charge of the metal ion, thus making the Co^{3+}-O_2^{2-} formalism attractive. This effect led to the early use of this formalism by the author's research group.

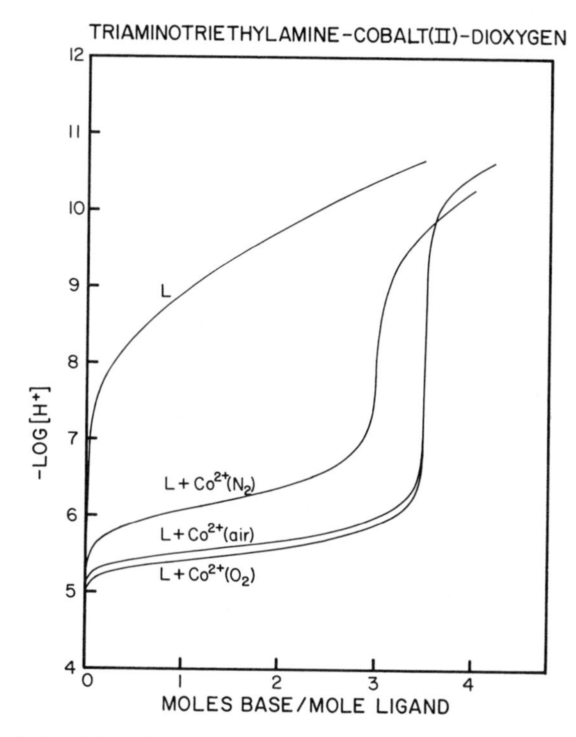

Figure 1. Potentiometric equilibrium curves for the system triaminotriethylamine (TREN) trihydrochloride-cobalt(II)-dioxygen. L = ligand under nitrogen in absence of metal ion; $L+Co^{2+}(N_2)$ = equimolar ratio of Co(II) and TREN under nitrogen; $L+Co^{2+}$(air) = equimolar ratio of Co(II) and TREN under air at 1.00 atm; $L+Co^{2+}(O_2)$ = equimolar ratio of Co(II) and TREN under oxygen at 1.00 atm; $T_M = T_L = 1.000 \times 10^{-3}$ M; t = 25°C; μ = 0.100 M (KNO_3).

A large number of equilibrium constants of dibridged dioxygen complexes have now been determined[4,20-28] and those published through 1983 are listed in reference 4. Figure 2 illustrates the type of correlation reported for the magnitudes of the oxygenation constants as a function of ligand basicity It is noted that the linear correlation observed is not with the stability constants of the cobalt(II) complexes, but with the total basicities of the ligand donor groups as measured by the sum of the corresponding pK_a's. Thus the strength of sigma coordinate bonding between the ligand donor groups and the cobalt(II) ion seems to be the major factor influencing the extent of charge transfer from the metal ion to dioxygen, which similarly influences the strength of metal-dioxygen coordinate bonding.

Analogous studies of the stabilities of monobridged cobalt dioxygen complexes have been reported by Harris et al.[25,26] and Timmons et al.[27,28]

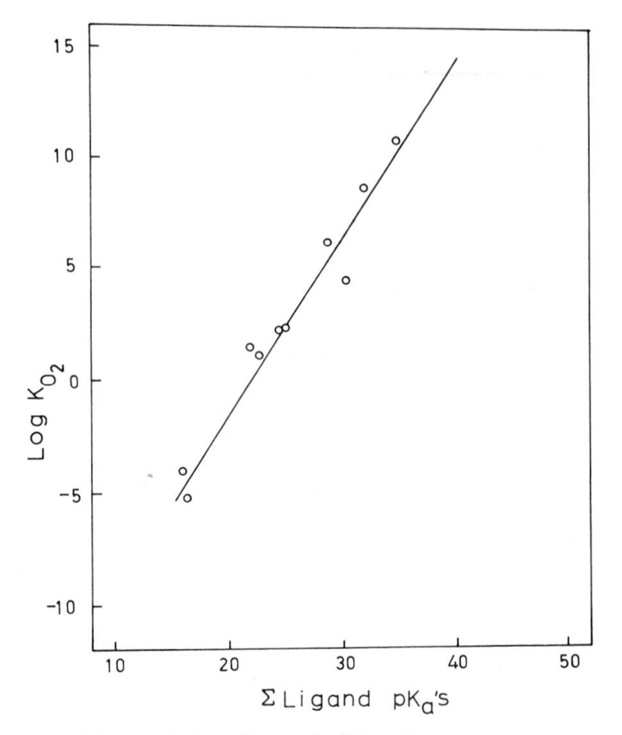

Figure 2. Correlation of log K_{O_2} of dibridged dioxygen complexes with the sum of the log protonation constants of the ligands.

Because of a shortage of the appropriate pentadentate ligands a number of polyamines varying in basicities of their donor groups, formulas **3-6**, were synthesized for comparison with the parent compound, TETREN, **7**, the most basic of the series. The correlation obtained for these and some additional cobalt(II) complexes is illustrated by Figure 3. Here also it was found that the oxygenation constants seem to be linear functions of the sums of the ligand pK_a's rather than the stability constants of the cobalt(II) complexes. The oxygenation constant data are separated into two groups, one series involving complexes containing only five membered chelate rings, with a separate correlation at lower stability for ligands which form two 5- and two 6-membered chelate rings. The separation into two groups shows that ligand conformation and steric factors may have measurable effects on the strength of metal-dioxygen binding.

The equilibrium constants[4] for the formation of a wide variety of dioxygen complexes are listed in Table 1. In order to make possible comparisons of dioxygen complexes under widely differing conditions, the constants in Table 1 are expressed as the equilibrium oxygen pressure required for half conversion of the metal complex in its lower valent state to the corresponding dioxygen complex. For dibridged dioxygen complexes $P_{1/2}$ is pH dependent, so that the p[H] is specified in such cases. The most striking aspect of

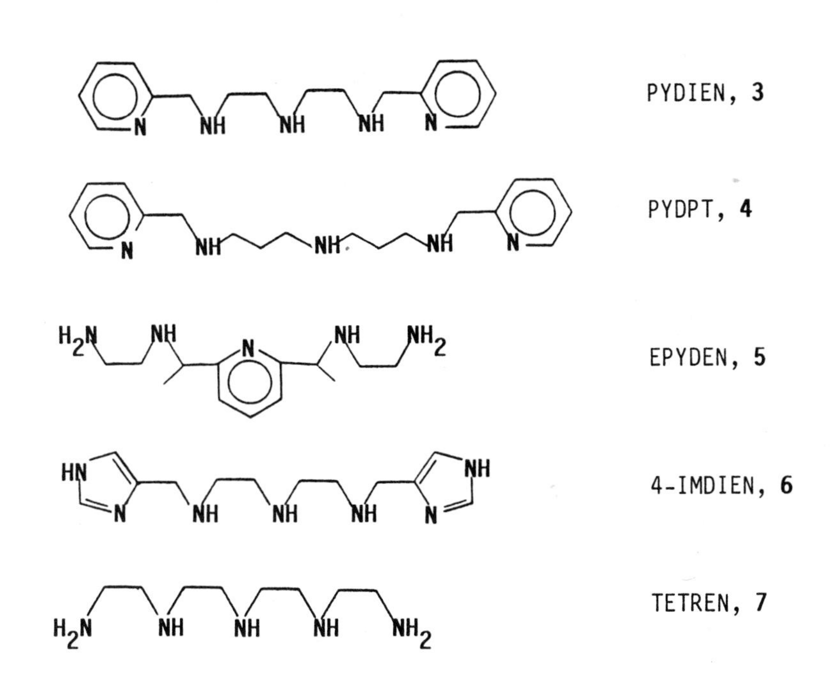

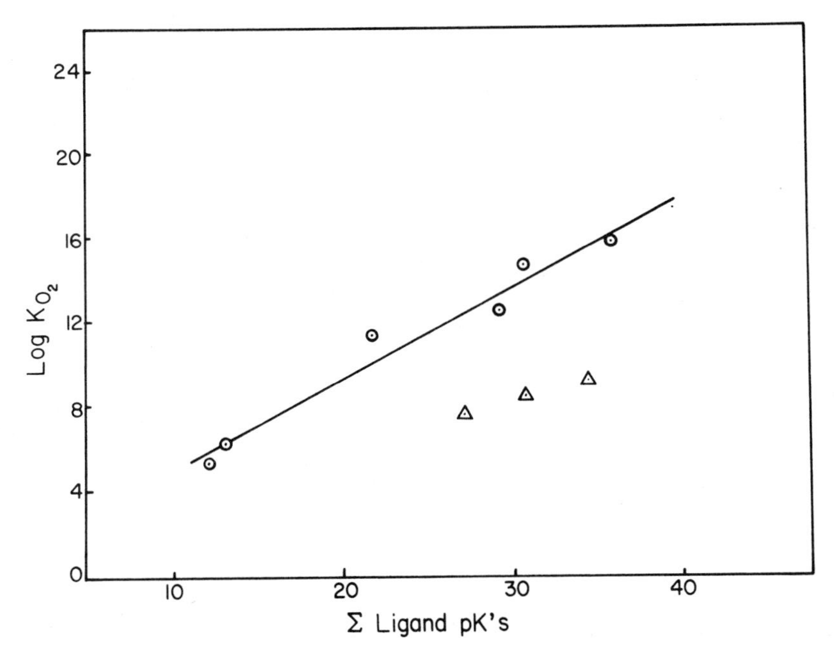

Figure 3. Correlation of log K_{O_2} of monobridged dioxygen complexes with the sums of log protonation constants of the ligands: (Δ) complexes with five-membered chelate rings, (⊙)complexes with five- and six-membered chelate rings.

Table 1. Comparison of Stabilities of Various Types of Dioxygen Complexes

No.	Oxygen-free complex	Dioxygen Complex	$P_{1/2}^{-1}$ (atm^{-1})	Conditions
1	human hemoglobin A, Hb	HbO_2	4.0×10^2	25°C, pH 7.4 (tris buffer)
2	FeTPivPP(Me_2Im), MLL'	MLL'O_2	2.0×10^1	25°C, toluene
3	CoTPivPP(Me_2Im), MLL'	MLL'O_2	8.4×10^{-1}	25°C, toluene
4	Co(SALEN), ML	$(ML)_2O_2$	2.3	25°C, Me_2SO
5	Co(ACACEN)PY, MLL'	MLL'O_2	4.0×10^2	-31°C, μ= 0.10 M
6	$Co(TEP)^{2+}$, ML	$(ML)_2O_2$	3.4×10^{13}	25°C, μ= 0.10 M
7	$Co(TREN)^{2+}$, ML	$(ML)_2(O_2)(OH)$	1.4×10^9	25°C, μ= 0.10 M pH 7
8	$Co(BPY)(TERPY)^{2+}$, MLL'	(MLL')O_2	1.4×10^3	25°C, μ= 0.10 M
9	$Co_2(BISTREN)OH^{3+}$, M_2LL'	(M_2LL')O_2	3.5×10^1	25°C, μ= 0.10 M pH 7
10	$Co_2PXBDE(EN)_2^{4+}$, $M_2LL'_2$	$(M_2LL'_2)O_2$	5.1×10^6	25°C, μ= 0.10 M
11	$Co_2BISDIEN^{4+}$, M_2L	$(M_2L)(O_2)(OH)$	5.1×10^4	25°C, μ= 0.10 M pH 7

the data in Table 1 is the wide variation of the oxygenation constants, involving a span of nearly fourteen orders of magnitude. It seems that hemoglobin and a number of synthetic cobalt(II) and iron(II) complexes form dioxygen adducts having relatively low stability constants. Such complexes can be readily decomposed into their components by small changes in pressure or temperature, or both. Such systems would be good candidates for oxygen separation or transport. While the polyamine and amino acid ligands produce cobalt complexes having very high dioxygen affinities, which under most conditions would dissociate with difficulty, they would probably serve as interesting models for oxygen activation and as catalysts for oxidation of organic substrates. In addition, their apparent very high stabilities may be reversed in aqueous solution by lowering the pH. This requirement would be rather cumbersome, however, for the development of systems for oxygen production or facilitated transport through membranes.

Two of the dioxygen complexes listed in Table 1, which are of special interest to this investigator, are BISDIEN[29] and BISTREN[30], which form macrocyclic and cryptate dicobalt complexes respectively. The dioxygen complex of dicobalt BISTREN, **8**, involves four basic nitrogen donors for each metal ion and should have a much higher oxygenation constant than the corresponding dixoygen complex formed from dicobalt BISDIEN, **9**, which has three basic nitrogen donors per metal ion. The fact that the

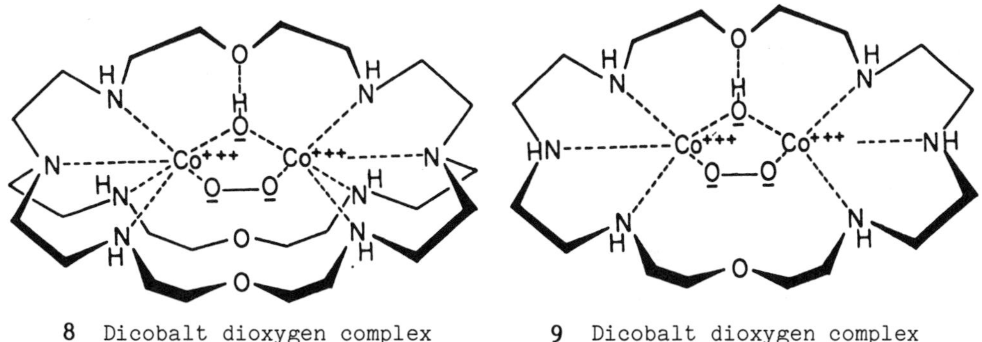

8 Dicobalt dioxygen complex of BISTREN

9 Dicobalt dioxygen complex of BISDIEN

latter is over three orders of magnitude more stable indicates that dioxygen complex formation must be severely restricted in the BISTREN cryptate complex, probably because of the less flexible nature of the cage (i.e., for steric reasons).

Of considerable interest with respect to the reactivities of cobalt dioxygen complexes to be discussed below is the fact that each metal ion in **9** has an additional coordination site which is not indicated in the formula, and may be occupied by a coordinated water molecule. That this is probably the case is indicated by the fact that **9** was reported[30] to undergo two additional dissociation reactions to form the complex BISDIEN-$Co_2(OH)_3O_2^{1+}$.

DEGRADATION REACTIONS OF COBALT DIOXYGEN COMPLEXES

All dioxygen complexes undergo irreversible degradation reactions to form inert complexes that have no affinity for dioxygen. This type of reaction has been found to follow one of the several pathways indicated in Table 2 at rates which may be very slow, or so rapid that it is difficult to isolate or work with the dioxygen complex. The factors that control the tendencies of oxygen complexes to undergo such reactions are not well understood, and it is still not possible to predict whether a proposed dioxygen complex would be stable or labile toward degradation.

In any case the rates of such reactions must be controlled and be sufficiently slow if the dioxygen complex is to be employed as a reactant (i.e., an oxidizing agent) or as a carrier for separation processes. The ability to predict or suppress such reactions would be invaluable for the design of dioxygen complexes for these applications, or for study as enzyme models. The remainder of this paper is concerned with the results of a series of investigations of the mechanisms of the irreversible degradation reactions of cobalt dioxygen complexes. As examples of the patterns of the oxidative degradation reactions that may take place, the

Table 2. Reactions of Binuclear Cobalt Dioxygen Complexes ($LCo^{3+}\text{-}\bar{O}\text{-}\bar{O}\text{-}Co^{3+}L$)

1. Spontaneous irreversible conversion to inert complexes
   a. Metal-centered oxidation
   Conversion to the inert Co(III) complexes of the original ligand with release of H_2O_2

$$LCo^{3+}\text{-}\bar{O}\text{-}\bar{O}\text{-}Co^{3+}L + 2H^+ \longrightarrow 2CoL^{3+} + H_2O_2$$

   b. Oxidative dehydrogenation of the coordinated ligand

$$LCo^{3+}\text{-}\bar{O}\text{-}\bar{O}\text{-}Co^{3+}L \longrightarrow 2H_2O + 2CoH_{-2}L^{2+}$$

   c. Oxygen insertion into coordinated ligand

$$LCo^{3+}\text{-}\bar{O}\text{-}\bar{O}\text{-}Co^{3+}L \longrightarrow 2CoLO^{2+}$$

$$LCo^{3+}\text{-}\bar{O}\text{-}\bar{O}\text{-}Co^{3+}L + H^+ \longrightarrow CoLO^{3+} + Co(OH)L^{2+}$$

<u>or</u> $$Co_2LO_2^{4+} + H^+ \longrightarrow CoLOCoOH^{5+}$$

reaction pathways for the dioxygen complexes of **3**, **5**, and **10** are described in detail.

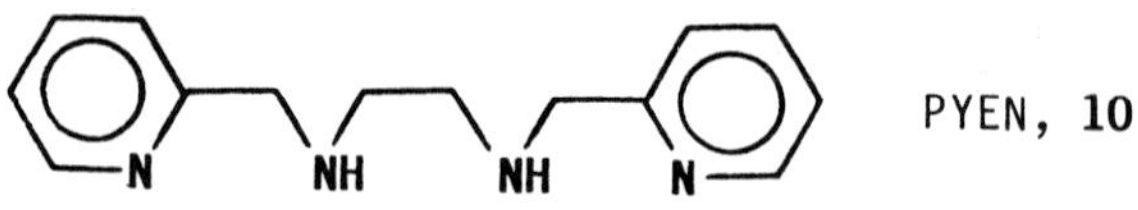

PYEN, **10**

<u>Degradation of the Cobalt(II)-PYDIEN-Dioxygen Complex</u>

The binuclear cobalt dioxygen complex derived from PYDIEN, **3**, is rapidly converted to the Co(II) complex of the dehydrogenated polyamine containing the imine double bond conjugated with a pyridine ring.[31,32] Experimentally, kinetics of reaction were measured by preparing the dioxygen complex in solution at pH values where it has high thermodynamic stability, removing excess oxygen, and raising the temperature to 35°C. The monoimine formed was isolated and identified by its IR spectrum, and by quantitative hydrolysis to 2-pyridinecarboxaldehyde. The evidence obtained in determining the nature of the reaction pathway and the stoichiometry of the oxidative dehydrogenation reaction is summarized in Table 3 and the reaction sequence is shown in Table 4. The cobalt(II) complex of the monoimine formed in the first oxidative dehydrogenation step forms a dioxygen complex which is thermodynamically somewhat less stable than the original complex, but is sufficiently stable to work with in the same manner. This complex in turn undergoes a second oxidative dehydrogenation to form a Co(II) complex of a coordinated diimine, which hydrolyzes spontaneously to a complex of diethylenetriamine and 2-pyridinecarboxaldehyde. When the reaction is carried out in the presence of excess oxygen the sequential

Table 3. Evidence for Formation of Co(II) Complex of Oxidatively Dehydrogenated Ligand

1. DC sampled polarograms indicate Co(II) complex formed.
2. Monoimine determined by hydrolysis to aldehyde and by TLC separation.
3. Uv-visible and IR spectra similar to those of synthetic imine complex.
4. No H_2O_2 formed in reaction mixture.
5. Product formed after anaerobic dehydrogenation forms a new dioxygen complex with O_2.

Table 4. Reaction Pathways for the Autoxidation of PYDIEN through Cobalt Dioxygen Complex Formation

$$2CoL^{2+} + O_2 \longrightarrow (CoL)_2O_2^{4+}$$
$$(CoL)_2O_2^{4+} \longrightarrow 2(CoL')^{2+} + 2H_2O$$
$$2(CoL')^{2+} + O_2 \longrightarrow (CoL')_2O_2^{4+}$$
$$(CoL')_2O_2^{4+} \longrightarrow 2(CoL'')^{2+} + 2H_2O$$
$$2(CoL'')^{2+} + O_2 \longrightarrow (CoL'')_2O_2^{4+}$$
$$(CoL'')_2O_2^{4+} + 2H^+ \longrightarrow 2(CoL'')^{3+} + H_2O_2$$

L = (structure: N NH NH NH N)

L' = (structure: N N NH NH N)

L" = (structure: N N NH N N)

dehydrogenation reactions overlap and only the final products can be isolated from the reaction mixture. These reactions are second order, first order in dioxygen complex and first order in hydroxide ion. Deprotonation of the aliphatic amino group undergoing dehydrogenation is therefore considered to be an essential part of the reaction mechanism.

Degradation of the Cobalt(II)-EPYDEN-Dioxygen Complex

The fact tht the cobalt dioxygen complex involving the coordinated pentamine EPYDEN, **5**, does not undergo oxidative dehydrogenation throws considerable light on the constitutional requirements of this reaction. In this case a so-called metal-centered oxidative degradation takes place, for which the products are the inert Co(III) complex of the original polyamine, and hydrogen peroxide. The evidence obtained for the nature of this reaction is summarized in Table 5.

The change in reaction pathway for the cobalt dioxygen complex containing **5**, relative to that of the dioxygen complex containing **3**, was interpreted in terms of the probable conformation of the EPYDEN ligand in the cobalt dioxygen complex formed[33]. The folded arrangement of the ethylenediamine moieties in **11** prevents the formation of an

Table 5. Evidence for Metal Centered Oxidation of the Cobalt EPYDEN Dioxygen Complex

1. H_2O_2 formed
   a. Detected polarographically
   b. Determined semiquantitatively by iodide titration
2. Co(III) complex of unchanged ligand formed
   a. Identified by measurement of electronic absorption spectrum
   b. Polarographic reduction of reaction product gives the same half-wave potential as that of the initial Co(II) EPYDEN complex
   c. The complex formed does not combine with dioxygen

imine double bond conjugated to the pyridine ring. This interpretation has more recently been supported by the crystal structure determination of the iodide obtained from a concentrated solution of **11** in the presence of excess iodide ion[34], as indicated by Figure 4.

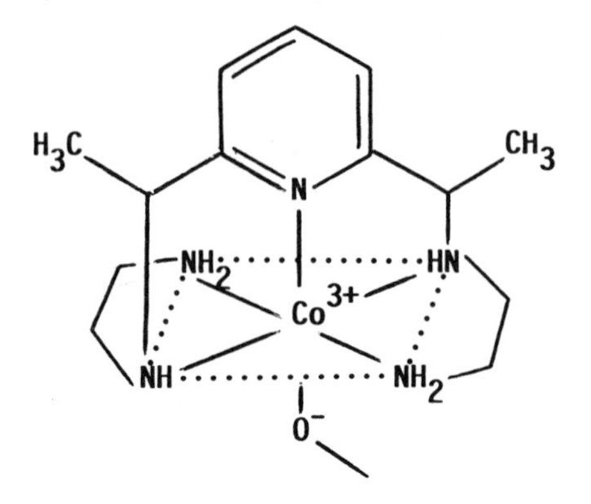

11 Cobalt dioxygen complex of EPYDEN

This new evidence led to a re-inspection of the crystal structure of the dioxygen complex containing coordinated ligand **3**, which had been reported earlier.[35] This structure clearly showed that the coordinated ligand in this dioxygen complex has a conformation of the aliphatic triamine bridge that is fairly close to the conformation required for the imine formed as the result of oxidative dehydrogenation.

The experimental work described above on the oxidative degradation of the cobalt dioxygen complexes containing pentamine ligands establishes the sensitivity of the dehydrogenation reaction to the conformation of the coordinated ligand. The kinetic studies described below on the cobalt dioxygen complex of **10** reveal still another requirement for oxidative dehydrogenation - that the amino group undergoing oxidative dehydrogenation must be proximal to the coordinated dioxygen.

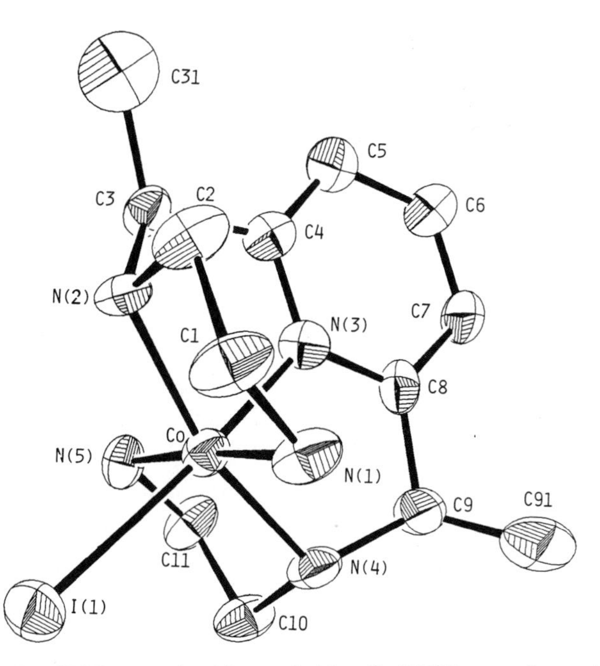

Figure 4. An ORTEP projection of the Co(III) complex obtained from **11**. Probability ellipsoids are at the 20% level.

Degradation of the Dibridged Dioxygen Complex of 1,6-Bis(2-pyridyl-2,5-diazahexanecobalt(II)

The dibridged (μ-hydroxo-μ-peroxo) complex, **12**, of the tetradentate polyamine (PYEN), **10**, analogous to PYDIEN was selected for oxidative degradation study. Apparently because of the stabilizing effect of the second μ-hydroxo bridge, the corresponding dioxygen complex was found to be more resistant to degradation than the corresponding monobridged complexes formed with the pentadentate ligands PYDIEN, **3**, and EPYDEN, **5**, described above. However, the reaction progressed smoothly with conveniently measurable kinetics between 45 and 60°C.[36]

The chelate formation constants of Co(II)-PYEN and its oxygenation constant were determined potentiometrically, and the dibridged dioxygen complex **12** was found to be the only complex species in solution between p[H] 4 and 11. Degradation of the dioxygen complex at elevated temperature resulted in dehydrogenation of the ligand to form a single reaction product, the Schiff base complex of pyridine-2-aldehyde and N-(2-pyridylmethyl)ethylenediamine. The reaction kinetics, on the other hand, were found to be complex, with two distinct second order rate constants leading to the same reaction product, which hydrolyzed in solution to the aldehyde and the diamine. These rate constants, $3.4 \times 10^{-4}\ M^{-1}\ s^{-1}$ and $9.3 \times 10^{-4}\ M^{-1}\ s^{-1}$, were found to be first order in dioxygen complex concentration and in hydroxide ion concentration. Because only a single product was obtained, it was concluded that the original dioxygen complex undergoing degradation exists in more than one form in solution.

Consideration of the possible conformations of the dibridged dioxygen complex **12** in solution led to the possibility of several complexes differing in the arrangement of the coordinated ligand. Three such substances are illustrated by formulas **13**, **14**, and **15**. The three conformations illustrated by formulas **13-15** show the ligands within each complex in equivalent positions. Other conformations are possible with ligands in different orientations. Thus there is ample opportunity for the existence of dioxygen complex species with differing rates of dehydrogenation. Such differences may be due to differences in proximity of the ligand groups undergoing oxidation to the coordinated dioxygen, as is apparent in formulas **13-15**, or may be due to the slow change in conformation of one (or more) species to a more reactive conformation.

The dehydrogenation reactions for which rate constants are indicated above were both found to have large deuterium isotope effects (6.4 for the slower reaction and 8.5 for the faster reaction). Consequently both the slow and fast reactions must involve direct transfer of a proton from the alpha carbon atom to the coordinated dioxygen. Because of the fact that conformation **13** provides the closest approach of the groups involved in such a proton transfer, it is suggested that it is the conformer responsible for the more rapid dehydrogenation reaction illustrated

POSSIBLE TOPOLOGIES OF DIOXYGEN COMPLEX OF
1,6-BIS(2-PYRIDYL)-2,5-DIAZAHEXANE (PYEN, L) $[Co_2L_2(OH)(O_2)]^{3+}$

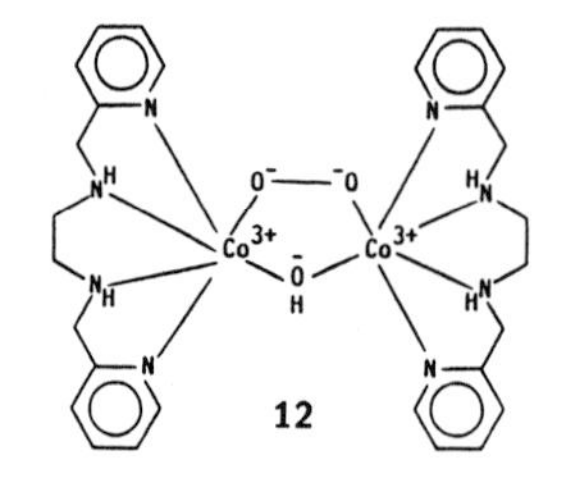

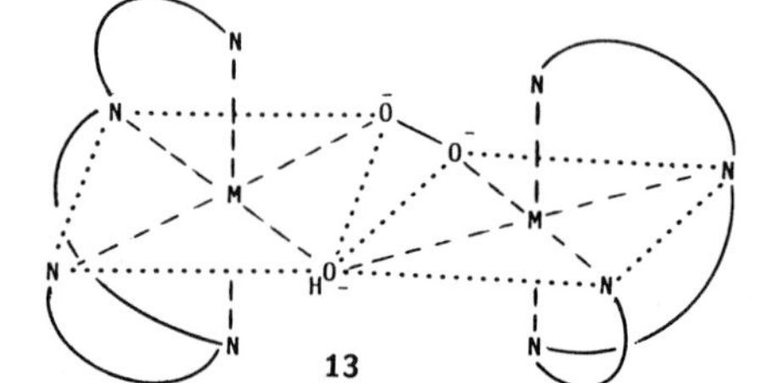

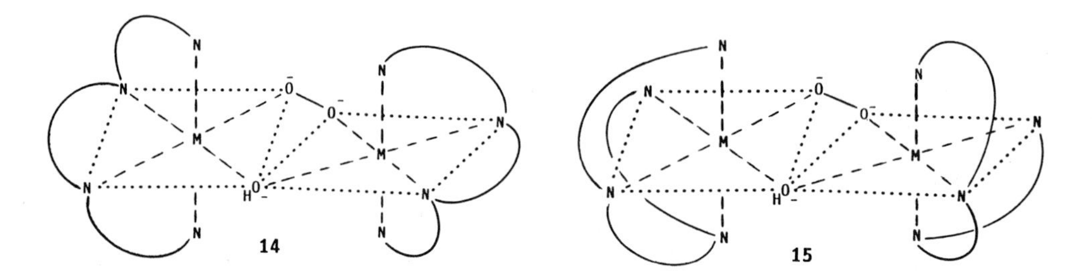

SCHEME I

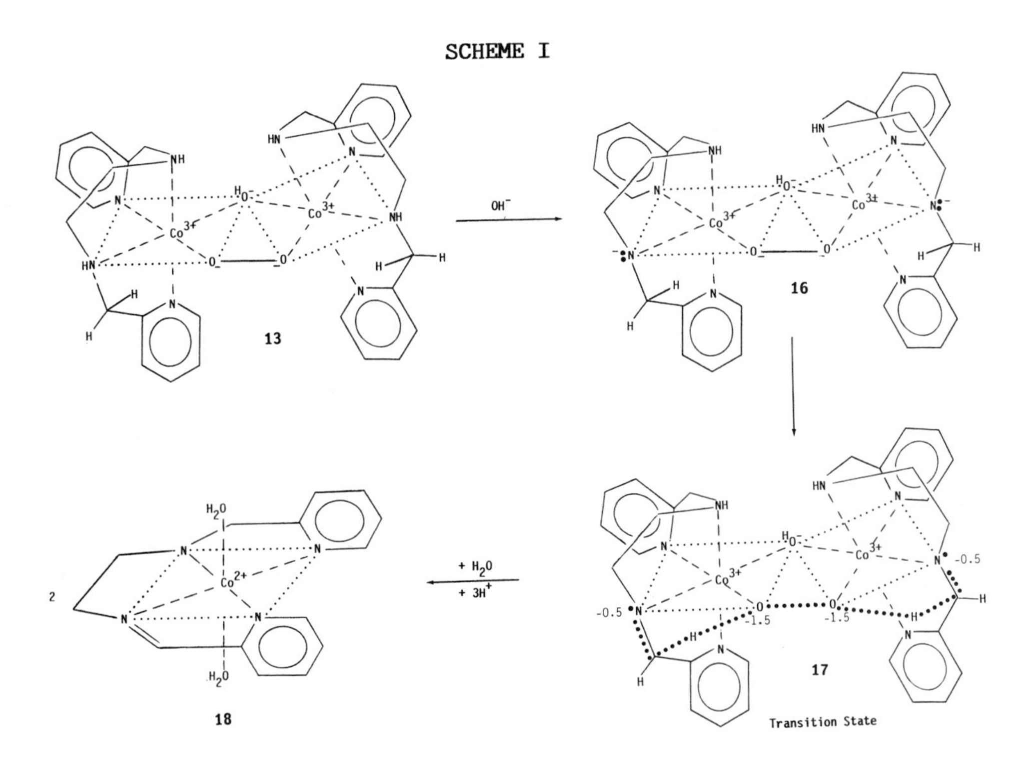

by Scheme I. The first order dependence on hydroxide ion concentration suggests an intermediate such as **16**, which then goes through a transition state **17** to form the reaction product **18**, which was identified as described above. The suggested transition state **17** illustrates the essential features of the proposed reaction mechanism: 1, direct proton transfer from the alpha carbon to the coordinated dioxygen; 2, homolytic O-O bond fission; 3, pre-equilibrium dissociation of coordinated amino nitrogen; and 4, electron transfer from the deprotonated amino group through the metal ion to the coordinated dioxygen. Although the kinetics indicates that more than one dioxygen complex species is involved in this dehydrogenation reaction, conformer **13** is selected as the most reactive form because it involves the shortest distance between the alpha carbon atoms and the coordinated dioxygen, and is therefore assigned as the species associated with the larger rate constant. One or more of the conformers with longer alpha carbon-dioxygen distances may account for the observed slower rate of dehydrogenation. Although the reaction mechanism indicated in the Scheme applies to the more rapidly-reactive species, a similar mechanism is suggested for the slow degradtion reaction because here also direct hydrogen ion transfer is considered to occur.

To summarize, the factors involved in ligand dehdyrogenation of cobalt dioxygen complexes are the following: 1, all dehydrogenation reactions are based catalyzed; 2, the imine double bond formed is conjugated with an aromatic ring (no dehydrogenation with purely aliphatic polyamines); 3, conformation of polyamine in dioxygen complex must be compatible with formation of trigonal R-C=N-R' (R = aromatic); 4, when isomers possible only configurations with a -CH- adjacent to the coordinated dioxygen undergo dehydrogenation; 5, large deuterium isotope effect; 6, concerted O-O bond scission, imine double bond formation, and proton transfer to coordinated peroxo oxygen.

DIOXYGEN COMPLEX REACTIONS IN PROGRESS

Although the studies of kinetics and mechanisms of the degradation reactions of the cobalt dioxygen complexes containing the ligands **3**, **5** and **10** have provided interesting guidelines for oxidative dehydrogenation of the ligand, and metal-centered oxidation with the formation of hydrogen peroxide, there are still many unanswered questions regarding the reaction pathways that require further investigation. One of the main complications has been the number of possible conformers of the dioxygen complexes, which complicates the interpretation of the degradation reactions.

In order to restrict the number of dioxygen complex conformers while maintaining proximity of the oxidation-sensitive positions to the coordinated dioxygen, it has been decided to study the cobalt dioxygen complexes containing macrocyclic ligands, as in **19**, **20**, and **21**. One of these ligands, the dipyridyl $[30]N_6O_4$ macrocycle in **19** has been reported by Nelson[37]. The three macrocyclic ligands **19-21** have been synthesized and studies of their cobalt dioxygen complexes are now underway. A predicted reaction sequence for the oxidative dehydrogenation for the $[24]N_6O_2$ analog of **19** is illustrated in Scheme II. The successive dehydrogenations **22** → **23** and **24** → **25** are similar to those postulated by Nelson[37] for the degradation of the Cu(I) complex of **19**, with the exception that the dioxygen complex intermediates may be isolated and identified for the cobalt complexes, whereas they were not stable enough to be detected in the Cu(I)/Cu(II) systems.

The binuclear dibridged dioxygen complex **26** (R = H) has been synthesized and characterized by R. Menif in this laboratory. Its oxidative degradation reactions (Scheme III) now under investigation may take one or more of three possible pathways: 1, reaction A resulting in oxidative dehydrogenation; 2, metal-centered oxidation B to give an inert binuclear cobalt(III) complex, **29**, and hydrogen peroxide; or oxygen insertion, C, into the aromatic ring to give a phenolate-bridged binuclear cobalt(III) complex. Evidence suggesting the elimination of pathways A and B have now been obtained, and further experimental work is underway to identify a possible oxygen insertion reaction product such as **28**.

SCHEME II

PREDICTED REDOX CYCLE IN THE OXYGENATION AND OXIDATIVE DEHYDROGENATION OF A BINUCLEAR MACROCYCLIC COBALT COMPLEX

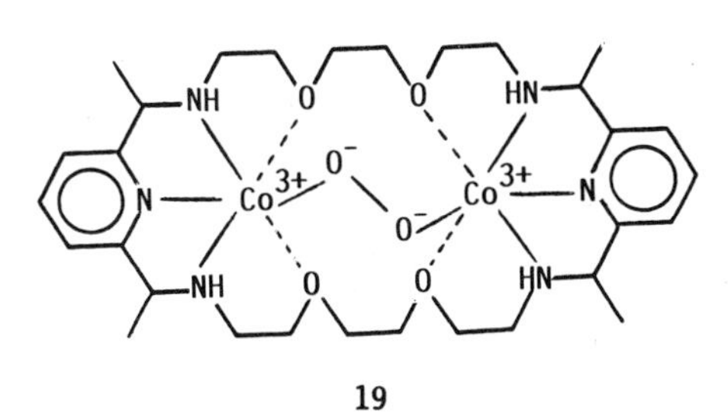

19

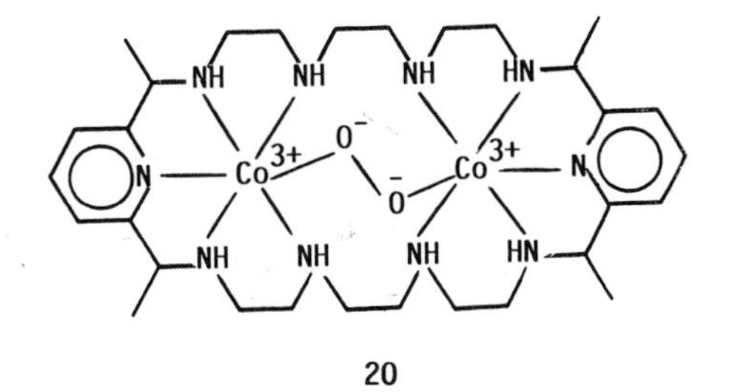

20

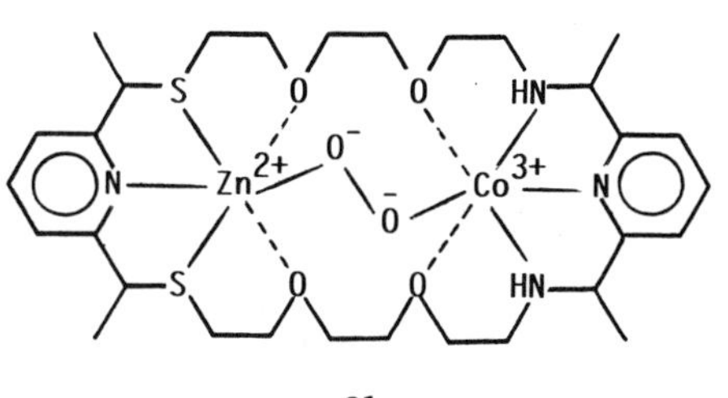

21

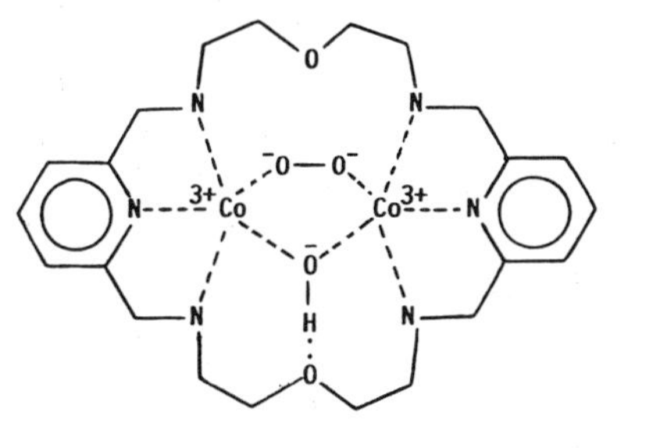

22

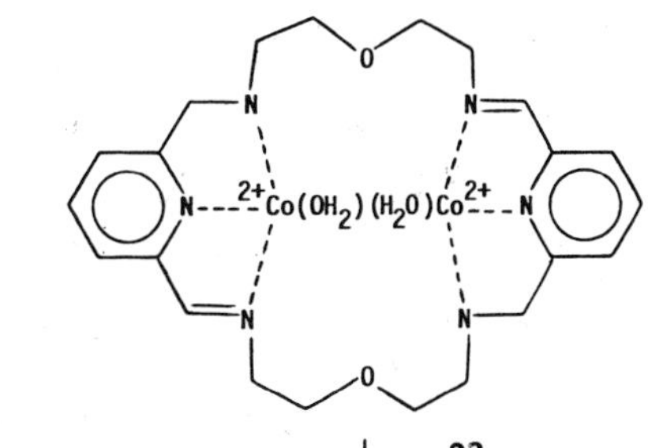

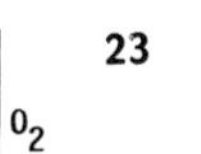

23

O_2

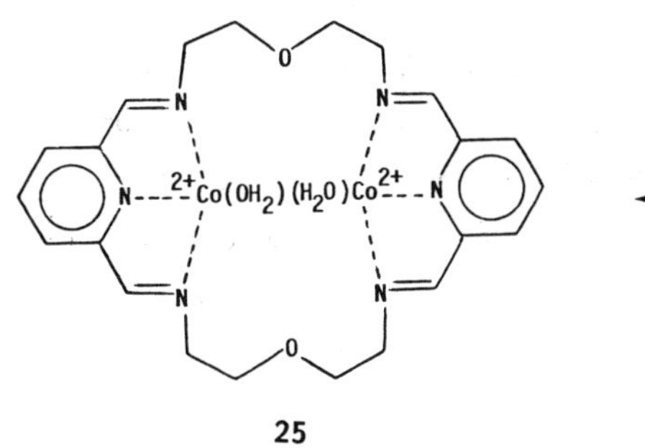

25

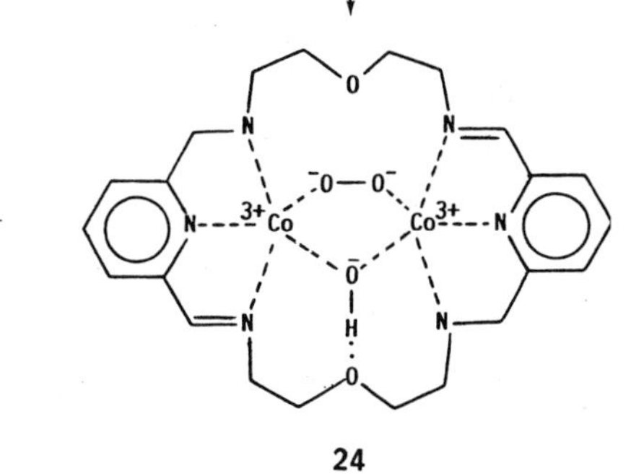

24

SCHEME III

POSSIBLE PATHWAYS OF THE IRREVERSIBLE DEGRATION OF DIBRIDGED DICOBALT DIOXYGEN COMPLEXES OF α,α'-m-XYLYLBIS(TRIETHYLENETETRAAMINE)

R = H, CH_3, OCH_3, OH

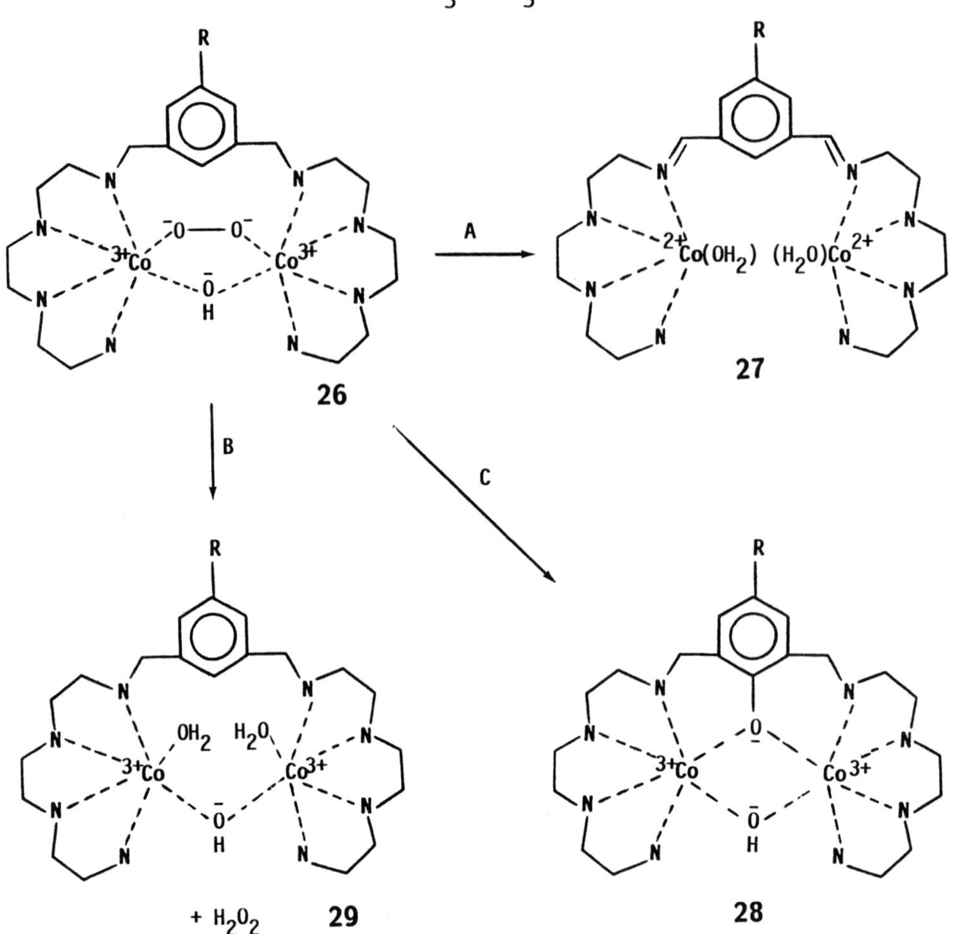

It has been noted above that the dicobalt BISDIEN dibridged dioxygen complex 9 has an additional coordination site on each metal ion that may bind an additional ligand. It is therefore planned to use these coordination postions to bind a bridging reducting ligand such as oxalate or catechol. Scheme IV illustrates a possible reaction pathway for the oxidation of a bridging oxalate ligand by a coordinated μ-peroxo group. Synthetic work leading to the formation of dibridged macrocyclic dicobalt complexes of this type is now in progress.

ACKNOWLEDGEMENT

Acknowledgement is made to the Donors of the Petroleum Research Fund, administered by the American Chemical Society and The Robert A. Welch Foundation (Grant No. A-259) for support of this research.

SCHEME IV

PROPOSED REDOX CATALYSIS IN BINUCLEAR MACROCYCLIC COBALT DIOXYGEN COMPLEXES. OXALATE OXIDATION

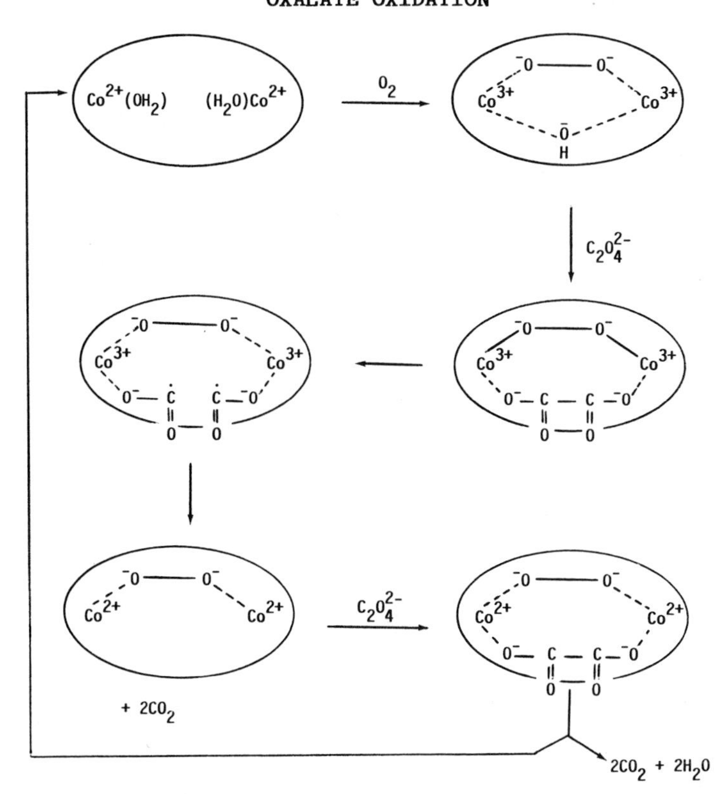

REFERENCES

1. A.E. Martell and M. Calvin, "Chemistry of the Metal Chelate Compounds", Prentice Hall, New York (1952).
2. G. McLendon, and A. E. Martell, Coord. Chem. Revs. 19:1 (1976).
3. R. D. Jones, D. A. Summerville, and F. Basolo, Chem. Rev. 79:139 (1979).
4. E. C. Niederhoffer, J. H. Timmons, and A. E. Martell, Chem. Rev. 84:137 (1984).
5. A. E. Martell, Acc. Chem. Res. 15:155 (1982).
6. A. E. Martell, Proc. Welch Chem. Conf. XXVIII:7 (1984).
7. A. Werner and A. Myelius, Z. Anorg. Chem. 16:245 (1898).
8. T. Tsumaki, Bull. Chem. Soc. Jpn. 13:252 (1938).
9. M. Calvin, R.H. Bailes, and W. K. Wilmarth, J. Am. Chem. Soc. 68:2254 (1946).
10. G. H. Barkelew and M. Calvin, J. Am. Chem. Soc. 68:2257 (1946).
11. W. K. Wilmarth, S. Aranoff and M. Calvin, J. Am. Chem. Soc. 68:2263 (1946).
12. M. Calvin and C. H. Barkelew, J. Am. Chem. Soc. 68:2267 (1946).

13. O. L. Harle and M. Calvin, J. Am. Chem. Soc. 68:2612 (1946).
14. R. H. Bailes and M. Calvin, J. Am. Chem. Soc. 69:1886 (1947).
15. H. Diehl, L. M. Liggett, G. Harrison, C. Hach and R. Curtis, Iowa State Coll. J. Sci., 22:165 (1948); and 13 preceding papers.
16. P. Burk, J. Hearon and A. Schade, J. Biol. Chem. 165:723 (1946).
17. P. Burk, J. Hearon, H. Levy and A. Schade, Fed. Proc., Fed. Am. Soc. Exp. Biol. 6:242 (1947).
18. J. Hearon, Fed. Proc., Fed. Am. Soc. Exp. Biol. 6:259 (1947).
19. J. Hearon, A. Schade, H. Levy and P. Burk, Cancer Res., 7:713 (1947).
20. R. Nakon and A. E. Martell, Inorg. Chem. 11:1002 (1972).
21. R. Nakon and A. E. Martell, J. Inorg. Nucl. Chem. 94:1365 (1972).
22. R. Nakon and A. E. Martell, J. Am. Chem. Soc. 94:3026 (1972).
23. G. McLendon and A. E. Martell, J. Coord. Chem. 4:235 (1975).
24. G. McLendon and A. E. Martell, J. Chem. Soc. Chem. Commun. 223 (1975).
25. W. R. Harris, I. Murase, J. H. Timmons, and A. E. Martell Inorg. Chem. 17:889 (1978).
26. W. R. Harris, J. H. Timmons, and A. E. Martell J. Coord. Chem. 8:251 (1979).
27. J. H. Timmons, W. R. Harris, I. Murase and A. E. Martell Inorg. Chem. 17:2192 (1978).
28. J. H. Timmons, A. E. Martell, W. R. Harris, and I. Murase Inorg. Chem. 21:1525 (1982).
29. R. J. Motekaitis, A. E. Martell, J. M. Lehn, and E. Watanabe Inorg. Chem. 21:4253 (1982).
30. R. J. Motekaitis, A. E. Martell, E. Watanabe, and J. M. Lehn Inorg. Chem. 22:609 (1983).
31. C. J. Raleigh and A. E. Martell, Inorg. Chem. 24:142 (1985).
32. C. J. Raleigh and A. E. Martell, Inorg. Chem. 25:1182 (1986).
33. C. J. Raleigh and A. E. Martell, J. Chem. Soc. Chem. Commun. 335 (1984).
34. C. J. Raleigh and A. E. Martell, J. Coord. Chem. 14:139 (1985).
35. J. H. Timmons, R. H. Niswander, A. Clearfield, and A. E. Martell, Inorg. Chem. 18:2977 (1979).
36. A. Basak and A. E. Martell, Inorg. Chem., submitted.
37. S. M. Nelson, C. V. Knox, M. McCann, and M. G. B. Drew, J. Chem. Soc. Dalton 1669 (1981).

REVERSIBLE COMPLEXES FOR THE RECOVERY OF DIOXYGEN

John A. T. Norman, Guido P. Pez,* and David A. Roberts

Corporate Science and Technology Center
Air Products and Chemicals, Inc.
P.O. Box 538, Allentown, Pennsylvania 18105

ABSTRACT

Dioxygen is produced in tonnage quantities by the distillation of air at cryogenic temperatures. In recent years, alternative technologies have emerged that employ O_2- or N_2-selective sorbents or O_2-permselective polymer membranes. New transition metal complexes that can bind O_2 reversibly and with high specificity may provide the basis for even better processes for dioxygen recovery. One of the more promising approaches is the use of such complexes as O_2 carriers in facilitated transport immobilized liquid membranes. The performance of the cyclidene lacunar "protected site" dioxygen complexes developed by D. Busch et al. has been evaluated in such membranes operating at ca. 0°C. The complexes facilitate the transport of dioxygen and result in O_2 permeabilities and O_2/N_2 selectivities that have been related in a preliminary manner to the complex concentration, equilibrium O_2 binding, reaction kinetics, and carrier and O_2 diffusivities. While the cyclidene complexes proved to be useful in these experimental studies, for practical membranes new carriers would have to be devised that are much more stable toward oxidative degradation. The synthesis and structure of a new "protected-site" reversible cobalt dioxygen complex are described.

INTRODUCTION

Dioxygen is a large-volume commodity chemical, with production in the United States currently exceeding 18 million tons per year. Most pure dioxygen is produced in large-scale cryogenic plants, in which incoming air is compressed, liquified, and separated into its components by distillation. Even the most modern cryogenic plants, however, operate at only ~20% thermodynamic efficiency, and foreseeable engineering innovations are unlikely to greatly improve their performance.

---

*Presenter

Although cryogenic methods continue to dominate, especially for higher purity, large-tonnage oxygen and nitrogen production, several separation technologies based on carbon molecular sieve[1] and zeolite[2] adsorbents and polymer membranes have recently emerged as cost-competitive alternatives for certain lower volume applications. Other methods that, in principle, have the potential to separate dioxygen from air more efficiently than any of these techniques are based on the highly specific but reversible binding of dioxygen by certain metal coordination complexes.

During the past forty years, considerable research has been devoted to dioxygen binding by transition metal complexes. Dioxygen complexes representing a wide variety of structural types are now known for many of the transition metals, and their synthesis and chemistry have been extensively reviewed.[3] Most work has been stimulated by attempts to model important biological dioxygen carriers such as hemoglobin, myoglobin, hemerythrin, and hemocyanin,[4] as well as the enzymes and components of biological energy transduction systems believed to involve dioxygen.[5] However, until recently only a relatively modest effort has been directed toward the development of air separation and oxygen storage technologies based upon coordination compounds that reversibly bind dioxygen.

Absorption Processes

In principle, the separation of air can be effected by utilizing reversible dioxygen complexes in either absorption or membrane processes. In an absorption process, dioxygen is taken up by either a porous solid complex or a solution. After saturation is achieved, the solid or solution carrier is isolated from the air feed stream, and the dioxygen is recovered by reducing the O_2 pressure or by thermal desorption. Because the system is selective for dioxygen, it avoids the need for compression of the accompanying dinitrogen as is practiced in the cryogenic process. An absorption process can also be scaled down to lower production levels, where air liquefaction plants become less efficient.

For a metal complex to function as a useful O_2 sorbent in an air separation process, it must have high capacity, adequate dioxygen binding thermodynamics and kinetics, acceptable cost, and chemical stability. A number of potential solid and solution reversible O_2 absorbents have been examined. The most notable is the family of cobalt(II) bis(salicylal)ethylenediamine, "salcomine," and similar O_2-reactive solids pioneered by Calvin[6] and Diehl,[7] which were later developed for on-board oxygen support systems for the U.S. Air Force.[8] Some more recent examples include the sterically protected porphyrins,[9] complexes anchored to solid supports such as organic polymers,[10] silica,[11] and zeolites.[12] Solutions of traditional salcomine and other known O_2 complexes have been claimed to be useful absorbents in dioxygen separations.[13]

Despite the wide variety of complexes studied, no synthetic carriers have all the properties required for dioxygen separation. Their most serious drawback is chemical instability caused by auto-oxidation reactions.

Membrane Processes

Several emerging technologies for air separation employ semipermeable organic polymer membranes, consisting essentially of a very thin film of an organic polymer on a suitable hollow fiber or flat sheet support. Separations are achieved on the basis of differential solubility and diffusivity of the gases in the organic polymer film.[14]

Because of their physical similarities, however, only limited resolution can be expected for gaseous dioxygen and dinitrogen using such polymer membranes. In addition, because of the generally relatively low diffusion rates of gases in polymers, the membranes must be exceptionally thin (<0.1 μm) in order to achieve high gas permeation rates. However, despite these limitations, silicone rubber and other ultra-thin-film polymer membranes having O_2:N_2 permeability ratios or selectivities of 2-3:1 are finding uses for the separation of air.[15]

Facilitated Transport Membranes

A recent approach to a potentially much more effective separation of air and of other gas mixtures uses, as the membrane, a thin film of a liquid in which a chemically reactive, reversible gas carrier is dissolved. The carrier facilitates transport of the gas across the membrane, resulting in a higher flux and selectivity of the desired reactive gas than would be achievable with the liquid alone.

The mechanism of transport is illustrated in Figure 1, which shows a membrane cross-section. At the feed side (higher pressure) of the membrane, gas A reacts selectively with carrier B, forming the gas/carrier complex AB (Equation 1), which diffuses in the direction of the concentration gradient, toward the permeate side. Here the reaction described by Equation 1 is reversed; gas A is liberated and the carrier returns to the feed side of the membrane.

$$\text{A (gas)} + \text{B (carrier)} \underset{k_{-1}}{\overset{k_1}{\rightleftharpoons}} \text{AB (complex)} \qquad (1)$$

A is also transported by a conventional solution/diffusion mechanism through the solvent.

An actual membrane usually consists of a solution, containing the carrier, immobilized in the pores of a thin porous polymer. The liquid is held within the relatively small pores (<0.4 μm) by capillary forces. Way et al.[16] have published an excellent review of such facilitated transport liquid membranes.

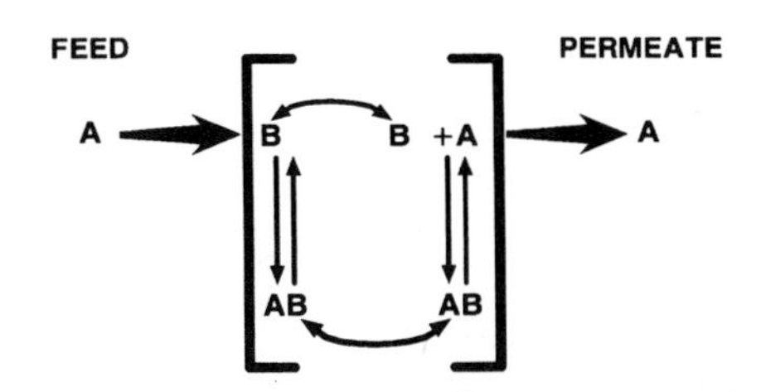

Fig. 1. Schematic representation of processes occurring in a carrier-mediated facilitated transport membrane.

The facilitated transport of dioxygen through an aqueous membrane containing hemoglobin as the carrier was first demonstrated by Scholander.[17] He reported increases in dioxygen flux up to eight times those observed for simple diffusion. This phenomenon was confirmed by later workers who modeled the transport process on the basis of a reversible, chemically reactive diffusing carrier, as outlined above.

The most comprehensive study on facilitated dioxygen transport was carried out by Bassett and Schultz,[18] who used bis(histidinato)-cobalt(II) as the reversible dioxygen carrier. Oxygen fluxes were enhanced over those arising from simple diffusion and their values were related to critical fundamental parameters of the system, namely, membrane thickness, the concentration and rate of diffusion of the carrier, and its oxygenation and deoxygenation kinetics. Unfortunately, their studies were limited by the low solubility (~0.05 M) and chemical instability of the bis(histidinato)cobalt(II) complex. A relatively recent report by Bend Research Inc. to the U.S. Department of Energy[19] describes research wherein relatively high dioxygen fluxes were obtained with immobilized liquid membranes containing cobalt(II) complexes with Schiff base ligands of the bis(salicylal)ethylenediamine (salen) type and also with examples of the cyclidene lacunar ligands described below.

From even the above qualitative mechanism of facilitated dioxygen transport through liquid membranes, one can project the properties of a reversible dioxygen carrier needed to yield an effective O_2-separation membrane. The carrier (B) should be present at a high concentration in the liquid medium; both clearly need to be relatively involatile. The dioxygen affinity, as measured by $K_{eq} = k_1/k_{-1}$ of Equation 1, should be chosen to yield the maximum O_2 loading on the feed side, with a concomitant minimum O_2 loading on the permeate side, for the specific membrane operating conditions. For thin membranes, the kinetics of dioxygen binding and desorption will also determine the O_2 loading at points within the membrane and, hence, the oxygen flux. To be effective, the dioxygen carrier also must be relatively mobile in the liquid medium. Finally, transport properties of the carrier solutions must remain essentially unaltered in the presence of dioxygen for long periods--most likely several years in a commercial operation.

This set of properties is very demanding and not even approached by the behavior of any known dioxygen carriers. The most difficult requirement to meet is that of long-term chemical stability, because all such carriers, even hemoglobin, degrade by auto-oxidation. Thus, although facilitated transport membranes could theoretically provide an elegant method for separating dioxygen from air, the synthesis of a truly effective carrier remains a very challenging problem for inorganic coordination chemists.

In this paper, we describe our preliminary studies of O_2-facilitated transport membranes using as reversible carriers the class of "protected site" or "lacunar," 1:1-binding cobalt(II) dioxygen complexes developed by Busch and co-workers at Ohio State University.[20] Also described is a new reversible oxygen complex which was prepared as part of a continuing effort to arrive at optimum carrier species for transporting dioxygen in liquid membranes.

MEMBRANE STUDIES

Background and Mathematical Models

As discussed above, the simplest mechanism for facilitated transport involves a reversible reaction between a gas A at the feed side of a membrane and a carrier solute B to yield the gas/carrier complex AB, which dissociates at the permeate side (Equation 1).

Figure 2 shows the idealized concentration profiles for each of these species along the cross-section of a liquid membrane, operating in essentially a "diffusion-controlled" mode (vide infra). In an analysis of the system, both diffusion of the species and characteristics of the chemical reactions between them must be considered. The following fundamental parameters determine the flux of gas A through the membrane:

- Concentration of A (C_A) at the feed (C_A^0) and permeate sides (C_A^L).
- Total concentration of carrier in the membrane ($C_T = C_A + C_{AB}$).
- Forward (k_1) and reverse (k_{-1}) reaction rate constants.
- Diffusivities of components (D_A, D_B, D_{AB}).

Kemena et al.[21] have reviewed the various models used to describe facilitated transport and present a mathematical method for calculating the gas flux for a flat plate membrane in terms of the above parameters. The salient feature of the model is that diffusion is assumed to be governed by Fick's law (Equation 2) and the steady-state mass balance of each reacting component (Equation 3):

$$\text{flux } (J) = -D_i dC_i/dx \qquad (2)$$

$$-dJ/dx = D_i d^2C_i/dx^2 = r_i \qquad (3)$$

D_i is the diffusion coefficient of species i, dC_i/dx represents the concentration gradient or driving force for diffusion, and r_i is the rate of disappearance of species i.

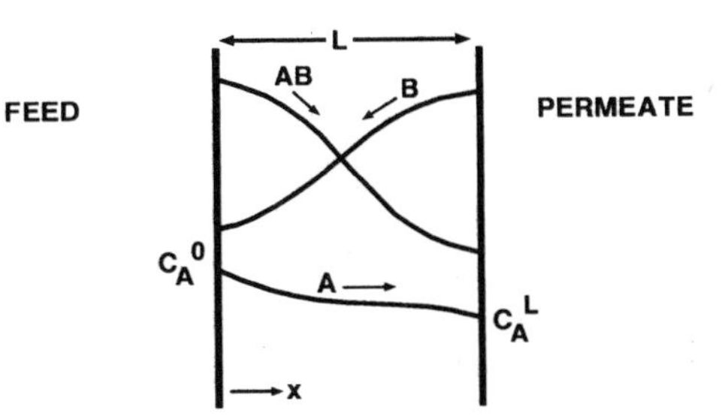

Fig. 2. Concentration profiles for the species in a liquid membrane. A represents the dissolved gas, B the unbound carrier, and AB the gas/carrier complex.

For one-dimensional flow in a flat membrane system, a mass balance for each species yields the differential Equations 4, 5, and 6:

$$D_A \frac{d^2C_A}{dx^2} = k_1C_AC_B - k_{-1}C_{AB} \tag{4}$$

$$D_B \frac{d^2C_B}{dx^2} = k_1C_AC_B - k_{-1}C_{AB} \tag{5}$$

$$D_{AB} \frac{d^2C_{AB}}{dx^2} = k_{-1}C_{AB} - k_1C_AC_B \tag{6}$$

The boundary conditions are as follows: at $x = 0$ and $x = L$, $dC_B/dx = 0$, since the carrier cannot leave the confines of the membrane; at $x = 0$, $C_A = C_A^0$; at $x = L$, $C_A = C_A^L$.

Since carrier B is usually a much larger molecule than the gas to be transported, it is assumed that $D_B = D_{AB}$. The flux of A is obtained by solving the equations for dC_A/dx, dC_B/dx, etc., and expressing the total flux of A as follows:

$$\text{flux } (J_A) = -D_AdC_A/dx - D_{AB}dC_{AB}/dx \tag{7}$$

It is only possible to solve these differential equations by numerical methods. For the special case where the system is at equilibrium throughout the membrane, and transport is thereby diffusion-limited, the flux can be calculated from Equation 8, where $K_{eq} = k_1/k_{-1}$ is the equilibrium constant for the reaction.

$$J_A = \frac{D_A(C_A^0 - C_A^L)}{L} + \frac{D_{AB}K_{eq}C_T(C_A^0 - C_A^L)}{L(1 + K_{eq}C_A^0)\,(1 + K_{eq}C_A^L)} \tag{8}$$

A useful quantity for describing the effectiveness of the carrier is the facilitation factor (F):

$$F = \frac{\text{total flux of A with carrier-mediated facilitation}}{\text{unfacilitated flux of A}} \tag{9}$$

Since the unfacilitated flux is given by the first term of Equation 8, the facilitation factor for the diffusion-limited case is given by the expression:

$$F = 1 + \frac{D_{AB}K_{eq}C_T}{D_A(1 + K_{eq}C_A^0)(1 + K_{eq}C_A^L)} \tag{10}$$

The optimum K_{eq} for diffusion-limited facilitation depends on the feed C_A^0 and permeate C_A^L gas concentrations and is given by Equation 11.

$$K_{eq}(\text{optimum}) = (C_A^0C_A^L)^{-1/2} \tag{11}$$

At each membrane interface, the concentration of A is related to the partial pressure of the gas (p) in contact with the membrane and is given by:

$$C_A^{\,0} = Hp_A^{\,0} \text{ and } C_A^{\,L} = Hp_A^{\,L} \qquad (12)$$

where H is Henry's law constant.

Kemena et al.[21] have also mathematically explored the extent of facilitated transport for various useful regimes of the fundamental parameters C_T, k_{-1}, K_{eq}, D_A, and D_{AB}, and the membrane thickness L. In many cases, especially with thin membranes and low reverse reaction rates, the flux is determined by both the reaction kinetics and rates of diffusion. This is nicely illustrated in the recent studies of Koval et al.[22] on the facilitated transport of CO in a liquid membrane mediated by an iron complex carrier.

Experimental Membrane Design and Evaluation

We prepared and evaluated several experimental carrier-mediated facilitated transport membranes for the separation of O_2 from air. Immobilized liquid membranes were prepared by impregnating a disc of a porous polymer, generally poly(vinylidene fluoride) (PVDF), ca. 5 cm in diameter and 25 to 130 μm in thickness, with a filtered solution of the metal complex at a known concentration under dinitrogen. The membrane disc was then loaded into a cell holder (see Figure 3), which was subsequently mounted in a membrane testing apparatus (Figure 4).

The entire system was flushed with solvent-saturated helium. When constant temperature was reached, the upstream side was switched to a solvent-saturated feed of dry air at slightly above ambient pressure. The downstream or permeate side of the membrane was swept with a similarly saturated helium stream leading to a gas chromatograph used for O_2 and N_2 analyses. The permeate O_2 and N_2 levels were monitored at 10-20-min intervals; data gathered over several hours of steady-state operation were used to calculate the permeability of the respective gases. Experimental rates for permeation of the gases are expressed in terms of Barrer units, where 1 Barrer = (10^{-10} sccm·cm/cm^2·sec·cmHg).

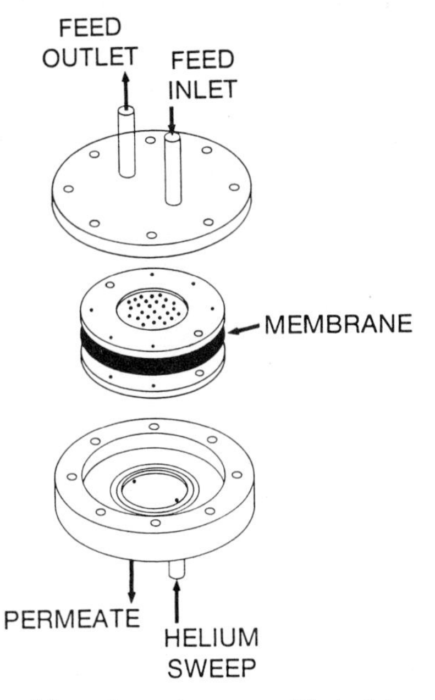

Fig. 3. Schematic of membrane cell holder for membrane testing apparatus.

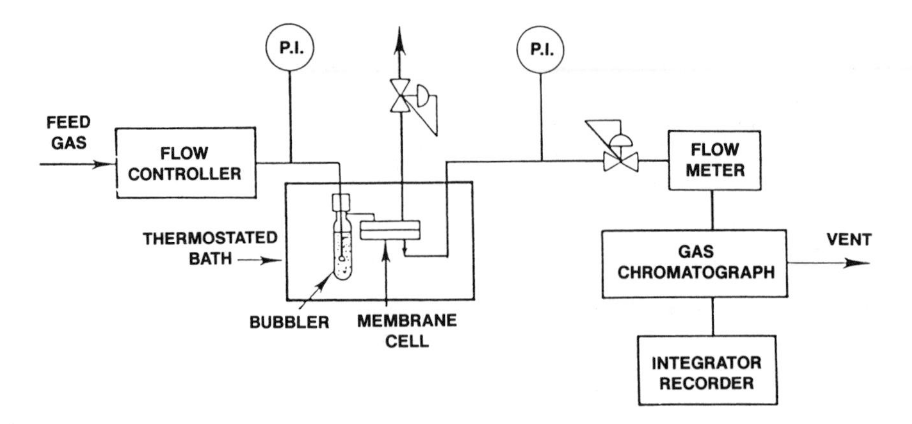

Fig. 4. Membrane testing apparatus

Equilibrium constants for dioxygen binding (K_{eq}) and for axial base coordination (K_B) of the metal complex were determined spectrophotometrically at typical concentrations of 10^{-4} M. Details of this technique are described elsewhere.[20c]

As explained earlier, the major challenge in arriving at an effective air-separation membrane is that of finding a suitable metal complex dioxygen carrier. For the most favorable case of a diffusion-controlled membrane, according to Equation 10 the maximum facilitation factor (and hence the highest O_2 flux and O_2/N_2 selectivity) will be obtained with carriers which can be used at a high concentration (C_T), are highly mobile (D_{AB}), and have a favorable K_{eq}. The optimum K_{eq} depends on the feed (C_A^0) and permeate (C_A^L) dioxygen concentrations and hence the partial O_2 pressure at these interfaces (Equations 11 and 12).

While there is no known carrier having this combination of properties, the cyclidene lacunar complexes[20] cited earlier have a number of attractive characteristics and thus seemed to be particularly appropriate for use in these membrane studies. The complexes consist of a cobalt(II) ion that is coordinated by a macrocycle containing four nitrogen atoms (N). This N_4-Co system is in an approximately planar environment, but the superstructure "folds" out of this plane by the effect of the bridging polymethylene chain R^1, as shown in the following structure.

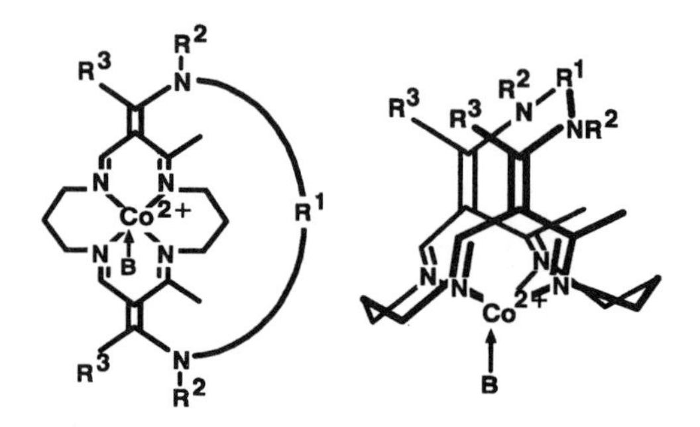

Table 1. Equilibrium O_2 Binding Data for C_6 and C_8 Cyclidene Lacunar Cobalt(II) Complexes at 0°C in DCB

Complex	Axial Base	$P_{1/2}(O_2)$ (Torr)
$C_6(Co)$	NMeIm, 1.5 M in DCB	2.7
$C_6(Co)$	DCB only	241
$C_8(Co)$	NMeIm, 1.5 M in DCB	0.2
$C_8(Co)$	DCB only	147

This molecular configuration results in the formation of a "protected pocket" or lacuna for coordination of the dioxygen, which precludes the formation of peroxy-bridged dimers. A basic ligand (B), usually N-methylimidazole (NMeIm), occupies the axial coordination site as shown.

The complexes are quite soluble in organic solvents. In fact, they show a 1:1 cobalt-to-dioxygen binding stoichiometry and their K_{eq} can be easily tuned to useful ranges of oxygen half-saturation pressure [$P_{1/2}(O_2) = 1/K_{eq}$ of ca. 50 to 200 Torr], by altering the length of the R^1 polymethylene chain and by adjusting the basicity of the axial base, B. Samples of the complexes with $R^2 = R^3 = CH_3$ and $R^1 = -(CH_2)_6-$ were initially made in our laboratory according to ref. 20b. Subsequently, this material and compounds with $R_3 = -(CH_2)_6-$ and $-(CH_2)_8-$ (hereinafter referred to as the C_6, C_8, etc. metal cyclidene lacunar complexes) were provided for these studies by Busch.

Our initial attempts at using solutions of the $C_6(Co)$ complex containing (as is customary) a large excess of the NMeIm axial base as liquid membranes were not successful because of poor stability of the systems toward auto-oxidation. Work by Busch et al. had shown that there is a strong dependence of the decomposition rate on the concentration of the axial base. Since the affinity of the complexes for even the relatively strong axial base NMeIm is quite low (K_B for the C_6 cobalt(II) complex in $CH_3CN = [C_6.NMeIm]/[NMeIm][C_6] = 79.4 \pm 0.9$ M^{-1}), high concentrations are needed for maximum O_2 binding. We reasoned, however, that an adequate O_2 coordination could be achieved by using polar organic solvents, in the absence of an additional axial base. Equilibrium oxygen binding data [expressed in terms of $P_{1/2}(O_2)$] are listed in Table 1 for the C_6- and C_8-bridged complexes in 1,4-dicyanobutane (DCB) in the presence and absence of NMeIm, respectively.

In a liquid membrane containing the $C_6(Co)$ complex in DCB, ca. 40% dioxygen loading would be expected at the feed interface for an O_2 pressure of 160 Torr. While this loading level is not ideal, we considered it at least adequate. Hence, the first membrane studies were performed using the system of cyclidene lacunar complexes in DCB.

Results and Discussion

Initial Studies. Performance data for membranes consisting of DCB solutions of the cyclidene complexes immobilized in porous poly(vinylidene fluoride) (PVDF) and operating at ca. 0°C are shown in Table 2. The first

Table 2. Facilitated Transport of Dioxygen in Liquid Membranes[a] Containing C_6- and C_8-Bridged Cyclidene Lacunar Complexes in DCB

#	Temp (°C)	Complex	Concn (M)	$P(O_2)$ (Barrers)	$\frac{P(O_2)}{P(N_2)}$	F	Permeate O_2 (%)
1.	-0.5	Neat DCB	0	36.2	2.22		37
2.	-1.0	$C_6(Ni)$	0.088	31.4	2.21		
3.	-0.6	$C_6(Ni)$	0.143	26.7	2.21		
4.	-0.2	$C_6(Co)$	0.090	45	3.5	1.6	48
5.	-0.2	$C_6(Co)$	0.143	48	3.6	1.7	49
6.	-0.8	$C_6(Co)$	0.192	76[b]	5.3	2.4	59
7.	-0.5	$C_6(Co)$	0.324	43[c]	5.0	2.2	57
8.	-0.4	$C_8(Co)$	0.185	68	5.3	2.4	58

[a]Liquid membrane solutions supported in porous PVDF (Millipore Corp.); thickness, 125 μm; porosity, 0.75; tortuosity, 1.25; area 10.7 cm^2. Conditions: solvent-saturated dry air feed with $P(O_2)$ ranging from 164 to 168 torr; likewise, solvent-saturated permeate He sweep, each at ca. 15 scc/min, with <30 Torr transmembrane pressure differential. Values were corrected for the porosity and tortuosity of the support. Note that the data are for the membrane operating under conditions of essentially "zero recovery," i.e., where the feed stream composition as it passes across the membrane is not significantly changed by the permeating gas.
[b]Mean $P(O_2)$, $\sigma = 0.1$, for period of 8-30 hr on stream.
[c]Mean $P(O_2)$, $\sigma = 0.3$, for period of 1.5-5 hr; decreased by 20% after 20 hr (see text).

entry gives the oxygen permeability [$P(O_2)$] and the O_2/N_2 permeability ratio or selectivity for the passage of air through the membrane, when no complex is present. Enriched air (37% O_2) is seen in the helium-swept permeate stream because of the apparently greater permeability of O_2 than N_2 in DCB. This is probably because O_2 is more soluble than N_2 in this liquid.

A solution of the cyclidene lacunar nickel(II) complex (which is unreactive toward dioxygen) in DCB gave lower O_2 and N_2 permeabilities, which was expected in view of the greater viscosity and consequently lower rates of gas diffusion in comparison to those in the neat solvent. The $P(O_2)/P(N_2)$ ratio was found to be remarkably constant at 2.21 ± 0.02 for these and other solutions of O_2-inert cyclidene complexes in DCB at ca. 0°C.

Membranes containing the C_6 cobalt(II) complex had significantly higher O_2 fluxes than the above control $C_6(Ni)$ systems. The facilitation factor (F) for dioxygen permeation given in Tables 2 and 3 is the ratio of $P(O_2)/P(N_2)$ for the Co(II)-containing membrane to $P(O_2)/P(N_2)$ for the control. The latter is taken to be 2.21, as explained above. This method of calculation gives slightly different F values than would be obtained using the $P(O_2)$ facilitated and unfacilitated values alone, but is considered to be a better representation of the system since the dinitrogen permeability essentially acts as an internal standard to correct for any small differences in membrane porosity, thickness, etc.

Table 3. Facilitated Transport of Dioxygen in Liquid Membranes[a] Containing C_5- and C_6-Bridged Lacunar Complexes. Effect of Solvent, Axial Base, and Membrane Thickness

#	Temp (°C)	Complex Concn (M)	Solvent Axial Base (equiv/Co)	Membrane Thickness (μm)	$P(O_2)$ (Barrer)	$\frac{P(O_2)}{P(N_2)}$	F	Permeate O_2 (%)
1.	0.2	C_6(Ni) 0.088	CH_3CN -	128	677	1.75		
2.	0.2	C_6(Co) 0.088	CH_3CN -	128	1290	3.2	1.8	
3.	0.0	C_6(Co) 0.192	DCB BzIm, 2equiv	128	89[b]	9.8	4.4	72
4.	0.3	C_5(Co) 0.192	DCB BzIm, 1equiv	128	33[c]	5.1	2.3	57
5.	0.0	C_5(Co) 0.192	DCB BzIm, 2equiv	128	113[d]	13	5.8	78
6.	0.0	C_5(Co) 0.192	DCB BzIm, 2equiv	63	96	11	4.9	
7.	0.0	C_5(Co) 0.192	DCB BzIm, 2equiv	25	86	10	4.4	

[a]Liquid membrane solutions supported in porous polymers: PVDF (128 μm), polypropylene (Celgard[R], Celanese Corp., 25 μm). Experimental conditions otherwise as described previously (see text and footnotes to Table 2).
[b]Mean $P(O_2)$, $\sigma = 1$ for period 2-15 hr on stream; decreased by 15% after 48 hr.
[c]Mean $P(O_2)$, $\sigma = 1.2$ for 3.3-24 hr; decreased by 23% after 100 hr.
[d]Mean $P(O_2)$, $\sigma = 0.4$ for 2.7-14 hr; decreased by 47% after 100 hr.

The data in Table 2 clearly show that the cobalt(II) cyclidene complexes facilitate the transport of dioxygen across the membrane. In the best cases, enriched air containing up to 57-59% O_2 was obtained in the permeate stream. Monitoring the dioxygen permeability of a membrane with time permitted an evaluation of the carrier lifetime (see footnotes, Tables 2 and 3). The C_6 and C_8 cobalt(II) complex carriers in DCB at ca. 0°C proved to be sufficiently stable toward auto-oxidation for recording meaningful experimental membrane data. The most stable system studied was the membrane containing a 0.192 M solution of the C_6(Co) complex in DCB where the dioxygen permeability was essentially constant for the period of 8 to 30 hr on stream at -0.8°C (#6, Table 2). This is significantly longer than lifetimes observed in our preliminary experiments done in the presence of excess NMeIm.

However, the oxygen permeability for a membrane containing a more concentrated solution of the same complex, although initially constant for the first 1.5 to 5 hr on stream, subsequently decreased steadily under otherwise similar conditions (#7, Table 2). The mechanism of auto-oxidation of these cyclidene complexes is poorly understood. In this situation, the decomposition rate may also depend on such extrinsic factors as trace impurities in the solvent or characteristics of the porous polymer support material.

Facilitated Transport Model. In the membrane experiments described above, the concentration of dioxygen in the helium-swept permeate stream was very low (100-300 ppm), so that the term C_A^L in Equation 10, which represents the concentration of dissolved O_2 at the permeate interface, was essentially zero. Furthermore, by expressing K_{eq} in reciprocal O_2 pressure units ($Torr^{-1}$) and C_A^L, the O_2 concentration at the feed interface in terms of Henry's law constant (H) (moles of O_2/liter·Torr) for the solubility of oxygen, Equation 10 can be rewritten in the form:

$$F = 1 + \frac{D_{CoO_2}}{H \cdot D_{O_2}} \left[\frac{C_T K_{eq}}{1 + K_{eq}\, p(O_2)}\right] \qquad (13)$$

where $p(O_2)$ is the oxygen partial pressure in the feed stream.

If this diffusion-limited equilibrium transport model is applicable, the ratio of diffusivities can be estimated from the slope of a line derived by plotting F - 1 versus the term in square brackets, for a range of C_T, K_{eq} and $p(O_2)$ values.

Based on the data in Table 2 for C_T ranging from 0 to 0.192 M (including the point of F = 1 where C_T = 0), a linear correlation of the five points [correlation coefficient (R^2) = 0.93] together with an O_2 solubility assumed to be equal to that in acetonitrile [8.1 mM/(liter·atm) at 25°C[23]] gives a value of D_{CoO_2}/D_{O_2} ~ 0.03. This estimate is instructive because it points to an important limitation of liquid membranes with this particular carrier system. Thus while incorporating a dioxygen complex in the membrane greatly increases the overall O_2 concentration, the transport of this gas is only slightly enhanced (F_{max} ~ 2.4 up to [Co] = 0.192 M), in part because of the relatively low rate of diffusion of the bulky complex carrier.* By using

* Membrane performance depends strongly on the ratio of diffusion constants. For this example, an order of magnitude improvement in D_{CoO_2}/D_{O_2} from 0.03 to 0.3 would increase the product stream concentration from 59 to 90% dioxygen.

a membrane containing a solution that is 0.324 M in the cobalt C_6 complex, a lower O_2 permeability and a facilitation factor of only ~2.2 are realized, probably indicative of an even smaller relative diffusivity of the carrier at this higher concentration.

Since low carrier diffusivities in 1,4-dicyanobutane appeared to be a problem, some experiments were performed using a less viscous solvent, acetonitrile. Membrane performance data for a 0.088 M solution of the $Co(C_6)$ complex and of the corresponding Ni(II) control in CH_3CN are listed as the first two entries in Table 3. Although the dioxygen permeability is about one order of magnitude greater than in DCB, the dioxygen enrichment level and facilitation factor are only slightly enhanced, suggesting that the ratio of carrier-to-dioxygen diffusivities does not increase markedly. While the overall performance of the membrane is improved by using less viscous lower molecular weight solvents, their concomitantly greater volatility makes it difficult to maintain the liquid membrane.

Since in the experiments using the $C_6(Co)$ complex in DCB the carrier is only about 40% oxygen-loaded at the feed interface, we felt that the facilitation factor could be improved by adding small amounts of N-benzylimidazole (BzIm), which is a stronger base than the DCB solvent. For a membrane where 2 equiv of this base is added per mole of the $C_6(Co)$ complex, the facilitation factor improved significantly, from 2.4 to 4.4 (cf. entries #6 and #3 in Tables 2 and 3, respectively). Also (Table 3, entries #4 and #5), for the C_5-bridged cobalt complex [which under similar conditions has a lower O_2 affinity than the $C_6(Co)$ complex], F depends strongly on the amount of added base. The best performance in these experiments was achieved with a carrier solution consisting of a $C_5(Co)$ complex in DCB to which 2 equiv of axial base was added (#5, Table 3). The membrane functioned with a dioxygen facilitation factor of 5.8 and afforded 78% O_2-enriched air. However, the auto-oxidation problem seemed to be exacerbated by the use of this level of axial base.

The facilitated transport data in Table 2 were discussed above in terms of the diffusion controlled facilitated transport model, Equation 13. While this relation provided an estimate of the ratio of carrier-to-oxygen diffusion coefficients, the data are unfortunately insufficient to provide an adequate test for the validity of this model. The model assumes that the reactions of O_2 with the complex proceed at rates that are much faster than the time of transit of the unbound dioxygen through the membrane. If this is not the case, the facilitation factor will be a function of the membrane thickness. To test this sensitivity to the membrane diffusional path length, experiments were conducted using a solution of the $C_5(Co)$ carrier in porous supports of different thicknesses. Results are shown in Table 3, examples #5 to #7. There is an apparent decrease in F in going from the standard 128-μm PVDF to thinner membranes, which implies that, at least for the latter, there are kinetic limitations to the facilitated transport.

DIOXYGEN METAL COMPLEX SYNTHESIS

With a knowledge of the chemistry, performance, and limitations of the Busch cyclidene lacunar complexes in immobilized liquid facilitated-transport membranes, we sought to prepare new O_2 carriers that would offer improved resistance toward auto-oxidation. We felt this might be achieved by synthesizing cyclidene lacunar complexes having additional steric protection about the cobalt center. A typical cyclidene complex 1 (which exists as a dipositive cation) is shown in the following structure. Since the polymethylene chain "strap" passes on one side of the molecule, we felt that a complex with a more effectively protected O_2 binding site could be prepared, in principle, by joining carbons 1 and 1' to carbons 2 and 2', respectively, in the form of two cyclohexene rings (structure 2).

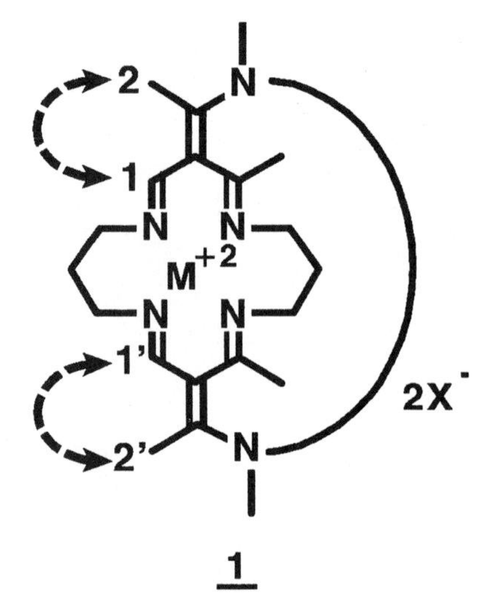

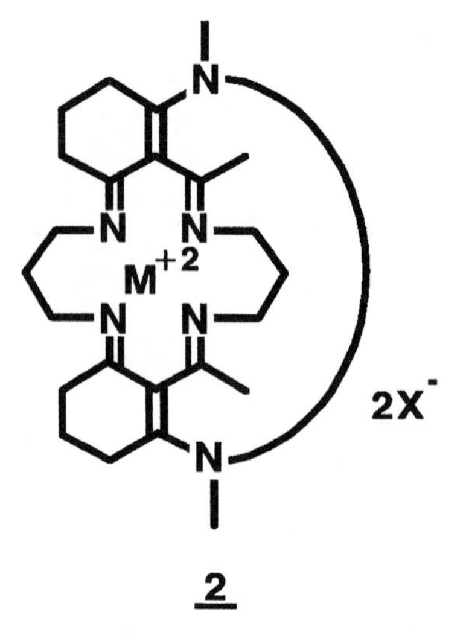

Our initial synthesis efforts toward the desired "cyclohexyl" complex 2 are shown in Scheme 1. In this sequence, 1,3-cyclohexanedione was acylated at the 2 position via treatment with acetyl chloride/pyridine and then $AlCl_3$, to form 2-acetyl-1,3-cyclohexanedione (3). This trione was then refluxed with oxalyl chloride in 1,2-dichloroethane to yield the previously reported[24] chlorovinyldione (4). Reaction of this compound

Scheme 1: Preparation of Macrocyclic Intermediates 7 and 7'

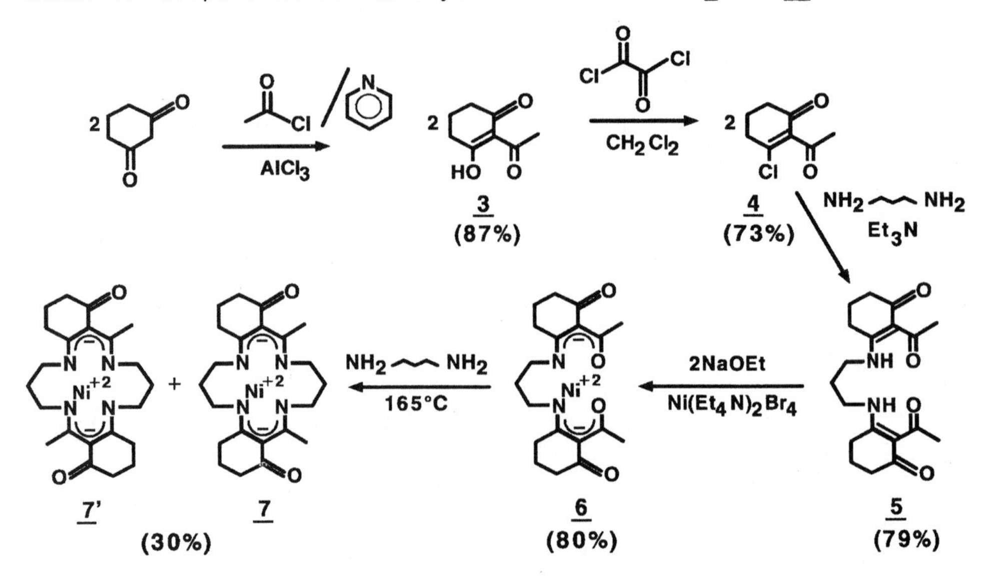

with 1,3-propanediamine in the presence of triethylamine afforded the ligand 5.[25] Its nickel(II) complex 6 was prepared as a bright red solid by deprotonation with sodium ethoxide, followed by addition of bis(tetraethylammonium)tetrabromonickel(II)ate. Refluxing this complex with 1,3-diaminopropane at 165°C resulted in ring closure to yield the macrocyclic N_4 complex, surprisingly in both the cis and trans forms, 7 and 7', respectively. Fractional crystallization of this mixture from methanol allowed the separation of pure 7', which was identified by its unique ^{13}C NMR spectrum. The C_{2h} symmetry of 7' dictates that there be only three chemically nonequivalent methylene carbons in the amine bridging unit, whereas there are four in the cis isomer 7.

At this point we realized that if this imprint of trans symmetry could be carried through the last three synthesis steps, we would arrive at a lacunar complex with a bridging moiety running directly across the center of the molecule. We felt that this novel trans bridge could provide an unusually high degree of steric protection for the metal center, and our subsequent efforts were thus directed at preparing such a trans bridged structure (Scheme 2).

The trans diketone complex 7' was dialkylated via reaction with triethyloxonium tetrafluoroborate in CH_2Cl_2; then monomethylamine was added directly to yield the diamine complex 8 as a yellow crystalline solid. This compound was then "bridged" by reaction with 1,7-ditosylheptane under conditions of high dilution following the general procedures developed by Busch et al.[20b] This gave the lacunar nickel(II) complex 10, which we intended to convert to the dioxygen reactive cobalt(II) form. However, all attempts to remove Ni^{+2} from 10 failed, and we were forced to prepare the desired compound 12 by bridging the Co(II) complex 11. The latter was easily made by demetalation of 8 with HCl/$NaPF_6$, followed by reaction of the free protonated ligand 9 with cobalt(II) acetate. The final product (12) was obtained after recrystallization from methanol as a yellow microcrystalline solid, which was characterized from a combination of elemental analyses and a single-crystal X-ray structure (vide infra).

Scheme 2: Synthesis of Trans-Bridged Co(II) "Cyclohexyl" Complex 12

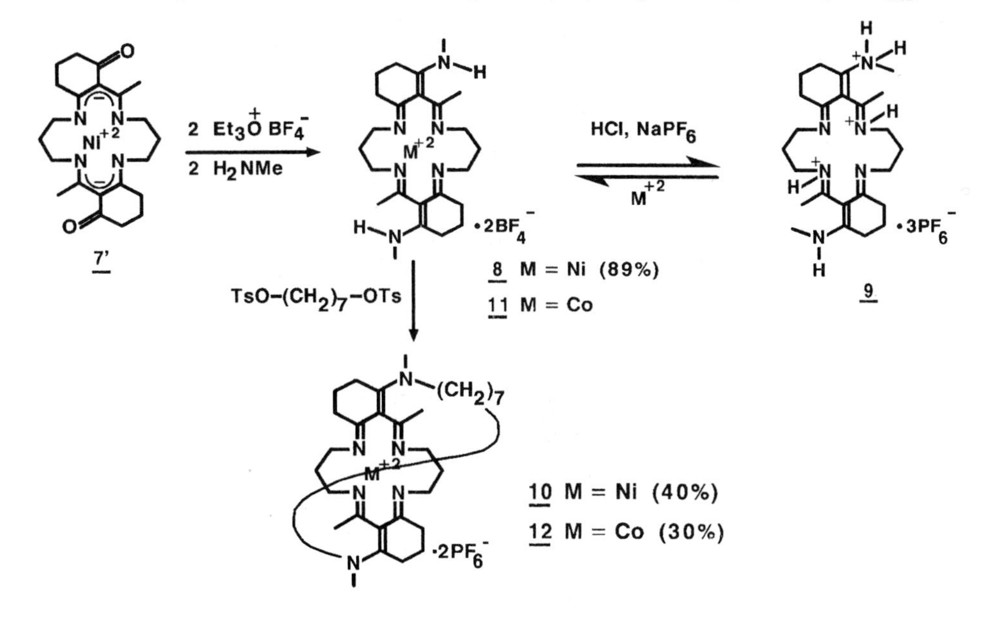

Table 4. Dioxygen Half-Saturation Pressure for Complex 12

T(°C)	$P_{1/2}(O_2)$ (Torr)
-16.5	418
-23	151
-35	65

The dioxygen binding at a concentration of 5×10^{-4} M in acetonitrile containing 1.5 M N-methylimidazole was determined by the spectrophotometric method[20c] cited earlier. Exposure of the solutions to O_2 at room temperature did not give any significant spectral changes; however, upon cooling, reversible dioxygen reactivity was seen. Table 4 lists preliminary estimates obtained for the equilibrium O_2 binding.

Surprisingly, the dioxygen affinity is much less than that of the analogous C_6- and C_8-bridged lacunar cyclidene complexes prepared by Busch et al. (cf. Table 1). An X-ray structural determination (performed by C. Day of Crystallitics Inc.) confirmed the expected structure of compound 12 (see Scheme 2), with the polymethylene chain straddling across the molecule from the two trans-positioned bridgehead nitrogen atoms. An ORTEP diagram of a side view of the molecule is shown in Figure 5.

A cavity or lacuna (in the terminology developed by Busch) is clearly seen above the Co(II) center, where O_2 binding occurs. However, a space filling model using the van der Waals radii for the respective atoms shows that there is indeed a very tight fit for O_2 in this cavity, which is presumably the basis for the unexpectedly low dioxygen affinity (Figure 6). Structural modifications of the complex to increase the size of the lacuna may be expected to enhance O_2 binding.

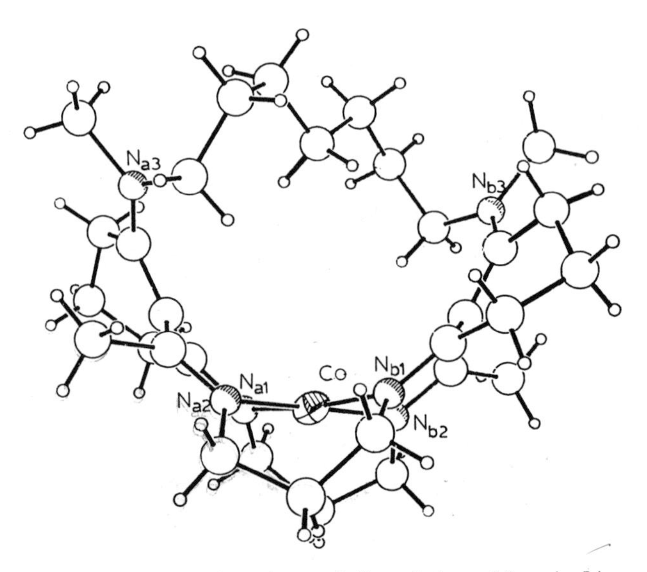

Fig. 5. ORTEP drawing of "cyclohexyl" cobalt complex 12 (n = 7)

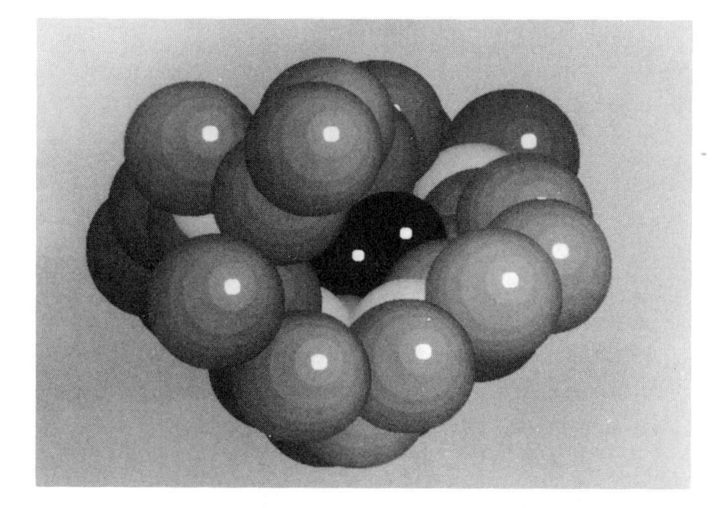

Fig. 6. Space filling model of the "cyclohexyl" complex 12 showing a cobalt-coordinated dioxygen (as two dark spheres) contained within the lacuna.*

CONCLUSION

In this work we have shown that cyclidene lacunar complexes can function as effective O_2 carriers for transporting dioxygen in immobilized liquid membranes. The performance of the complexes has been related in only a preliminary manner to such factors as concentration, O_2 affinity, reaction kinetics, and the ratio of dioxygen and dioxygen carrier diffusivities. The complexes provide an excellent system for a fundamental study of facilitated transport where, by using the mathematical model described, the facilitation factor could be accurately predicted from a knowledge of the above independently determinable basic carrier properties.

From a practical viewpoint, facilitated transport liquid membranes employing the cyclidene complex carriers have a number of inherent limitations. Although the $P(O_2)/P(N_2)$ permeability ratios of 3-13 we observed are higher than corresponding values of 2-4 for organic polymer membranes, the magnitude of our observed dioxygen permeabilities are such that very thin membranes of the order of 1-5 μm or less would be required to yield adequate dioxygen fluxes. However, as we have seen in these preliminary studies, when the membrane thickness is reduced, the dioxygen flux does not increase correspondingly, because at very short diffusional path lengths the facilitated transport becomes increasingly limited by the complex-O_2 reaction kinetics. It appears that complex carriers having an appropriate O_2 binding constant with much faster dioxygen off-rates may be needed to be effective with very thin membranes. Also, maintaining carrier solutions of finite volatility in such thin films represents a considerable technical challenge.

---

* Figure 6 was derived using Chem-X computational software (Chemical Design Ltd., Oxford, England) on the basis of the crystal structure parameters for 12 and assuming Co-O_2 bond distances and angles to be as found for Co(t-Bsalten)(bzImid)(O_2).[26] Standard van der Waals radii for C, N, and O, and the covalent radius for Co(II) (1.16 A) were employed. Hydrogen atoms were not included.

To achieve high O_2 permeabilities, the carrier must be present at a high concentration in the solution and still be quite mobile. We found, surprisingly, that the ratio of carrier to dioxygen diffusivities for the cyclidene complexes in DCB ($^{D}CoO_2/^{D}O_2$) is only about 0.03, and appears to worsen upon increasing the concentration. To attain the best O_2 fluxes, an optimal combination of complex/solvent and temperature would have to be reached.

The lifetime of the membrane is currently limited by the decomposition rate of the carrier complex, although we have shown that one can extend this lifetime significantly by varying the nature of the solvent and the concentration and strength of the axial base. This tendency of dioxygen complexes to decompose via auto-oxidation reactions is the major problem limiting their utility in separation processes. Useful metal complex O_2 absorbents and facilitated transport carriers will need to be far more resistant to oxidative degradation.

The design of dioxygen complexes that meet absorption process and membrane requirements presents a very challenging goal for synthetic chemists. One of the most promising approaches seems to be the synthesis of 1:1 cobalt-to-dioxygen binding complexes, where the bound O_2 is held in a sterically protected site, which in principle precludes dimerization and other unwanted reactions. We have prepared a new dioxygen complex that further illustrates this concept and also serves to emphasize the need for synthesis efforts to arrive at new dioxygen carriers that are tailored for use in industrial applications.

ACKNOWLEDGEMENTS

Contributions of K. Venkataraman and M. Morris to the membrane testing work are gratefully acknowledged. We also thank S. Auvil, F. Herman, and P. Burban for very helpful discussions, and W. L. Barrow for computation of the space filling model.

REFERENCES

(1)(a) Knoblauch, K. Chem. Eng. 1978 (Nov 6), 87. (b) Nandi, S.P.; Walker, Jr. P.L. Sep. Sci. 1976, 11, 441. (c) Juntgen, H.; Knoblauch, K.; Hardner, K. Fuel 1981, 60, 817.

(2) Breck, D.W. "Zeolite Molecular Sieves"; John Wiley & Sons: N.Y., 1974, 709. See also ref 1b.

(3)(a) Gubelmann, M.H.; Williams, A.F. Struct. Bonding 1983, 55, 1. (b) Niederhoffer, E.C.; Timmons, J.H.; Martell, A.E. Chem. Rev. 1984, 84, 137.

(4)(a) Spiro, T.G. "Metal Ion Activation of Dioxygen"; John Wiley: New York, 1980. (b) Royal Society of Chemistry. "Oxygen and Life", Special Publication #39; Royal Society of Chemistry, ISBN 0 85186 285 8, 1981.

(5)(a) Chang, C.K.; Dolphin, D. J. Am. Chem. Soc. 1976, 98, 1607. (b) Ricard, L.; Schappacher, M.; Weiss, R.; Montiel-Montoya, R.; Bill, E.; Gonser, U.; Trautwein, A. Nouv. J. Chim. 1983, 7, 405.

(6) Martell, A.E.; Calvin, M. "Chemistry of the Metal Chelate Compounds"; Prentice-Hall Inc.: Englewood Cliffs, N.J., 1952, pp 336-357.

(7) Diehl, H. Iowa State Coll. J. Sci. 1947, 21, 271.

(8) Adduci, A.J. Presented at the 107th National Meeting of the American Chemical Society, Chicago, Ill., August, 1975.

(9) Collman, J.P.; Brauman, J.I.; Doxsee, K.M.; Halbert, T.R.; Hayes, S.E.; Suslick, K.S. J. Am. Chem. Soc. 1978, 100, 2761.

(10)(a) Wöhrle, D.; Bohlen, H.; Blum, J.K. Makromol. Chem. 1986, 187, 2081. (b) Nishide, H.; Yoshioka, H.; Wang, S.; Tsuchida, E. Ibid. 1985, 186, 1513.

(11)(a) Tashkova, K.A.; Andreev, A.J. Mol. Struct. 1984, 115, 55. (b) Basolo, F.; Hoffman, B.M.; Iken, J.A. Acc. Chem. Res. 1975, 8, 392.

(12)(a) Herron, N. Inorg. Chem. 1986, 25, 4714. (b) Schoonheydt, R.A.; Pelgrims, J. J. Chem. Soc. Dalton 1981, 914. (c) Lunsford, J.H.; Camara, M.J. Inorg. Chem. 1983, 22, 2498.

(13) Roman, I.C. U.S. Patent 4,451,270 (1984). See also Ref 19.

(14)(a) Lonsdale, H. K. J. Membr. Sci. 1982, 10, 81. (b) Harga, K. Chem. Econ. Eng. Rev. 1981, 13, 20.

(15)(a) Kimura, S.G.; Browall, W.R. J. Membr. Sci. 1986, 29, 69. (b) Haraya, K. Chem. Econ. Eng. Rev. 1981, 13, 20.

(16) Way, J.D.; Noble, R.D.; Flynn, T.M.; Sloan, F.D. J. Membr. Sci. 1982, 12, 239.

(17) Scholander, P.F. Science 1960, 131, 585.

(18) Bassett, R.J.; Schultz, J.S. Biochim. Biophys. Acta 1970, 211, 194.

(19) Baker, R.F.; Lonsdale, H.K.; Matson, S.L. (Bend Research Inc.). "Liquid Membranes for the Production of Oxygen-Enriched Air", Contract No. DE-AC06-79ER10337; DOE/ER/10337-1, DE85 006056.

(20)(a) Stevens, J.C.; Busch, D.H. J. Am. Chem. Soc. 1980, 102, 3285. (b) Busch, D.H.; Olszanski, D.J.; Stevens, J.C.; Schammel, W.P.; Kojima, M.; Herron, N.; Zimmer, L.L.; Holter, K.A.; Mocak, J. Ibid. 1981, 103, 1472. (c) Stevens, J.C. PhD Thesis, Ohio State University, 1979.

(21) Kemena, L.L.; Noble, R.D.; Kemp, N.J. J. Membr. Sci. 1983, 15, 259, and references cited therein.

(22) Koval, C.A.; Noble, R.D.; Way, J.D.; Louie, B.; Reyes, Z.E.; Bateman, B.R.; Horn, G.M.; Reed, D.L. Inorg. Chem. 1985, 24, 1147.

(23) Achord, J.M.; Hussey, C.L.; Anal. Chem. 1980, 52, 601.

(24) Akhrem, A.A.; Moisenkov, A.M.; Lakhvich, F.A. Izv. Akad. Nauk. SSSR, Ser. Khim. 1971, 12, 2786; Chem. Abstr. 1972, 71, 47914F.

(25) Bamfield, P. J. Chem. Soc. A 1969, 2021.

(26) Gall, R.S.; Rogers, J.F.; Schaefer, W.P.; Christoph, G.G. J. Am. Chem. Soc. 1976, 98, 5135.

SUMMARY - DIOXYGEN COMPLEXES WITH TRANSITION METAL - F. BASOLO

I will give you my take-home message from each talk, and each of you will have your own take-home message, which does not necessarily have to be the same as mine. First, if you remember, Mike Hall did a beautiful job of telling us all about the virtues of molecular orbital theory. My take home message here is that molecular orbital theory is here to stay; it explains "everything" providing one uses the theory properly. The thing that I was personally interested in was the fact that apparently the people at Strasbourg who did the ab initio calculations on the manganese porphyrin dioxygen complex may have arrived at the wrong assignment of structure because of not having included some parameter in their calculations. The calculations done more recently by Mike support our structure, which was based on experiment. This is why as an experimental chemist I always tell students to do experiments and not be turned off by theory.

The talks by Tom Loehr and Ralph Wilkins I have lumped together. They both talked about the three respiratory natural proteins. My take-home message from these two talks is that where natural proteins of this size are concerned one does not have a handle on them as far as definitive X-ray structure. However, one can get a lot of information by applying spectroscopy judicially as did Tom, particularly as we saw in his dioxygen complexes where he used ^{16}O, ^{18}O labeling of the dioxygen moiety so effectively. Ralph showed us it is possible to do kinetic and mechanism studies on complicated biological systems and obtain meaningful results. Ordinarily kinetic and mechanism studies are done on substrates of known structure in reactions of known stoichiometry. Apparently it is possible to focus on the active site of a complicated biological reaction and get useful information on the mechanism of reaction.

I now come to Tom Meyer's talk and I have in my notes the comment "where, oh where would Tom Meyer be if it weren't for ruthenium?" Ruthenium has been very good to inorganic chemists; helping Henry Taube get the Nobel Prize and the basis of thousands of papers on $[Ru(bipy)_3]^{n+}$. I will further take home with me the presumption that this need not be just ruthenium but any oxo-platinum metal complex is potentially a better oxygen atom transfer agent than are oxo-metal complexes of oxophilic early transition metal complexes. I make this assumption in spite of knowing that oxo-Mo, -Fe, -Mn, and -Cr complexes do transfer the oxo group.

The talks of Daryle Busch and Art Martell I lump together also because they both told us about the elegant ligands used in their attempt to generate synthetic oxygen carriers which did not degrade in solution over a period of time. My take-home message here is that very good coordination chemistry is being done that provides much more stable oxygen carriers, but it is doubtful that some of the more extreme demands being asked of these systems will be met.

Last but not least in my group, is the talk by Guido Pez. The take-home message on Guido's talk in my notes is that there is money to be made in some of these synthetic oxygen carriers. Unfortunately the solution stability demands placed on the complexes for membrane separation of oxygen from air seem a bit excessive. Perhaps in order to prevent degradation, one may have to avoid organic ligands. Two suggestions are the use of appropriate metal ions in iso or heteropolyacid systems, or in zeolites.

It has been a pleasure and an educational experience for me to chair this opening session. I want to thank again all the speakers for their excellent talks, and I want now to turn the meeting over to Don Sawyer and to the speakers of the final session. Thank you.

SUMMARY - OXYGEN ACTIVATION BY TRANSITION METALS - D.T. SAWYER

The chapters in the preceding section are concerned with binding dioxygen reversibly and unactivated. In contrast the contributions in this section discuss the activation of dioxygen to become an active oxidant and oxygenase via transition-metal catalysts. The biological catalysts (oxidases, oxygenases, peroxidases, and cytochromes P-450) represent highly selective systems for the activation of oxygen for reaction with substrate molecules via a single pathway. Nature's design is such that O_2 is made an effective oxidizing agent for specific substrates, but does not attack the host ligands. The chemical characteristics of biological oxygen-activation catalysts are discussed in the initial chapters, and are followed by descriptions of four industrial processes (based on homogeneous transition-metal complexes) that utilize oxygen to enhance the value of organic substrates and of hydrogen sulfide. The final chapters discuss the development of transition-metal heterogeneous catalysts for the selective incorporation of oxygen atoms from dioxygen into organic substrates.

Several common "threads of thought" are present within this diverse group of biological and industrial activation systems for dioxygen:

(a) Ground state triplet dioxygen ($^3O_2$) is unreactive because it is a weak electron-transfer oxidant, has a large activation barrier for singlet substrates, and its two oxygen-atoms are highly stabilized by 119 kcal of bond energy.

(b) Activation is accomplished by electron-transfer reduction of O_2 to $O_2^{-\cdot}$ and HOOH, atomization of $^3O_2$ to ^1O, and spin-state conversion from $^3O_2$ to $^1O_2$.

(c) Lipid peroxidation, autoxidation of unsaturated fats and oils, and rancidification of foodstuffs require an oxy-radical initiator ($\cdot OH$, $HO_2^{\cdot}$, $RO\cdot$, or $MO\cdot$). Reducing equivalents plus O_2 and trace metals produce HOOH and reduced transition-metal ions [Fe(II), Mn(II), Cu(I), and V(II)], which in turn react to give $\cdot OH$ and $MO\cdot$.

(d) Catalyst geometry is critically important to the activation of substrate and dioxygen, to the selectivity of the chemistry, and to the stability of the catalyst against attack by activated oxygen.

(e) Oxygenase chemistry requires the formation of an electrophilic metal-oxene reactive intermediate (weakly stabilized oxygen atom via covalent bond formation). The common feature of the transition metal catalysts is the presence of unpaired electrons that can couple with the unpaired p electrons of $^3O_2$ and ^3O to form d-p covalent bonds. Selectivity is achieved via the relative bond energies for the oxygen in the intermediate and the substrate product.

In summary, this group of chapters provides insights to the common chemistry for the activition dioxygen by metalloproteins, organometallic catalysts, and transition-metal surfaces. The transformations are characteristic of electrophilic and biradical processes, rather than those associated with oxo or nucleophilic centers.

THE CHEMISTRY AND ACTIVATION OF DIOXYGEN SPECIES (O_2, $O_2^{-}\cdot$, AND HOOH) IN BIOLOGY

Donald T. Sawyer

Department of Chemistry
Texas A&M University
College Station, Texas 77843

ABSTRACT

Biological systems activate dioxygen(O_2) for controlled energy transduction and chemical syntheses. This is accomplished via electron-transfer reduction of O_2 to $O_2^{-}\cdot$ and HOOH, and its atomization with metalloproteins to accomplish atom-transfer chemistry. These reactive intermediates have been characterized by the use of (a) transition-metal complexes as models for metalloproteins and (b) model substrates.

Because dioxygen is the natural product from the dehydrogenation of water via photosystem II of green-plant photosynthesis (a process that began about 2.7 billion years ago with the appearance of blue-green algae),[1,2]

$$2\ H_2O \xrightarrow[\text{[Mn]}]{4h\upsilon} {}^3O_2 + (4H^+ + 4e^-) \qquad (1)$$

nature has had ample time to develop effect catalysts to activate and control its reactivity with organic substrates. The presence of copious quantities of reduced iron in the oceans consumed all of the O_2 that was produced for another 0.7 billion years. After its "titration" the earth's atmosphere rapidly changed from 1 per cent O_2 to the present 21 per cent concentration. With dioxygen in the atmosphere solar radiation transformed a small fraction of it to ozone, which provided a protective shield against short wave-length UV light and thereby made possible the evolution of terrestial life from marine organisms.

$$3\ O_2 \xrightarrow{h\nu} 2\ O_3 \quad (2)$$

The redox thermodynamics of O_2 are directly dependent upon proton activity,

$$O_2 + 4H^+ + 4e^- \rightarrow 2\ H_2O \quad E°' \quad (3)$$

which in turn depends upon the reaction matrix. Table 1 summarized the pK_a' values for a series of Brønsted acids in several aprotic solvents and water.[3] In acetonitrile the activity values for pK_a' range from -8.8 for $(H_3O)ClO_4$ to 30.4 for H_2O. This means that the formal potential (E°') for Reaction 3 in acetonitrile(MeCN) is +1.75 V vs. NHE in the presence of 1M $(H_3O)ClO_4$ and -0.56 V in the presence of 1M $(Bu_4N)OH$. Another limiting factor with respect to chemical-energy flux for oxidative metabolism and respiration is the solubility of O_2. Because of its non-polar character dioxygen is much more soluble in organic solvents than in H_2O (Table 2).[4] The reduction potentials for O_2 and various intermediate species in H_2O at pH 0, 7, and 14 are summarized in Figure 1;[5,7] similar data for O_2 in MeCN at pH -8.7, 10.0, and 30.4 are presented in Figure 2.[4,8,9]

The reduction manifolds for O_2 (Figures 1 and 2) indicate that the limiting step (in terms of reduction potential) is the first election transfer to O_2, and that an electron source adequate for the reduction of O_2 will produce all of the other reduced forms of di-

Table 1. Effective pK_a' Vaues for Brønsted Acids in Aprotic Solvents and Water. [a]

Brønsted Acid	Solvent [0.5M $(Et_4N)ClO_4$]				
	MeCN	DMF	Me_2SO	pyr	H_2O
$(H_3O)ClO_4$	-8.8	0.7	2.6	4.6	0.0
$MePhSO_3H$	-3.8	-	-	-	-
$(pyrH)ClO_4$	1.8	3.1	4.4	5.7	4.9
$2,4\text{-}(NO_2)_2PhOH$	4.3	4.5	4.9	5.5	-
$(NH_4)ClO_4$	5.9	9.6	11.7	7.3	8.7
$(Et_3NH)Cl$	10.0	9.9	12.7	7.6	10.1
PhC(O)OH	7.9	11.5	13.6	11.6	3.2
2-pyr-C(O)OH	8.6	-	-	-	-
PhOH	16.0	19.4	20.8	20.1	9.2
p-EtOPhOH	19.3	21.5	23.8	21.8	9.6
H_2O	30.4	34.7	36.7	30.5	14.8

[a] Ref. 3.

Table 2. Solubilities of O_2 (1 atm) in various Solvents [a]

Solvent	$[O_2]_{1\ atm}$, mM
H_2O	1.0
Me_2SO	2.1
DMF	4.8
pyr	4.9
MeCN	8.1
hydrocarbons	~10.
fluorocarbons	~25.

[a] Ref. 4

oxygen ($O_2^{-}\cdot$, $HO_2\cdot$, H_2O_2, HO_2^{-}, $\cdot OH$) via reduction, hydrolysis, and disproportionation steps (Scheme 1).[10,11] Thus, the most direct means to activate O_2 is the addition of an electron (or hydrogen atom), which results in significant fluxes of several reactive oxygen species.

Reactivity of $O_2^{-}\cdot$ and $HO_2\cdot$ The dominant characteristic of $O_2^{-}\cdot$ in any medium is its ability to act as a strong Brønsted base via formation of $HO_2\cdot$,[12,13] which reacts with allylic hydrogens, itself, or a second $O_2^{-}\cdot$ (Scheme 1). Within water superoxide ion is rapidly

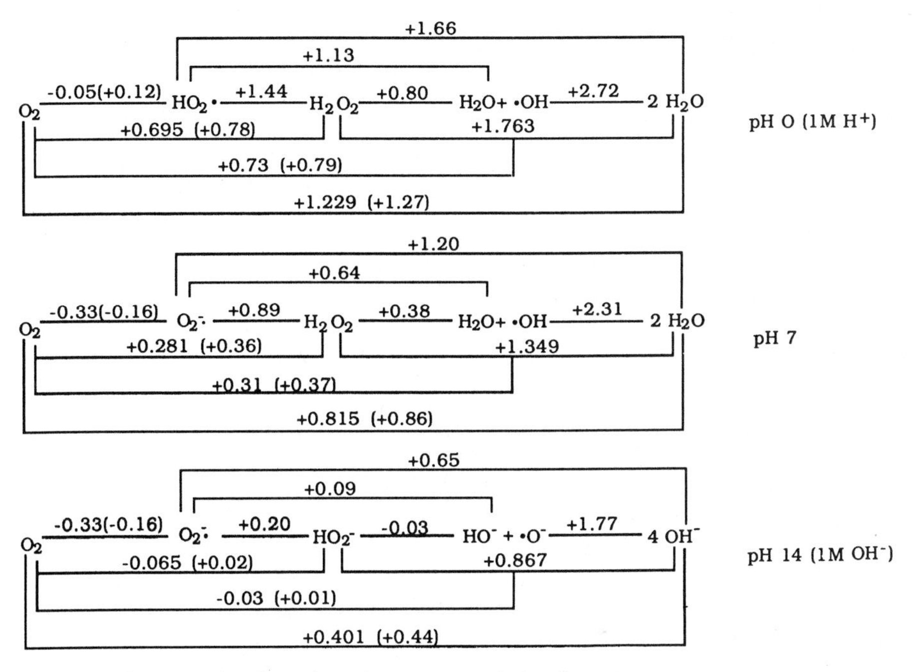

Figure 1. Standard reduction potentials for dioxygen species in water (O_2 at 1 atm). Formal potentials for O_2 at unit activity.

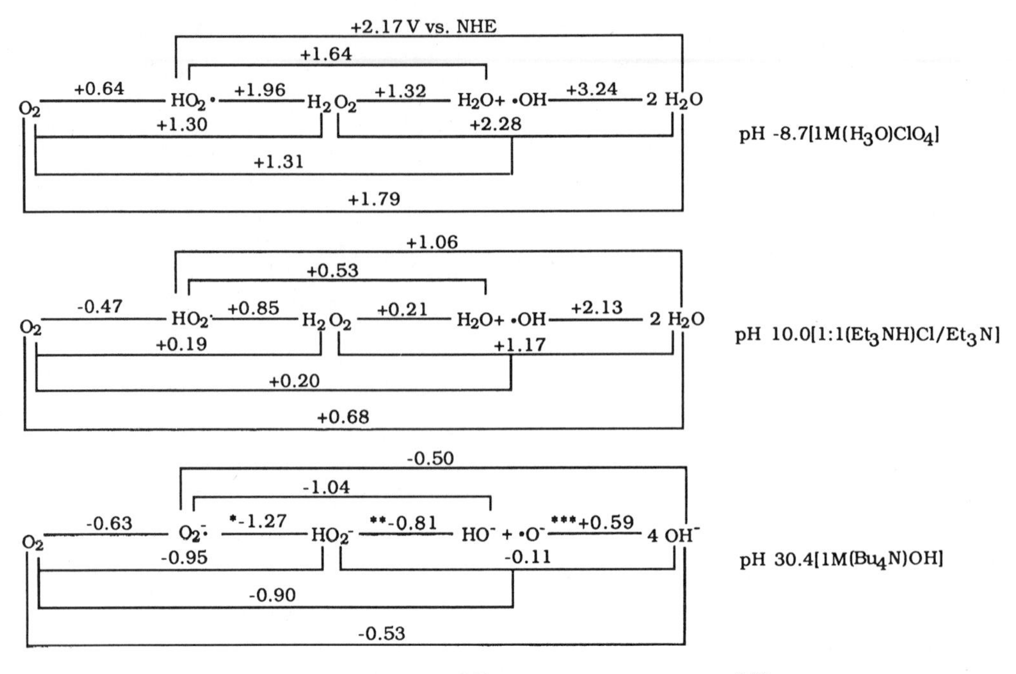

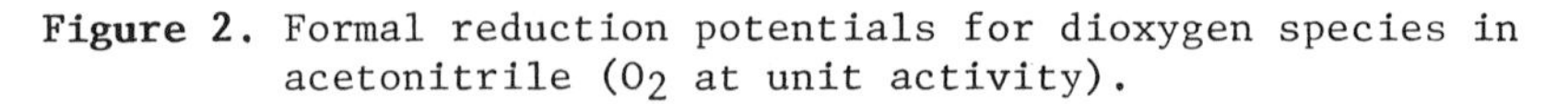

Figure 2. Formal reduction potentials for dioxygen species in acetonitrile (O_2 at unit activity).

converted to dioxygen and peroxide

$$2\ O_2^{-}\cdot + H_2O \rightarrow O_2 + HO_2^- + HO^- \qquad K,\ 2.5\times10^{8}\ M \qquad (4)$$

Such a proton-driven disproportionation process means that $O_2^{-}\cdot$ can deprotonate acids much weaker than water (up to $pK_a \approx 23$).[14]

Under aprotic conditions $O_2^{-}\cdot$ is a strong nucleophile that reacts with esters, acid halides, and halogenated hydrocarbons[12,15]

$O_2 + e^-$ ⇌ (-0.6V) $O_2^{-}\cdot$ —HA→ $HO_2\cdot + A^-$

- —1,4-CHD→ 1,3-CHD· + H_2O_2 k, $10^2 M^{-1}s^{-1}$ → 1/2 1,3-CHD + 1/2 PhH
- —$HO_2\cdot$→ $H_2O_2 + O_2$ k, $10^4 M^{-1}s^{-1}$
- —$O_2^{-}\cdot$→ $HO_2^- + O_2$
  - —HA→ $H_2O_2 + A^-$
  - —H_2O_2→ $O_2^{-}\cdot + H_2O + \cdot OH$

Scheme 1 (Me_2SO)

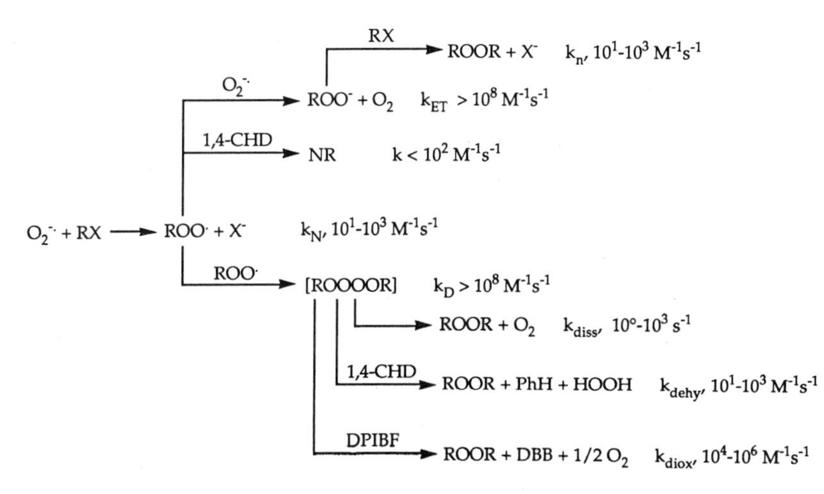

Scheme 2. (RX = CCl_4, F_3CCCl_3, $PhCCl_3$, BuBr, BuCl)

via displacement of alkoxide or halide ion, respectively, to give an organic peroxy radical (ROO·, Scheme 2). With benzil the initial reaction by $O_2^{-}\cdot$ is a nucleo-philic addition to a carbonyl carbon, which is followed by $O_2^{-}\cdot$ reduction of the oxy radical.[16]

$$PhC(O)C(O)Ph + O_2^{-}\cdot \rightarrow \left[Ph\overset{O-O\cdot}{\underset{O^-}{C}}C(O)Ph \rightarrow Ph\overset{O-O}{\underset{^-O\quad O\cdot}{C-C}}-Ph \right]^{O_2^{-}\cdot} \rightarrow 2PhC(O)O^- + O_2 \quad (5)$$

The data of Figures 1 and 2 indicate that $O_2^{-}\cdot$ is a moderate one-electron reducing agent [cytochrome c(Fe^{III}) is reduced in H_2O[17] and iron(III) porphyrins in dimethylformamide].

$$Fe^{III}TPP^+ + O_2^{-}\cdot \rightarrow Fe^{II}TPP + O_2 \quad \Delta E,\ +0.7\ V \qquad (6)$$

Superoxide is an effective hydrogen-atom oxidant for substrates with coupled hetero-atom (O or N) dihydrogroups such as catechols, ascorbic acid, 1,2-disubstituted hydrazines, dihydrophenazine, and dihydrolumiflavin.[18,19] The general mechanism involves the rapid sequential transfer to $O_2^{-}\cdot$ of a proton and a hydrogen atom to form HOOH and the anion radical of the dehydrogenated substrate. With 1,2-diphenylhydrazine the azobenezene anion radical product is rapidly oxidized by dioxygen.

$$PhNHNHPh + O_2^{-}\cdot \rightarrow PhN^{-}\cdot NPh + HOOH \qquad (7)$$
$$PhN^{-}\cdot NPh \xrightarrow{O_2} PhN{=}NPh + O_2^{-}\cdot$$

Hence, $O_2^{-}\cdot$ serves as the initiator for the autooxidation of such dihydrosubstrates and the chemical generation of HOOH under biological conditions.

Superoxide ion reacts with proton sources to form $HO_2\cdot$, which disproportionates via a second $O_2^-\cdot$ or itself (Scheme 1). However, with limiting fluxes of protons to control the rate of $HO_2\cdot$ formation from $O_2^-\cdot$, the rate of decay of $HO_2\cdot$ is enhanced by reaction with the allylic hydrogens of excess 1,4-cyclohexadiene (1,4-CHD).[20] Because $HO_2\cdot$ disproportionation is a second-order process, low concentrations favor hydrogen-atom abstraction from 1,4-CHD. This is especially so for Me_2SO, in which the rate of disproportionation for $HO_2\cdot$ is the slowest ($PhCl>MeCN>H_2O>DMF>Me_2SO$).

The initial product from the reaction of RX($R=Cl_3C$, F_3CCCl_2, $PhCCl_2$, and Bu) with $O_2^-\cdot$ in acetonitrile is $ROO\cdot$, which (a) can be reduced by a second $O_2^-\cdot$ to form ROO^- (a reactive nucleophile) or (b) dimerize to form ROOOOR (Scheme 2). The latter has a half life that ranges from ~1s (R=Bu) to ~10^{-3}s($R=Cl_3C$), and homolytically dissociates to ROOR and O_2. The longer-lived forms of ROOOOR react with (a) 1,4-cyclohexadiene(1,4-CHD) via dehydrogenation to give PhH, ROOR, and HOOH, (b) diphenylisobenzofuran(DPIBF) via dioxygenation to give dibenzoylbenzene(DBB) and ROOR, and (c) rubrene via dioxygenation to give its endoperoxide and ROOR. Thus, the reactivity of ROOOOR parallels that of singlet dioxygen ($^1O_2$). Because of its diffusion-controlled dimerization, the primary product from the $CCl_4/O_2^-\cdot$ reaction, $Cl_3COO\cdot$, does not exhibit any reactivity with 1,4-CHD. Hence, at millimolar concentrations this peroxy radical (the suspected cytotoxin from the aerobic activation of CCl_4) does not exist long enough to react with allylic hydrogens. The biological hazard of CCl_4 may be due to the transient formation of $Cl_3COOOOCCl_3$.

Activation of HOOH by Lewis Acids. As with $O_2^-\cdot$, the reactivity of HOOH is dependent upon the solution matrix. In aqueous media its interaction with Fe(II) produces $\cdot OH$ and $Fe^{III}(^-OH)^{2+}$. The subsequent reactivity of $\cdot OH$ with organic substrates via H-atom abstraction yields carbon radicals, which propagate chain reactions and autooxidations in the presence of O_2.

However, the addition of HOOH in MeCN to a solution that contains $[Fe^{II}(MeCN)_4](ClO_4)$ in dry MeCN(<0.005% H_2O) catalyzes a rapid disproportionation of HOOH via the initial formation of an adduct $[Fe^{II}(HOOH) \leftrightarrow Fe(O)(OH_2)]^{2+}$, which oxidizes a second HOOH to

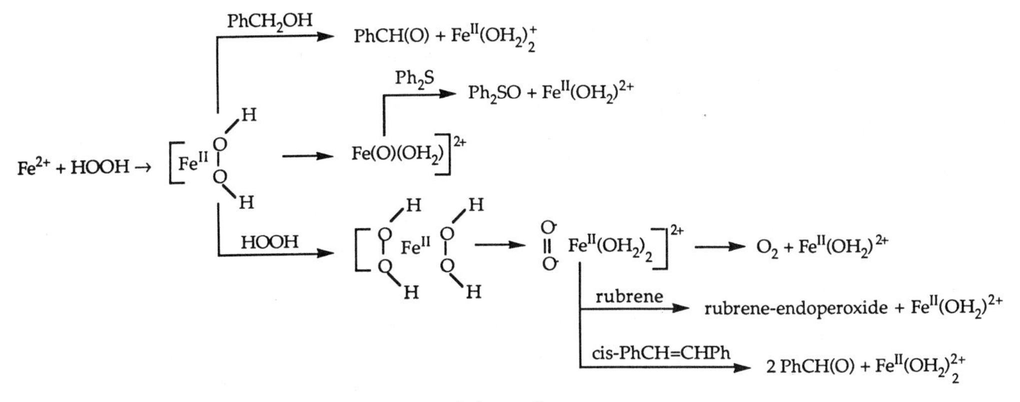

Scheme 3

O_2 (the iron catalyst remains as Fe(II) because its oxidation potential is +1.8 V vs. NHE in this base-free medium).[22] This same intermediate cleanly oxidizes alcohols, aldehydes, thioethers, and substituted hydrazines by a two-electron process (Scheme 3). The products for the Fe^{II}-HOOH oxidations are consistent with those that result from catalase- and some peroxidize-catalyzed processes.

In the presence of excess HOOH the $Fe^{II}(MeCN)_4{}^{2+}$ catalyst forms a reactive adduct, $Fe^{II}(^1O_2)(OH_2),^{2+}$ that reacts with diphenylbenzofuran, 9,10-diphenylanthracene, or rubrene to form exclusively dioxygenated products.[23] Such reactivities parallel those of $^1O_2$ with this group of substrates.

In the same base-free medium(dry MeCN) $Fe^{III}Cl_3$ activates HOOH to form a reactive intermediate that oxygenates alkanes, alkenes and thioethers, and dehydrogenates alcohols and aldehydes.[24] Such reactivity indicates that the intermediate is a highly electrophilic Fe^{III}-(O) species (formed by the strong Lewis acidity of $Fe^{III}Cl_3$ in MeCN relative to HOOH). Anhydrous $Fe^{III}Cl_3$ catalyzes the stereospecific epoxidation of norbornene, the demethylation of N,N-dimethylaniline, and the oxidative cleavage of PhCMe(OH)CMe(OH)Ph (and other α-diols) by hydrogen peroxide (Table 3 and Scheme 4).[25] For each class of substrate the products parallel those that result from their enzymatic oxidation by cytochrome P-450. The close congruence of the products indicates that the reactive oxygen in the $Fe^{III}Cl_3$/HOOH model system and the active form of cytochrome P-450 is essentially the same, with strong electrophilic oxene character (stabilized singlet atomic oxygen).

Table 3. Products and conversion efficiencies for the $Fe^{III}Cl_3$-catalyzed epoxidation of olefins, demethylation of $PhNMe_2$, and oxidative cleavage of 1,2-diols by H_2O_2 in dry acetonitrile.

Substrate (RH)	Reaction conversion efficiency, % [a]	Catalyst turnover number [b]	Products (yield)
norbornene	52	5	*exo*-epoxide (80%), other non-epoxide products (20%)
cyclohexene	37	4	epoxide (64%), dicyclohexyldioxane (13%)
1,4-cyclohexadiene	39	4	benzene (76%), epoxide (17%)
cis-PhCH=CHPh	63	6	PhCHO (50%), epoxides (50%) (*cis*-to-*trans* epoxide ratio, 2.5:1)
$PhNMe_2$	39	4	PhNHMe (95%), PhN(CHO)Me (5%)
PhCMe(OH)CMe(OH)Ph	30	3	PhC(O)Me (100%)

[a] Percentage of substrate converted to products.

[b] Millimoles of RH converted per mmol of $Fe^{III}Cl_3$ added.

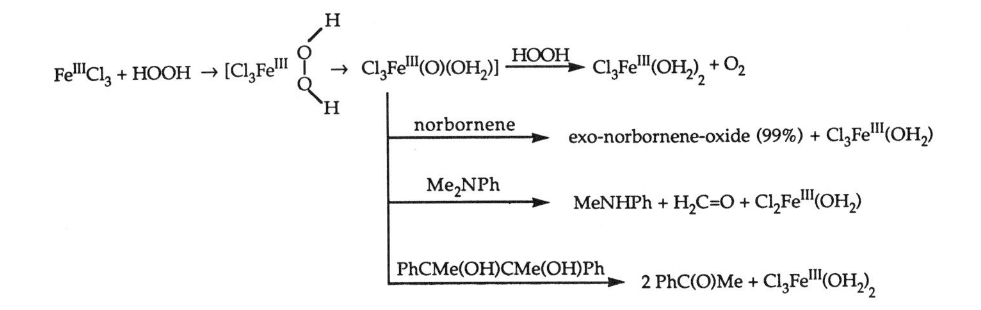

Scheme 4

Formation and Reactivity of Atomic Oxygen [O]. The function of peroxidase enzymes is the activation of HOOH to provide two oxidizing equivalents for the subsequent oxidation of a variety of substrates. The interaction of horseradish peroxidase (an iron(III) heme that has a proximal imidazole) with HOOH results in the formation of a green reactive intermediate known as Compound I. The latter is reduced by one electron to give a red reactive intermediate, Compound II.[26] Both of these intermediates contain a single oxygen atom from HOOH, and Compound I is two oxidizing equivalents above the iron(III)-heme state with a magnetic moment equivalent to three unpaired electrons (S=3/2). A recent EXAFS study[27] summarizes the physical data in support of formulations of $[(Por^{-}\cdot)Fe^{IV}(O^{2-})]^{+}$ for Compound I, and $[(Por^{2-})Fe^{IV}(O^{2-})]$ for Compound II; and concludes that both species contain an oxo-ferryl group (Fe=O) with a bond length of 1.64Å.

A recent summary[28] of the activation of O_2 by cytochrome P-450 (an iron(III)-heme protein with a proximal cysteine thiol) concludes that the reactive form of this monooxygenase also contains an oxo-ferryl group $[(RS^{-})(Por^{-}\cdot)Fe^{IV}(O^{2-})]$. The monooxygenase chemistry of cytochrome P-450 has been modeled via the use of $(TPP)Fe^{III}Cl$ (TPP=tetraphenylporphyrin dianion) and $(OEP)Fe^{III}Cl$ (OEP=octaethyl-porphyrin dianion) with peracids,[29,30] iodosobenzene,[29,30] 4-cyano-N,N-dimethyl aniline-N-oxide,[31] and hypochlorite[32] to oxygenate model substrates. On the basis of the close parallel with the products from the cytochrome P-450-catalyzed reactions and the net two-oxidizing equivalents of the catalytic cycles for cyt P-450/(O_2 + $2H^{+}$ + 2e) and HRP/H_2O_2, a general consensus has developed that the reactive intermediate of cytochrome P-450 is analogous to Compound I with a $Fe^{IV}(O^{2-})$ group.

All contemporary work indicates that the reactive intermediate for HRP-I and cytochrome P-450 is an oxygen-atom adduct of (imid)-$(Por^{2-})Fe^{III}$ and $(RS^-)(Por^{2-})Fe^{III}$.[27,33] The common belief is that atomic oxygen invariably removes two electrons from iron(III) and/or (Por^{2-}) to achieve an oxo(O^{2-}) state. Although this misconception is general for the oxygen compounds of transition metals, there is no thermodynamic, electronegativity, or theoretical basis to exclude stable $M(O^{-}\cdot)$ and M(O) species.[34] Thus, the atomic-oxygen adduct of $(Por^{2-})Fe^{III}(B)^+$ should be viewed as the resonance hybrid of several valence-bond formulations.

$$[(Por^{2-})Fe^{V}(O^{2-})^+ \leftrightarrow (Por^{-}\cdot)Fe^{IV}(O^{2-}) \leftrightarrow (Por^{\circ})Fe^{III}(O^{2-})^+ \leftrightarrow (Por^{2-})Fe^{IV}(O^{-}\cdot)^+ \leftrightarrow (Por^{-}\cdot)Fe^{III}(O^{-}\cdot)^+ \leftrightarrow (Por^{\circ})Fe^{II}(O^{-}\cdot)^+ \leftrightarrow (Por^{2-})Fe^{III}(O)^+ \leftrightarrow (Por^{-}\cdot)Fe^{II}(O)^+]$$

The last two of these are simple O-atom adducts without intramolecular electron transfer to oxygen, but stabilized by d-p orbital overlap [similar to the addition of [O] to CO to give O=C=O or O_2 to heme-Fe(II) or give heme-Fe(II)(O_2)].[35]

In recent discussion[34] I have argued that high-valent transition metal ions such as iron(IV) are thermodynamically incompatible with the strongly electronegative oxo dianion, particularly when electron transfer results in unpaired valence electrons that can stabilize each other via covalent bond formation. Such stabilization of the two unpaired electrons of ground-state atomic oxygen attenuate its redox potential and its reactivity to an extent proportional to the energy of the metal-oxygen covalent bond. Tables 4 and 5 summarize the reduction potentials for atomic oxygen (as the free atom and in various compounds) in aqueous solution and in MeCN.[5,6,36]

The results of a recent investigation[37] of model systems provide compelling evidence that stabilized atomic oxygen is present in Compound I and Compound II of horseradish peroxidase and the reactive form of cytochrome P-450. Thus, the combination of tetrakis(2,6-dichlorophenyl)-porphinato iron(III) perchlorate (**1**, Scheme 5) with pentafluoro-iodosobenzene, m-chloroperbenzoic acid, or ozone in acetonitrile at -35°C yields a green porphyrin-oxene adduct (**2**). This species, which has been characterized by spectroscopic, magnetic and electrochemical methods, cleanly and stereo-specifically epoxidizes olefins (>99% exo-norbornene-oxide).

Table 4. Redox Potentials for Oxygen Species in Aqueous Media

Reaction	E°, V vs. NHE
pH 0:	
$O_{(g)} + 2H^+ + 2e^- \rightarrow H_2O$	+2.42
$O_{(g)} + H^+ + e^- \rightarrow \cdot OH$	+2.12
$\cdot OH + H^+ + e^- \rightarrow H_2O$	+2.72
$O_{3(g)} + 2H^+ + 2e^- \rightarrow O_{2(g)} + H_2O$	+2.07
$O_3 + e^- \rightarrow O_3^{-}\cdot$	+0.95
$HOOH + 2H^+ + 2e^- \rightarrow 2H_2O$	+1.76
$HOIO_3 + H^+ + 2e^- \rightarrow IO_3^- + H_2O$	+1.6
$HOCl + 2H^+ + 2e^- \rightarrow HCl + H_2O$	+1.49
pH 7:	
$O_{(g)} + 2H^+ + 2e^- \rightarrow H_2O$	+2.01
$O_{(g)} + H^+ + e^- \rightarrow \cdot OH$	+1.71
$\cdot OH + H^+ + e^- \rightarrow H_2O$	+2.31
$O_{3(g)} + 2H^+ + 2e^- \rightarrow O_{2(g)} + H_2O$	+1.66
$O_{3(g)} + e^- \rightarrow O_3^{-}\cdot$	+0.95
$HOOH + 2H^+ + 2e^- \rightarrow 2H_2O$	+1.35
$IO_4^- + 2H^+ + 2e^- \rightarrow IO_3^- + H_2O$	+1.2
pH 14:	
$O_{(g)} + H_2O + 2e^- \rightarrow 2\ ^-OH$	+1.60
$O_{(g)} + H_2O + e^- \rightarrow \cdot OH + ^-OH$	+1.31
$O_{(g)} + e^- \rightarrow O^{-}\cdot$	+1.43
$\cdot OH + e^- \rightarrow ^-OH$	+1.89
$O^{-}\cdot + H_2O + e^- \rightarrow 2\ ^-OH$	+1.77
$O_{3(g)} + H_2O + 2e^- \rightarrow O_{2(g)} + 2\ ^-OH$	+1.25
$O_{3(g)} + e^- \rightarrow O_3^{-}\cdot$	+0.95
$O_3^{-}\cdot + H_2O + e^- \rightarrow O_{2(g)} + 2\ ^-OH$	+1.55
$HOO^- + H_2O + 2e^- \rightarrow 3\ ^-OH$	+0.87
$IO_4^- + H_2O + 2e^- \rightarrow IO_3^- + 2\ ^-OH$	+0.8
$ClO^- + H_2O + 2e^- \rightarrow Cl^- + 2\ ^-OH$	+0.89
$ClO^- + e^- \rightarrow Cl^- + O^{-}\cdot$	+0.02

The reaction chemistry and electronic characterization of the green adduct (**2**) are consistent with an oxygen atom covalently bound to an iron(II)-porphyrin radical center $[(P^{-}\cdot)Fe^{II}(O)^+]$. The latter has the spectral, magnetic, and redox characteristics of Compound I of horseradish peroxidase (HRP), and the selective stereospecific

Table 5. Redox Potentials for Oxygen Species in MeCN [H+ ≡ 1M $(H_3O)ClO_4$;-OH ≡ 1M $(Bu_4N)OH(MeOH)$].

Acid, pH(-8.8):	E°, V vs. NHE
$O_{(g)} + H^+ + e^- \rightarrow \cdot OH$	+2.64
$\cdot OH + H^+ + e^- \rightarrow H_2O$	+3.24
$O_{(g)} + 2H^+ + 2e^- \rightarrow H_2O$	+2.94
$O_{3(g)} + 2H^+ + 2e^- \rightarrow O_{2(g)} + H_2O$	+2.59
$HOOH + 2H^+ + 2e^- \rightarrow 2H_2O$	+2.28
$HOCl + 2H^+ + 2e^- \rightarrow HCl + H_2O$	+2.0
$F_5PhIO + 2H^+ + 2e^- \rightarrow F_5PhI + H_2O$	+1.02
$(Cl_8TPP^-\cdot)Fe^{II}(O)^+ + 2H^+ + 2e^- \rightarrow (Cl_8TPP^{2-})Fe^{III}(OH_2)^+$	+1.94
$(Cl_8TPP^{2-})Fe^{II}(O) + 2H^+ + 2e^- \rightarrow (Cl_8TPP^{2-})Fe^{II} + H_2O$	+1.50
Neutral, pH 9:	
$O_{(g)} + H_2O + e^- \rightarrow O^-\cdot (H_2O)$	+0.67
$O_{3(g)} + e^- \rightarrow O_3^-\cdot$	+0.35
$(Cl_8TPP^-\cdot)Fe^{III}(O)^{2+} + e^- \rightarrow (Cl_8TPP^-\cdot)Fe^{II}(O)^+$	+1.83
$(Cl_8TPP^-\cdot)Fe^{II}(O)^+ + e^- \rightarrow (Cl_8TPP^{2-})Fe^{II}(O)$	+1.51
$(Cl_8TPP^{2-})Fe^{II}(O) + m\text{-}ClPhC(O)OH + e^- \rightarrow$ $(Cl_8TPP^{2-})Fe^{III}(^-OH) + m\text{-}ClPhC(O)O^-$	+0.16
$(Cl_8TPP^{2-})Fe^{II}(O) + e^- \rightarrow (Cl_8TPP^{2-})Fe^{II}(O^-\cdot)$	-0.30
Base, pH 30.4:	
$O_{(g)} + H_2O + e^- \rightarrow \cdot OH + {}^-OH \rightarrow O^-\cdot (H_2O)$	+0.34
$O^-\cdot + H_2O + e^- \rightarrow 2\ {}^-OH$	+0.59
$\cdot OH + e^- \rightarrow {}^-OH$	+0.92
$O_{(g)} + H_2O + 2e^- \rightarrow 2\ {}^-OH$	+0.63
$O_3 + H_2O + 2e^- \rightarrow O_2 + 2\ {}^-OH$	+0.28
$ClO^- + H_2O + 2e^- \rightarrow Cl^- + 2\ {}^-OH$	-0.08
$HOO^- + H_2O + 2e^- \rightarrow 3\ {}^-OH$	-0.10

oxygenase character of the reactive intermediate for cytochrome P-450. Reduction of the green species by one-electron equivalent yields a red species (**3**, Scheme 5), which has the spectral characteristics and reactivity of Compound II of HRP. The iron(III) porphyrin(**1**) is an efficient catalyst for (a) the stereospecific epoxidation of olefins, (b) the dehydrogenation of alcohols, (c) the oxidative cleavage of α-diols and (d) the demethylation of dimethylaniline by F_5PhIO and m-ClPhC(O)OOH (Scheme 5). When HOOH is used as the source of oxygen extensive attack of the porphyrin ring occurs and there is no significant reaction with olefins or α-diols.

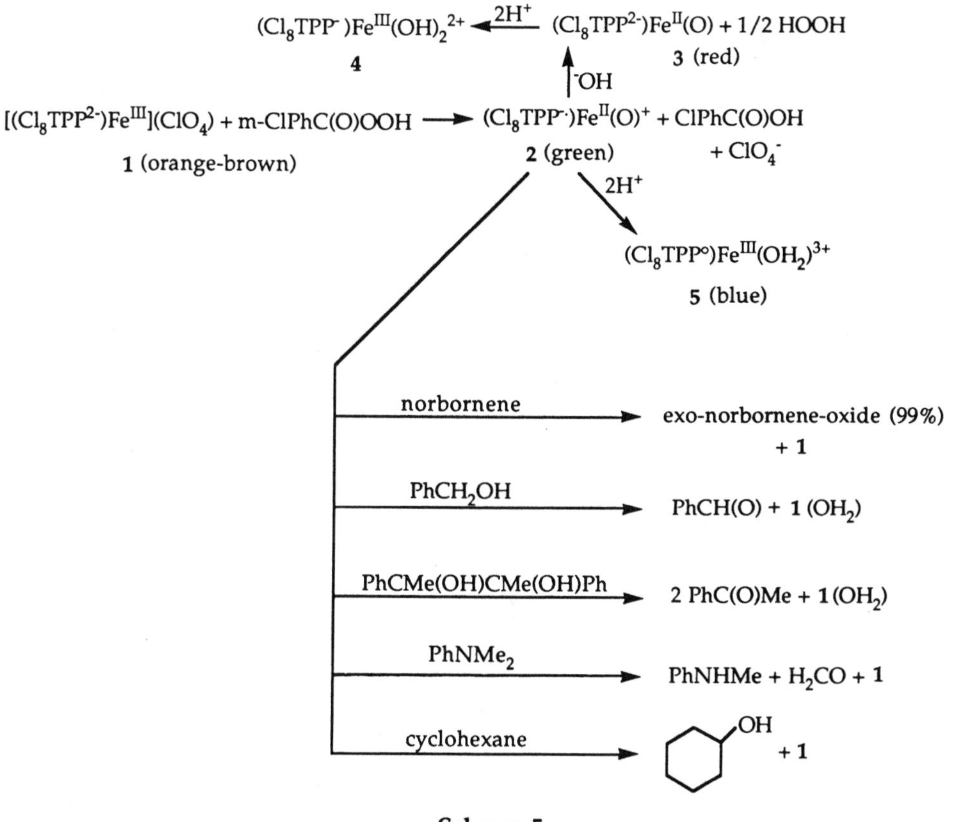

Scheme 5

Table 5 summarizes the redox thermodynamics for the various iron-oxene species of this model system in MeCN. The shift in the two-electron reduction potential for O(g) (from +0.63 V vs. NHE in a neutral unbuffed solution to +2.94 V in acidic media) is analogous to that observed when protons are added to the green iron oxene species (**2**, Scheme 5). Because the addition of an equivalent of ^{-}OH to **2** produces the same red species (3) as the addition of an electron to **2**, species **3** is formulated as an iron(II)-oxene. Production of this species by the combination $(Cl_8TPP)Fe^{III}(ClO_4)$, Me_3Py, and HOOH; and its limited epoxidation of olefins are consistent with an iron(II)-oxene (versus iron(III)-$O^{-}\cdot$) formulation. Addition of protons to **3** promotes an intramolecular two-electron transfer to the oxene oxygen [one from the porphyrin ring and one from iron(II)] to give $(Cl_8TPP^{-}\cdot)Fe^{III}(OH_2)^{2+}$ (**4**, Scheme 5). If species **3** contained hyper-valent iron or oxidized porphyrin, such a transformation with proton addition would not be expected.

The formation of **3** rather than **2** from the combination of $(Cl_8TPP)Fe^{III}(ClO_4$, 2,4,6-Me_3Py, and HOOH indicates that the latter is unable to transfer an O-atom to the iron(III) center. With HOOH alone there is rapid degradation of the porphyrin ring. Thus, the formation process for **3** requires a base and reducing agent (2,4,6-Me_3Py) to cause HOOH to act as an $(O^{-}\cdot)$ transfer agent to iron(III) with subsequent intramolecular electron transfer.

$$(Cl_8TPP^{2-})Fe^{III}(ClO_4) + 2\ Me_3Py + HOOH \rightarrow \underset{\mathbf{3}}{(Cl_8TPP^{2-})Fe^{II}(O)} + Me_3PyH^{+} + [Me_3Py(\cdot OH)] \quad (8)$$

Olefins are epoxidized by **3** to give the iron(II)-porphyrin,

$$(Cl_8TPP^{2-})Fe^{II}(O) + \text{norbornene} \rightarrow \text{exo-norbornene oxide} + (Cl_8TPP^{2-})Fe^{II} \quad (9)$$

which reacts with HOOH to give inactive catalyst

$$(Cl_8TPP^{2-})Fe^{II} + HOOH + Me_3Py \rightarrow (Cl_8TPP^{2-})Fe^{III}(^{-}OH) + [Me_3Py(\cdot OH)] \quad (10)$$

The reaction chemistry of Scheme 5 confirms that **2** acts as an oxygen-atom transfer agent towards olefins. The stereospecificity for the epoxidation of norbornene is consistent with the concerted insertion[38] of a singlet oxygen atom into the pi bond (analogous to the stereospecific transfer of a singlet oxygen atom from uncatalyzed m-ClPhC(O)OOH to norbornene). If **2** contained hypervalent iron, an electron-transfer mechanism would be favored, which results in a mixture of exo and endo epoxides.[30,39]

The magnetic moments for **2** (S=3/2) and for **3** (S=0) indicate extensive coupling between the ground state triplet p-orbitals of atomic oxygen and the half filled d-orbitals of iron(II). In terms of valence-bond considerations overlap by the metal-d and oxygen-p orbitals will result in the formation of a metal-oxygen s-bond and a metal-oxygen p-bond. The two-electron reduction potentials under acidic conditions (Table 5) for **2** (+1.94 V vs. NHE) and O(g)(+2.94 V) provide an approximate measure of the bond energy for the $(P^{-}\cdot)Fe^{II}{=}O$ covalent double bond; B.E. = ΔExnx23.1 kcal = 46.2 kcal (Table 5). Likewise, the two-electron reduction potential for **3** (+1.50 V vs. NHE) relative to that for O(g) (+2.94 V) provides an indication of the bond energy for the (P^{2-}) $Fe^{II}{=}O$ covalent double bond; B.E. = +1.44 x 2 x 23.1 = 67 kcal. Thus, the much lower reactivity of **3** with olefins is consistent with the greater stabilization of (O) by the iron(II) center. Also, for F_5PhIO the two-electron reduction potential under acidic conditions is +1.02 V vs. NHE (Table 3), which indicates an I-O bond energy of 89 kcal and accounts for its unreactive nature with olefins.

The spectroscopy, electrochemistry, and magnetic properties of **2** indicate that its iron center is equivalent to that of Compound I of HRP. Recent EXAFS studies[27,38] of Compound I confirm that it contains an Fe=O double bond (bond distance, 1.64Å), and that its conversion to Compound II (via one-electron reduction) gives a species with an Fe=O group that has the same iron-oxygen bond distance. Again, the spectroscopic and electrochemical properties of 3, and its reduced reactivity with olefins, indicate that the electronic structure of its iron-oxygen center is analogous to that of Compound II of HRP.

Catalase Redox Cycle:

$$(PhO^-)PFe^{III}(H_2O) + H_2O_2 \rightarrow (PhO\cdot)PFe^{II}(O) + H_2O$$

Compound I (catalase)
S = 1/2

$$(PhO\cdot)PFe^{II}(O) + H_2O_2 \rightarrow (PhO^-)PFe^{III}(H_2O) + O_2$$

Peroxidase Redox Cycle:

$$(Imid)(P^{2-})Fe^{III}(H_2O)^+ + H_2O_2 \rightarrow (Imid)(P^{-}\cdot)Fe^{II}(O)^+ + 2H_2O$$

S = 5/2 → Compound I (peroxidase), S = 3/2

$$(Imid)(P^{-}\cdot)Fe^{II}(O)^+ + RH \rightarrow [(ImidH)(P^{2-})Fe^{II}(O)]^+ + R\cdot$$

Compound II

↓ RH

$$[(Imid)(P^{2-})Fe^{III}(OH_2)] + R\cdot$$

Scheme 6

The present results indicate that **2** contains a stabilized oxygen atom, and the parallel chemistry with the active form of cytochrome P-450 prompts the conclusion that it also contains a stabilized atomic oxygen. We have argued elsewhere[25,34] that the most reasonable electronic formulation for the active form of cytochrome P-450 is $(RS\cdot)(Por^{2-})Fe^{II}(O)$ with an $(RS^{-}\cdot)$-Fe(II) covalent bond and an Fe(II)=O covalent double bond. The inability to form **2** with HOOH as the oxidant and the inefficient formation of the active form of cytochrome P-450 via the peroxide shunt[33,40] may mean that HOOH is not formed as an intermediate during the cytochrome P-450 activation

Cytochrome P-450 Redox Cycle:

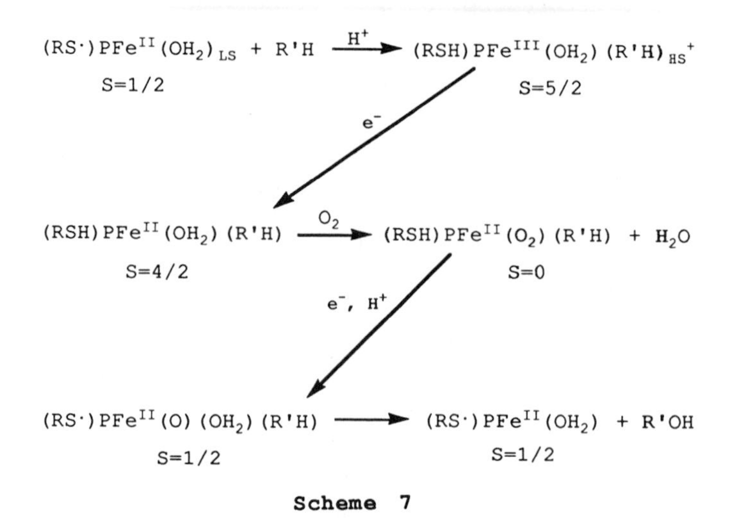

Scheme 7

cycle. A direct reduction cycle to give the active species seems likely

$$(RSH)(Por)Fe^{II} \xrightarrow{e^-,\ O_2} (RSH)(Por)Fe^{II}(O_2) \xrightarrow{e^-,\ H^+} [(RS\cdot)(Por)Fe^{II}(O) + H_2O] \tag{11}$$

Experiments with $(Cl_8TPP)Fe^{III}(ClO_4)$ and thiol ligands are in progress to test this proposition, and to achieve the formation and characterization of the reactive intermediate of cytochrome P-450.

On the basis of the preceding results and arguments reasonable reaction cycles are proposed for the activation of HOOH by the catalase and peroxidase proteins (Scheme 6), and for the activation of O_2 by the cytochrome P-450 protein (Scheme 7).

ACKNOWLEDGEMENT

This work was supported by the National Science Foundation under grant No. CHE-8516247.

REFERENCES

1. W. Day, "Genesis on Planet Earth," 2nd ed., Yale University Press, New Haven (1984).
2. H. Metzner, ed., "Photosynthetic Oxygen Evolution," Academic Press, New York (1978).
3. W.C. Barrette, Jr., H.W. Johnson, Jr., and D.T. Sawyer, Anal. Chem. 56, 1890-1898 (1984).

4. D.T. Sawyer, G. Chiericato, Jr., C.T. Angelis, E.J. Nanni, Jr., and T. Tsuchiya, Anal. Chem. 54, 1720-1724 (1982).
5. R. Parsons, "Handbook of Electrochemical Constants," Butterworths Scientific Publications, London, pp. 69-73 (1959).
6. H.A. Schwarz and R.W. Dodson, J. Phys. Chem. 88, 3643 (1984).
7. J. Wilshire and D.T. Sawyer, Acc. Chem. Res. 12, 105-110 (1979).
8. P. Cofre′ and D.T. Sawyer, Anal. Chem. 58, 1057-1062 (1986).
9. P. Cofre′ and D.T. Sawyer, Inorg. Chem. 25, 2089-2092 (1986).
10. D.T. Sawyer, J.L. Roberts, Jr., T. Tsuchiya, and G.S. Srivatsa, in"Oxygen Radicals in Chemistry and Biology," W. Bors, M. Saran and D. Tait, ed., Walter de Gruyter and Co., pp. 25-33. Berlin: (1984).
11. J.L. Roberts, Jr., M.M. Morrison, and D.T. Sawyer, J. Am. Chem. Soc. 100, 329 (1978).
12. J.L. Roberts, Jr. and D.T. Sawyer, Israel J. of Chem. 23, 430-438 (1983).
13. D.-H. Chin, G. Chiericato, Jr., E.J. Nanni, Jr., and D.T. Sawyer, J. Am. Chem. Soc. 104, 1296 (1982).
14. M.J. Gibian, D.T. Sawyer, T. Ungerman, R. Tangpoonpholvivat, and M.M. Morrison, J. Am. Chem. Soc. 101, 640-644 (1979).
15. J.L. Roberts, Jr., T.S. Calderwood, and D.T. Sawyer, J. Am. Chem. Soc. 105, 7691-7696 (1983).
16. D.T. Sawyer, J.J. Stamp, and K.A. Menton, J. Org. Chem. 48, 3733-3736 (1983).
17. (a) I. Fridovich, in "Oxygen and Oxy Radicals," M.A.J. Rodgers and E.L. Powers, eds., Academic Press, New York, p. 197 (1981) (b) J.A. Fee, Ibid., p. 205.
18. D.T. Sawyer, J.L. Roberts, Jr., T.S. Calderwood, H. Sugimoto, and M.S. McDowell, Phil. Trans. R. Soc. Lond. B311, 483-503 (1985).
19. T.S. Calderwood, C.L. Johlman, J.L. Roberts, Jr., C.L. Wilkins, and D.T. Sawyer, J. Am. Chem. Soc. 106, 4683-4687 (1984).
20. D.T. Sawyer, M.S. McDowell, and K.S. Yamaguchi, Chem. Res. Tox. submitted (1987).
21. S. Matsumoto, H. Sugimoto, and D.T. Sawyer , Chem. Res. Tox. submitted (1987).
22. H. Sugimoto and D.T. Sawyer, J. Am. Chem. Soc. 107, 5712 (1985).
23. H. Sugimoto and D.T. Sawyer, J. Am. Chem. Soc. 106, 4783 (1984).
24. H. Sugimoto and D.T. Sawyer, J. Org. Chem. 50, 1785 (1985).
25. H. Sugimoto, L. Spencer, and D.T. Sawyer, Proc. Natl. Acad. Sci. USA 84, 1731-1733 (1987).
26. (a) P. George, Adv. Catal. 4, 367 (1952); (b) P. George, Biochim. J. 54, 267 (1953); (c) P. George, Biochim. J. 55, 220 (1953).

27. J.E. Denner-Hahn, K.S. Eble, T.J. MacMurray, M. Renner, A.L. Balch, J.T. Groves, J.H. Dawson and K.O. Hodgson, J. Am. Chem. Soc. 108, 7819 (1986).
28. P.F. Guengerich and T.L. McDonald, Acc. Chem. Res. 17, 9 (1984).
29. (a) J.T. Groves, R.C. Haushalter, M. Nakamura, T.E. Nemo and B.J. Evans, J. Am. Chem. Soc. 103, 2884-2886 (1981); (b) J.T. Groves and Y. Watanabe, J. Am. Chem. Soc. 108, 7834-7836 (1986).
30. (a) P.S. Traylor, D. Dolphin, and T.G. Traylor, J. Chem. Soc. Chem. Commun. 279-280 (1984); (b) T.G. Traylor, T. Nakano, B.E. Dunlap, P.S. Traylor, and D. Dolphin, J. Am. Chem. Soc. 108, 2782-2784 (1986).
31. (a) C.M. Dicken, T.C. Woon, and T.C. Bruice, J. Am. Chem. Soc. 108, 1636-1643 (1986); (b) T.S. Calderwood and T.C. Bruice, Inorg. Chem. 25, 3722-3724 (1986); (c) T.S. Calderwood, W.A. Lee, and T.C. Bruice, J. Am. Chem. Soc. 107, 8272-8273 (1985).
32. J.P. Collman, T. Kodadek, and J.I. Brauman, J. Am. Chem. Soc. 108, 2588-2594 (1986).
33. Ortiz Montellano, Ed. "Cytochrome P-450," Plenum Press, New York, (1986).
34. D.T. Sawyer, Comments Inorg. Chem. VI, 103-121 (1987).
35. W.A. Goddard, III and B.D. Olafson, Proc. Nat. Acad. Sci., USA 72, 2335-2339 (1975).
36. P.K.S. Tsang, P. Cofre´, and D.T. Sawyer, Inorg. Chem. 26, 0000-0000 (1987).
37. H. Sugimoto, H.-C. Tung, and D.T. Sawyer, J. Am. Chem. Soc. submitted (1987).
38. M. Chance, L. Powers, T. Poulos, and B. Chance, Biochemistry 25, 1266 (1986).
39. E.G. Samsel, K. Srinivasan, and J.K. Kochi, J. Am. Chem. Soc. 107, 7606-7617 (1985).
40. M.B. MacCarthy and R.E. White, J. Biol. Chem. 258, 9153-9158 (1983).

OXYGEN ACTIVATION BY NEUTROPHILS

James K. Hurst

Department of Chemical and Biological Sciences
Oregon Graduate Center
Beaverton, Oregon

INTRODUCTION

Mobile, phagocytosing cells were first observed in starfish larvae in 1882 by Elie Metchnikoff, who was subsequently able to demonstrate their central role in host defense against infection in animals. In recognition of the significance of his discoveries, Metchnikoff was awarded the Nobel Prize in Physiology or Medicine in 1908. Since that time our understanding of the physiology and biochemistry of leukocytic cells has increased enormously; however, the microbicidal toxins produced by leukocytes and their disinfection mechanisms have remained poorly characterized, and are correspondingly the subject of increasing attention of medical researchers and biochemists.

This review is intended as an introduction to the field. As with many rapidly moving areas of science, there is considerable debate and conflicting viewpoints concerning central issues. I have attempted to present as much as possible the current consensus within a framework of unifying concepts, although this treatment is often speculative and incomplete, and the original literature should be consulted for details. The subject matter is restricted to the neutrophil (also frequently called polymorphonuclear leukocyte or granulocyte), which is the predominant white blood cell in our bodies, and whose primary function appears to be combating bacterial infection. Nonetheless, the underlying general biochemical principles should be applicable to other phagocytic cells in the peripheral circulation.

PHAGOCYTOSIS BY NEUTROPHILS [1]

Neutrophils migrate to sites of infection by responding to chemotactic factors, i.e., chemical signals, generated by reactions at these sites. Particles encountered that are recognized as foreign are bound tightly to the outer plasma membrane, eliciting a complex series of physiological and metabolic changes within the neutrophil leading to their encapsulation by phagocytosis. Once compartmented within the neutrophil, bacteria are rapidly killed and subsequently extensively digested. Neutrophilic recognition of foreign bodies is often aided by adsorption of glycoproteins derived from the host antibody and complement systems, a process termed opsonization.

The sequence of events comprising phagocytosis is illustrated stylistically in Figure 1. The process is initiated by binding of an opsonized microbe at specific cell surface receptor sites, stimulating oxygen consumption by activating a pyridine nucleotide-dependent oxidase located in the plasma membrane [2,3]. The triggering mechanism is not completely understood, but is thought to be indirect, involving activation of an intracellular phospholipase that generates "secondary messengers" which, in turn, activate a protein kinase, leading ultimately to kinase-catalyzed phosphorylation of a component of the oxidase enzyme complex [4]. The respiratory "burst," once initiated, lasts for 15-20 minutes and generates the one- and two-electron reduced products, O_2^- and H_2O_2, respectively. Coincidentally, the neutrophil plasma membrane invaginates in the region of binding, ultimately surrounding the particle and pinching off, isolating the particle within the neutrophil in a special lysosome called the phagosome. Granular lysosomes containing encapsulated biopolymers then migrate to the phagosome. Upon subsequent fusion of the lysosomal cell membranes, the granule contents are discharged into the phagosome. The

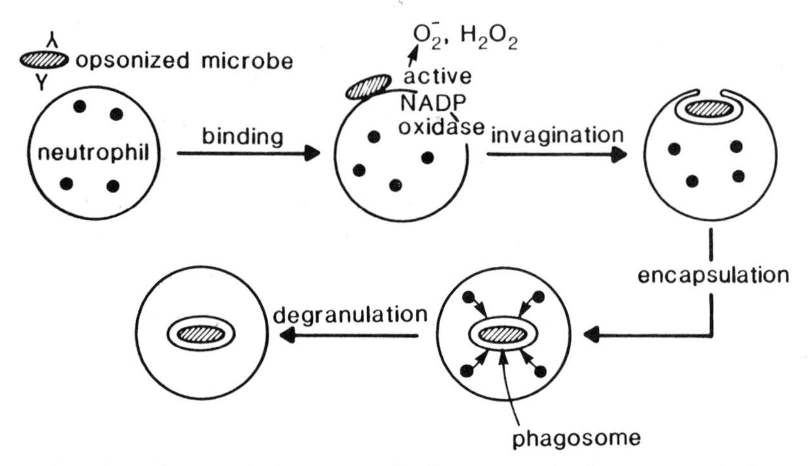

Fig. 1. Sequential steps of phagocytosis by neutrophils.

composition of the granules is diverse, including numerous digestive enzymes, cationic proteins and mucopolysaccharides, but they contain only two biopolymers that might be involved in oxidative microbicidal reactions, an unusual chlorin-containing [5] peroxidase called myeloperoxidase (MPO) and lactoferrin, which, however, appears to be predominantly demetalated. The amount of myeloperoxidase contained within the neutrophil is truly staggering, comprising 2-5% of the dry weight of the cell [6].

Phagocytosis is further illustrated in Figure 2, which is a tracing of an electron micrograph of a single neutrophil containing two bacteria within its phagosome. The cell was cytochemically stained for peroxidase activity and only those regions where positive response was observed have been indicated. These include the MPO-containing granules which have not yet fused with the phagosome, one granule which appears to be fusing (arrow), and the region immediately surrounding the bacterial cell walls. Since MPO is a cationic protein and the bacterial cell wall is negatively charged, electrostatic forces favor their association.

The entire process from recognition to degranulation requires at most a few minutes and leaves the microbe isolated in a highly inimical environment. Killing is quite rapid [7,8] and occurs on the same timescale as phagocytosis, although cellular disruption and depolymerization of microbial cellular [9,10] components continues for several hours afterwards. For Gram-negative bacteria, at least, completely inactivated cells can be recovered from the phagosome that are morphologically

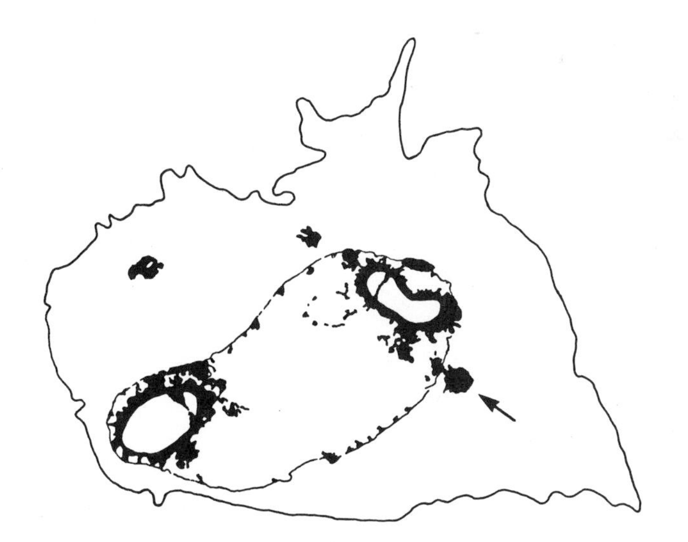

Fig. 2. Tracing of an electron micrograph of a human neutrophil containing two _Lactobacillus acidophilus_ within a phagocytic vacuole. Only the plasma and phagosomal membranes and regions of the cell staining for peroxidase are shown (Adapted from reference 1, p. 219).

indistinguishable from their viable counterparts [11]. Two observations underscore the importance of oxidative reactions to phagocytic disinfection. First, many organisms are killed much more effectively in the presence of oxygen than in anaerobic environments [12]. Second, this oxygen dependence is manifested in the congenital defect known as chronic granulomatous disease (CGD), which is characterized by an inability of one's neutrophils to mount a respiratory burst, although other aspects of phagocytosis appear normal. The consequences to the individual, which are often lethal, are the inability to combat certain types of infection, particularly involving pathogens that are catalase-positive and/or do not possess endogenous H_2O_2. These points are illustrated in Figure 3, where neutrophils from CGD patients are totally ineffective against S. aureus within a period of time that allows greater than 99% inactivation by normal neutrophils.

The respiratory burst can also be elicited by soluble stimuli, which bind at surface receptor sites without inducing extensive formation of phagosomes. Degranulation occurs in this instance at the plasma membrane, with attendant release of lysosomal components into the extracellular medium. This phenomenon is of considerable practical utility, since it provides a means for studying neutrophilic reactions without requiring isolation of or recovery from subcellular organelles. Stimuli often used to induce this cellular response include chemotactic peptides and tumor-forming surface-active agents, e.g., phorbol myristate acetate (PMA).

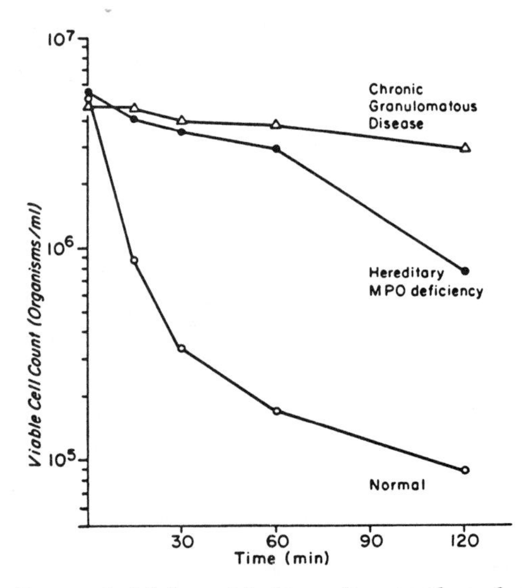

Fig. 3. Bactericidal activity of normal and deficient neutrophils. The test microorganism was Staphylococcus aureus (from reference 66; reproduced by permission of the Journal of the Reticuloendothelial Society).

THE NADPH-OXIDASE

The respiratory oxidase found in neutrophils and other phagocytic cells bears no resemblance to mitochondrial respiration. Oxygen reduction is tightly coupled to glucose oxidation via the hexose monophosphate pathway and is not inhibited by respiratory poisons such as N_3^- and CN^-. Both NADH and NADPH can act as immediate electron donors to the oxidase, although NADPH is thought to be the physiological electron donor, based upon its favorable binding constant. The major, if not sole [13-15], oxygen product is superoxide anion. Because oxidase-generated O_2^- and H_2O_2 react nearly quantitatively with membrane-impermeable scavengers added to the external medium, the site of oxygen reduction is thought to be the external surface of the plasma membrane [13]. The membrane everts during phagocytosis, so oxygen reduction should correspondingly occur within the phagosome. This interpretation is supported by cytochemical studies [16,17] which show O_2^- and H_2O_2 accumulation at the outer plasma and inner phagosomal membrane surfaces of stimulated neutrophils (Figure 4). In contrast, experiments comparing the effects of added NADPH and $NADP^+$ upon respiratory rates in intact and broken neutrophils [18] or plasma membrane vesicles [19] support the notion that the NADPH reduction site is located on the cytoplasmic side of the membrane. Specifically, stimulation of

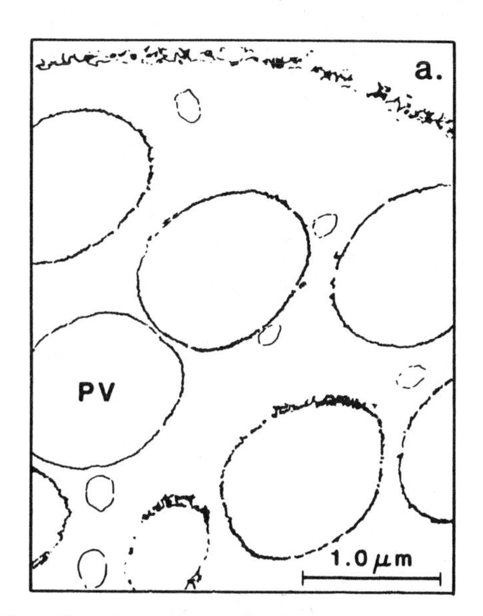

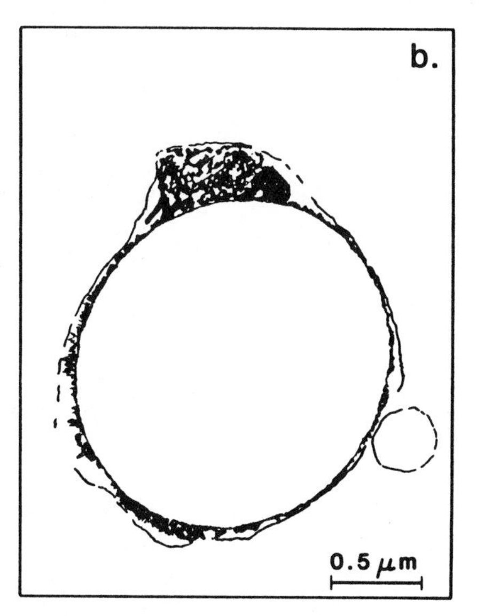

Fig. 4. Tracing of electron micrographs of neutrophils stimulated with polystyrene spheres. Only membranes and regions staining positively for H_2O_2 are shown. Panel a: phagocytic vacuoles (PV) and a portion of the plasma membrane; panel b: a single PV at higher magnification showing H_2O_2 accumulation is confined to the intraphagosomal space (Adapted from reference 16).

respiration by NADPH and its inhibition by $NADP^+$ is observed or enhanced only when access is provided to the inner membrane surface.

These results imply that the oxidase is transversely oriented across the membrane, which is conceptually appealing from the perspective of isolating the lethal reactions from the cytosolically-localized enzymatic processes that are driving them. The neutrophilic cytosol contains enzymes and metabolites, e.g., superoxide dismutase (O_2^-), catalase (H_2O_2) and a glutathione-glutathione peroxidase-glutathione reductase cycle (H_2O_2), that protect it from respiratory burst products that might escape the phagosome, whereas the phagosomal medium, being extracellularly derived, is devoid of these components. This topographic organization is depicted in Figure 5. Also included is a potentially protective sequence involving conversion of HOCl, a secondary oxidant formed by MPO-catalyzed [20] peroxidation of Cl^-, into a less-reactive [21] hydrophilic chloramine by reaction with taurine, a sulfonated amine present in high concentration in the cytosol.

If the description of the NADPH oxidase as a vectorial transmembrane redox enzyme is correct, it is likely by analogy with other membrane-bound electron-translocating systems to be a multicomponent particle. As with many membrane-localized redox systems, attempts to characterize the oxidase have been frustrated by difficulties in obtaining membrane-free soluble preparations that retain high O_2^--forming activity. Additionally, neutrophil stimulation prior to isolation is necessary to obtain oxidase

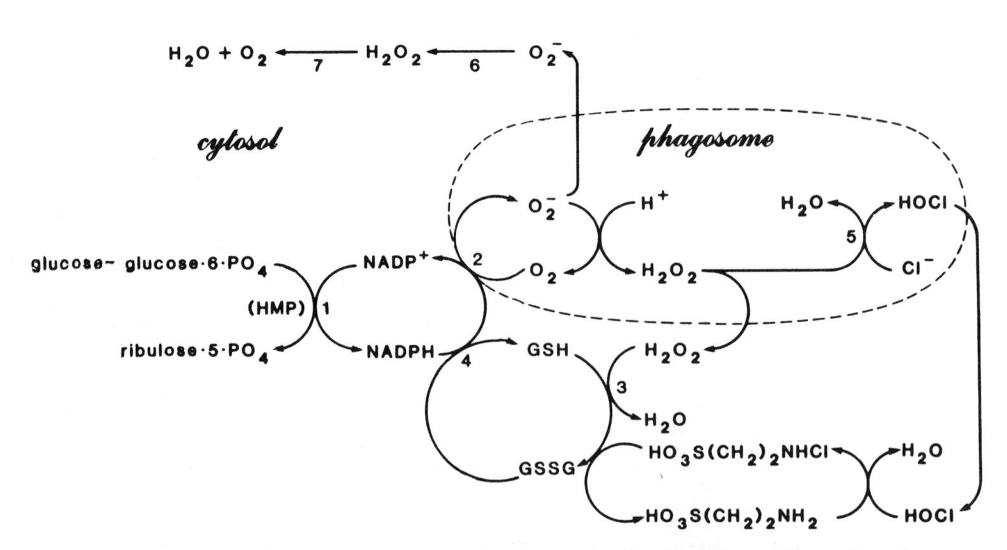

Fig. 5. Probable topographic arrangement of the NADPH oxidase in the neutrophil and cytosolic protection mechanisms. 1, dehydrogenases of the hexosemonophosphate shunt (HMP); 2, NADPH oxidase; 3, glutathione peroxidase; 4, glutathione reductase; 5, myeloperoxidase; 6, superoxide dismutase; 7, catalase.

activity. Consequently, many of the studies have been made on oxidase-containing plasma membrane fragments and other partially purified fractions, with widely differing results. There is now strong circumstantial evidence that one component of the oxidase is a low-potential heme-containing glycoprotein [22], however, which has alternately been designated as cytochrome b_{559} or cytochrome b_{-245}, based upon its ferroheme absorption spectrum [23] (λ_{α} = 559 nm) or midpoint reduction potential [24] ($(E_m)_{pH7}$ = -245 mV), respectively. This cytochrome is reduced by NADPH in plasma membranes from stimulated neutrophils [24] and oxidized by O_2 at rates which are kinetically competent [25] to account for the overall catalytic activity of the oxidase. It is usually found to copurify with the oxidase, is incorporated into the phagosomes of stimulated neutrophils [26], and appears to be absent or functionally abnormal in CGD neutrophils [27]. An FAD-containing flavoprotein is also usually found in partially purified NADPH oxidase preparations. The stimulated NADPH-dependent O_2^- production of solubilized extracts was also enhanced by FAD addition and inhibited by FAD analogues that are incapable of one-electron transfer to electron acceptors [28]. NADPH-dependent flavin semiquinone formation has been detected in stimulated plasma membrane fragments by EPR spectroscopy; the radical signal could not be elicited in membranes from unstimulated cells [29]. The measured midpoint reduction potential of the flavin was $(E_m)_{pH7}$ = -280 mV. Finally, separation of FAD and cytochrome b components of the oxidase appears to recently have been accomplished by cholate extraction of stimulated neutrophil plasma membranes [30].

Ubiquinone-10 (QH_2) has been proposed as a component of the oxidase respiratory chain [31-33], although this claim is highly controversial. Most recent studies failed to detect quinones in purified plasma membranes [34-36] and, assuming a normal midpoint potential [37] of $(E_m)_{pH7} \simeq$ +65 mV, it is difficult to imagine a functional role for quinone in an electron transport chain poised at -160 to -320 mV. Specifically, thermodynamic potentials favor only two-electron transfer from QH_2 to O_2 (Figure 6), but the trapping experiments previously mentioned indicate that nearly all H_2O_2 formed arises from O_2^- dismutation. Other plausible redox components such as ferredoxin-type iron-sulfur centers (FeS) have not been detected.

A minimal electron transport chain consistent with most observations is given in Figure 6. FAD is suggested as a secondary site of O_2 reduction by recent reports that a soluble NADPH oxidase preparation depleted in cyt b_{-245} retains O_2^- reductase activity [36] and direct two-equivalent reduction of O_2 occurs as a minor redox pathway [14,15], a reaction that is non-complementary for cyt b_{-245}. The scheme is not consistent with one recently described NADPH oxidase preparation which reputedly contained

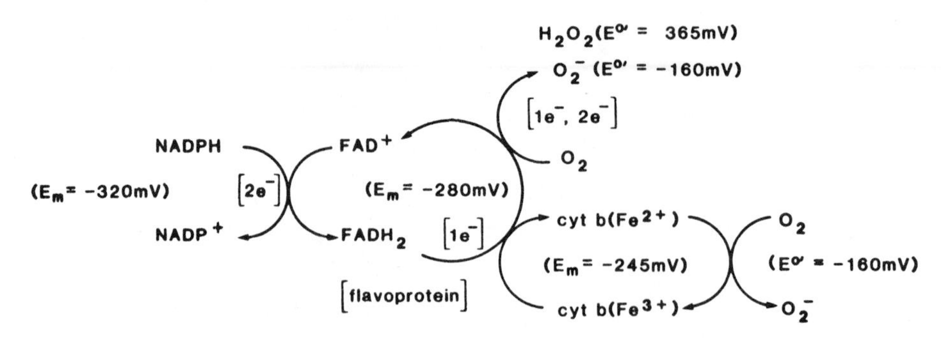

Fig. 6. Minimal composition and probable organization of the NADPH oxidase respiratory chain. Relevant thermodynamic properties are given in parentheses.

negligible FAD, yet retained strong O_2^--forming activity [38]. It is difficult to understand how noncomplementarity between the two-electron donor NADPH and one-electron acceptor cyt b_{-245} might be overcome in this case.

Activation of the NADPH oxidase appears to involve phosphorylation of a 47 kDa protein [39-41]. Recent studies have shown that this reaction correlates with the onset of the respiratory burst, that inhibition of protein kinase both blocks phosphorylation and abolishes the respiratory increase, and that phosphorylation of the protein does not occur in the autosomal recessive form of CGD, which is characterized by an apparently normal complement of cyt b_{-245}. The protein may therefore be the immediate trigger for activation of the burst in normal neutrophils. Its identity is unknown, but apparently is not cyt b_{-245}, the molecular weight for which exceeds 50 kDA [22]. The extreme lability of the isolated oxidase may be a consequence of dephosphorylation or loss of this component.

OXIDATIVE TOXINS FROM THE RESPIRATORY BURST

Various oxidants that have been proposed as the ultimate microbicidal agents derived from the respiratory burst are presented schematically in Figure 7. For purposes of discussion they can conveniently be divided into MPO-dependent and MPO-independent subgroups.

MPO-Dependent Toxins

Myeloperoxidase catalyzes the two-electron oxidation of Cl^-, Br^- and I^- by H_2O_2 in neutral and weakly acidic solutions; chloride ion is thought to be the physiological reductant because it is predominant in biological

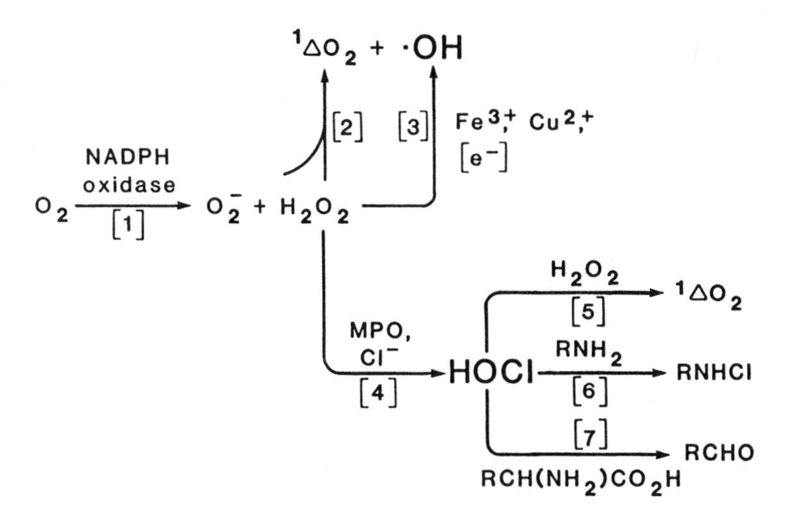

Fig. 7. Proposed neutrophil-generated toxins and their pathways for formation. Most likely ultimate toxins are indicated by the use of larger chemical formulae.

fluids. Hypochlorous acid, the immediate reaction product, is freely diffusible from the enzyme active site [20] and is potently microbicidal to virtually every cell type [1]. To determine if Cl^- peroxidation occurs in the phagosome, we probed the reaction environment using yeast cell wall fragments (zymosan) that had been covalently labeled with the fluorescent dye, fluorescein [42]. The dye is chlorinated progressively by HOCl in the 4'- and 5'-positions to form the corresponding mono- and dichlorofluorescein products (Eq. 1):

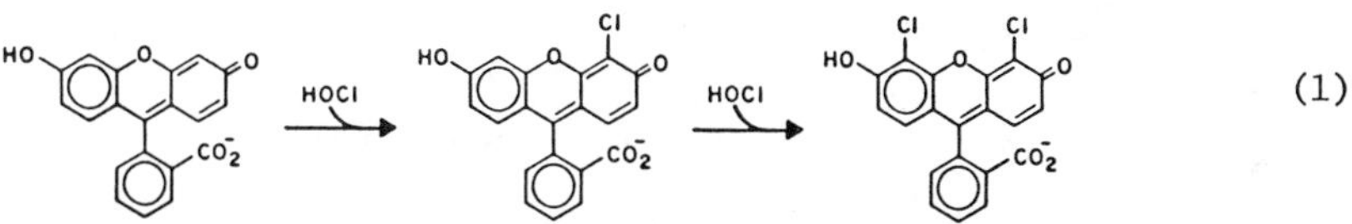

(1)

The same compounds are the predominant products of MPO-catalyzed Cl^- peroxidation. Increasing chlorination causes progressive red-shifting of the excitation and emission bands of the strongly fluorescing dianionic form of the dye as well as diminution of fluorescence quantum yields, presumably a consequence of heavy atom quenching. Fluorescence changes attributable to chlorination by neutrophils are shown in Figure 8. In this instance, the fluorescein-zymosan probe was unopsonized to minimize phagocytosis and the soluble stimulus PMA was used to activate the neutrophils. Under these conditions, the reaction occurs in the external medium, which was buffered at pH 7.4. The time course of fluorescence changes following stimulation was measured under conditions where chlorination would cause maximal reduction in intensity. The first four traces (a-d) in Figure 8

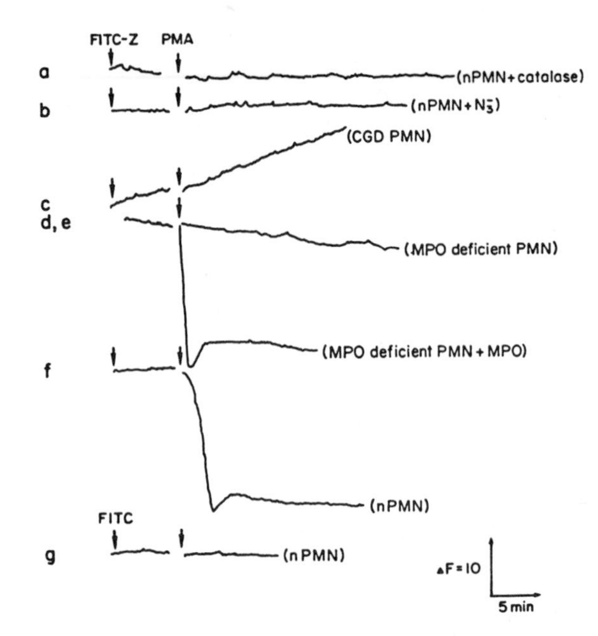

Fig. 8. Fluorescence changes with unopsonized fluoresceinated zymosan (FITC-Z) as an external probe of neutrophil (PMN) activation. Phorbol myristate acetate (PMA) was added as stimulus at the points indicated by the arrows. ΔF is the relative fluorescence intensity where 70 ΔF corresponds to total fluorescence from the particles; normal neutrophils (nPMN); chronic granulomatous disease (CGD) (from reference 42).

establish that the fluorescence losses observed in traces e and f are peroxidase-dependent. Specifically, the rapid changes following stimulation are not observed if H_2O_2 is removed by catalyzed disproportionation (a), if MPO is poisoned (b), or if the neutrophils are deficient in either H_2O_2-generating capability (c) or peroxidase activity (d). Monochlorination of the dye was confirmed by its recovery and HPLC analysis. Fluorescence changes observed with opsonized fluoresceinated zymosan are shown in Figure 9. In this instance, the particle was the stimulus and reaction was primarily intraphagosomal. Normal neutrophils exhibited marked fluorescence quenching (a) which was not observed with either MPO-deficient (b) or CGD neutrophils (c). The reaction could be prevented in normal neutrophils by adding N_3^- to inhibit MPO and could be elicited by addition of MPO to the medium containing MPO-deficient cells or a source of H_2O_2 to the CGD neutrophils. The results therefore strongly support MPO-catalyzed fluorescein chlorination as the explanation for fluorescence losses in phagocytosed fluoresceinated zymosan. Similar conclusions have been drawn in other laboratories from studies using a variety of HOCl-sensitive probes [43-46].

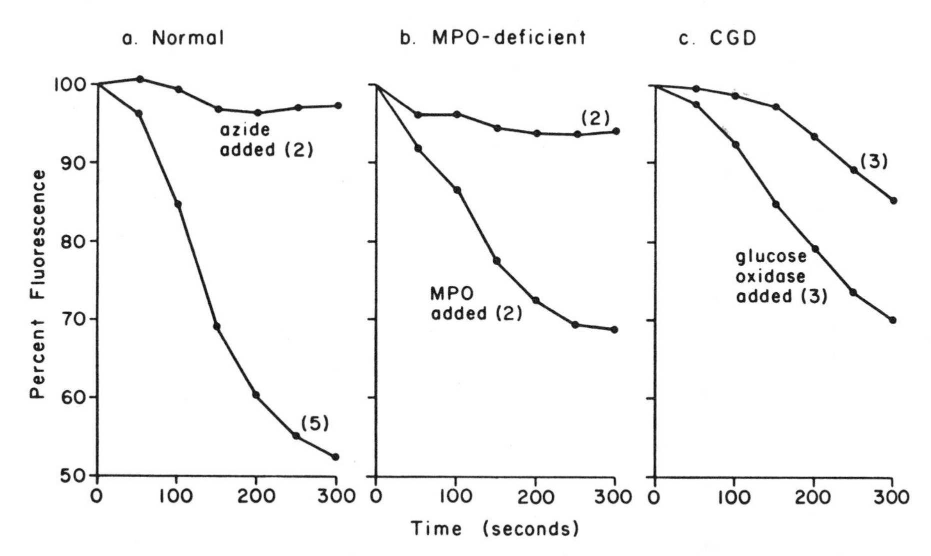

Fig. 9. Fluorescence changes upon ingestion of opsonized fluoresceinated zymosan by normal and deficient neutrophils. Abbreviations as in Fig. 8; fluorescence intensity changes are expressed as percent of initial values (from reference 42).

$^1\Delta O_2$ as a Neutrophil-Generated Toxin. Oxygenated solutions of many chromophoric dyes are microbicidal when exposed to light. At least part of this "photodynamic effect" is attributable to the maintenance of low steady-state concentration levels of $^1\Delta O_2$ formed by energy transfer from the dye triplet excited state [47]. A similar circumstance was suggested to occur in neutrophils, with $^1\Delta O_2$ arising from reaction of MPO-generated HOCl with excess H_2O_2 formed in the respiratory burst (Figure 7, reactions $1 \rightarrow 4 \rightarrow 5$). Evidence cited in support of this proposal included observations that stimulated neutrophils exhibit weak chemiluminescence [48], possibly attributable to $^1\Delta O_2$ dimol emission, and reagents which physically quench or react chemically with $^1\Delta O_2$ inhibited the reaction of 2,5-diphenylfuran, a $^1\Delta O_2$ trapping agent, with the HOCl or $MPO-H_2O_2-Cl^-$ system in the expected manner. Although reaction of HOCl with the hydroperoxide anion HO_2^- does give $^1\Delta O_2$ in 100% yield [50], these suggestions are not in accord with expectations based upon general reactivity principles of HOCl. The relative rates of chlorination of a wide variety of organic and inorganic compounds can be rationalized in terms of electrophile-nucleophile interactions between the electrophilic chlorine atom and nucleophilic centers on the other reactant [21,50]. An appropriate transition-state for HO_2^- reacting with unipositive chlorine compounds (X-Cl) is given in Figure 10. The reaction rate increases with electron-withdrawing character of the substituent X and reaction with the powerful nucleophile, HO_2^-, is effectively quenched by protonation to form

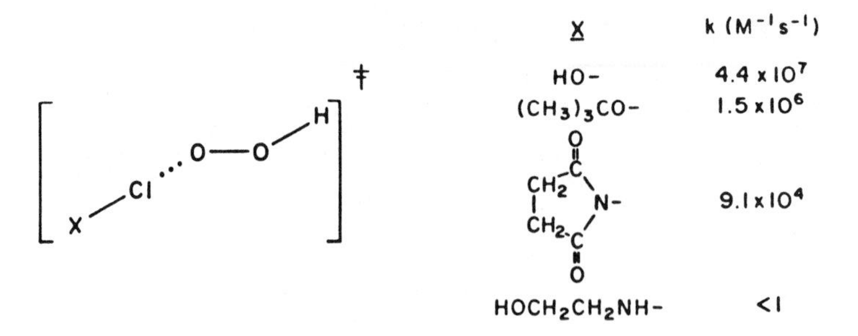

Fig. 10. Hypothetical transition state structure and kinetic data for reaction of the hydroperoxide anion with monovalent chlorine compounds. All reactions obey the rate law, $R = k[X\text{-}Cl][HO_2^-]$.

its very weakly nucleophilic [51] conjugate acid, H_2O_2. Under physiological conditions, hydrogen peroxide is nearly completely protonated and its rate of oxidation by HOCl is correspondingly low. The biological milieu provides many alternative nucleophilic sites for reaction with HOCl [52], so the question of $^1\Delta O_2$ toxicity will not arise simply because the opportunity for its formation is extremely limited. Consistent with this viewpoint, the inhibitory effects of various reagents upon reaction of diphenylfuran with $MPO\text{-}H_2O_2\text{-}Cl^-$ are now recognized to be due to their direct competitive reaction with HOCl [53] and chemiluminescence is not attributable to emission from $^1\Delta O_2$ [54]. There is presently no compelling evidence that $^1\Delta O_2$ plays a primary role in neutrophilic disinfection.

Chloramines and Aldehydes. Hypochlorous acid reacts rapidly with amines and amino acids to form the corresponding chloramine and N-chloramino acids (Figure 7, reactions 6,7). Consistent with the electrophile-nucleophile character of HOCl reactions, rate constants vary proportionately with nitrogen basicity [55]. However, because biological amines and amino groups exist predominantly in their unreactive protonated forms at neutral pH, their overall reaction rates are considerably attenuated under physiological conditions, permitting competitive reaction with other biological nucleophilic sites [52]. N-chloramino acids are unstable, and have been proposed to undergo decarboxylation-deamination to form the corresponding aldehydes [10,56] (Figure 7, reaction 8).

Chloramine (NH_2Cl) and lipophilic chloramines have been shown to be potently bactericidal [57] and, in titrimetric assay, are lethal to *Escherichia coli* at lower dose levels than HOCl [57-59]. Hydrophilic amines, however, protect bacterial cultures against disinfection by HOCl or the $MPO\text{-}H_2O_2\text{-}Cl^-$ system, presumably because the chloramine product is unable to penetrate the hydrocarbon barrier imposed by the bacterial plasma membrane [57]. Toxic chloramines formed by reaction of endogenous NH_3 or functional amino groups may act as intermediates in the lethal neutrophilic reactions [60] (Figure 7, reactions 1 → 4 → 6), but this role can be questioned on the grounds of kinetic competency. We have found by quench-flow kinetics that E. coli are inactivated upon exposure to bactericidal concentration levels of HOCl for periods shorter than 100 ms [61], whereas addition of NH_2Cl-reactive agents even several minutes after exposure to chloramine protects the cultures from inactivation [58,59]. These reactivity differences presumably reflect the slower chlorination and/or oxidation reactions of the less electrophilic NH_2Cl chlorine atom (Figure 10). Killing may occur more rapidly within the phagosome than is achievable with NH_2Cl, although identification of HOCl or chloramines as the ultimate intraphagosomal MPO-generated toxin is probably not crucial to understanding microbicidal mechanisms since their chemistries are undoubtedly quite similar. Participation of endogenous aldehydes in the set of toxic reactions is excluded by these experiments, however, since their formation rates are considerably slower [10,56].

Hypochlorous Acid. The quench-flow experiments [61] also establish that the lethal reactions in E. coli must involve biological compounds that are highly susceptible to oxidation by HOCl (or chloramines). HOCl displays a wide range of biochemical reactivity, rapidly oxidizing or chlorinating electron-rich biomolecules, but being virtually unreactive towards compounds not possessing nucleophilic sites [52]. Reactive molecules include ferredoxin-like FeS centers, purine and pyrimidine bases, conjugated polyenes and sulfhydryl groups in proteins (Figure 11). This high selectivity extends as well to bacterial cells [52], as illustrated in Figure 12 for HOCl bleaching of carotene in the bacterium, Sarcina lutea. Similar irreversible loss of chromophore has been demonstrated [52] in E. coli for b-type cytochromes, as well as destruction of FeS centers in membrane-localized respiratory dehydrogenases measured by ESR spectroscopy [62].

Microbicidal Mechanisms of HOCl. To discriminate among the various possible microbicidal mechanisms represented by these reactions, we have developed methods to correlate the extent of their occurrence with cellular death. Incremental addition of oxidant is made to bacterial cultures and viability, as measured by the ability of the organism to sustain colonial growth, is compared to metabolic capabilities and/or reaction at specific sites [58,59]. A typical titrimetric curve is given in Figure 13, where viability and respiratory rates of E. coli, as well as succinate

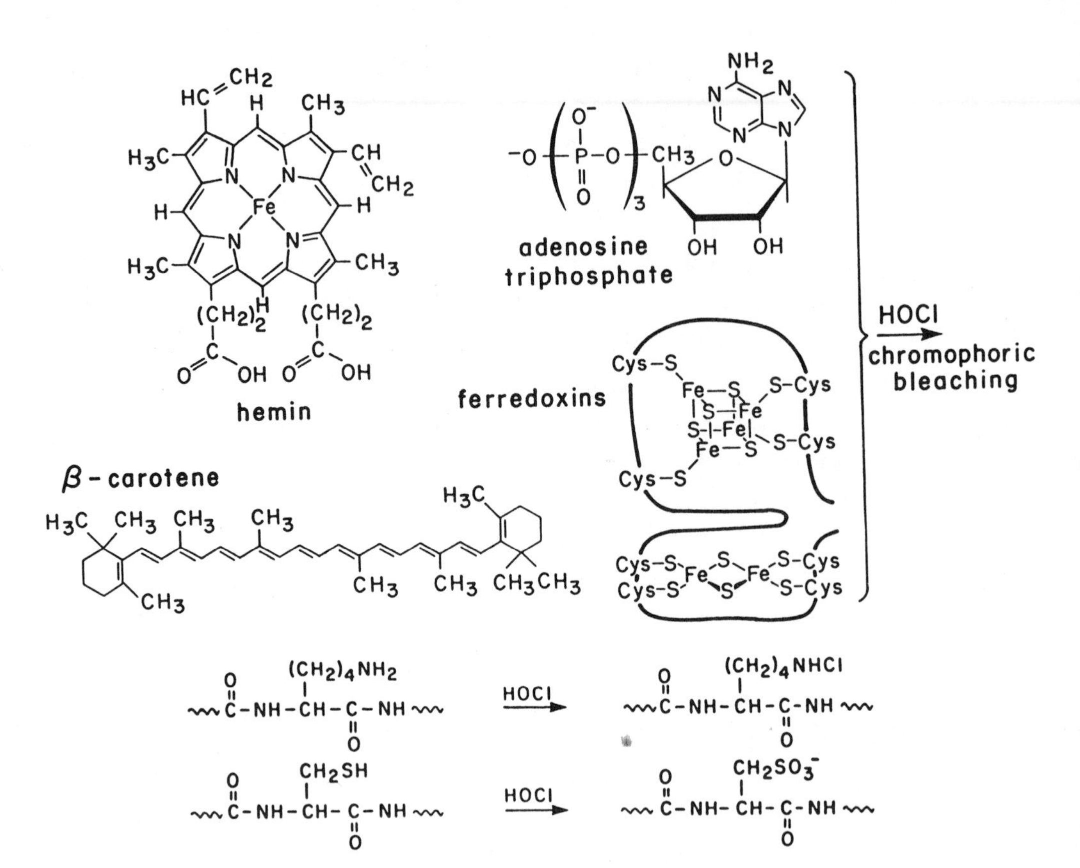

Fig. 11. Representative members of classes of biological compounds reactive towards HOCl.

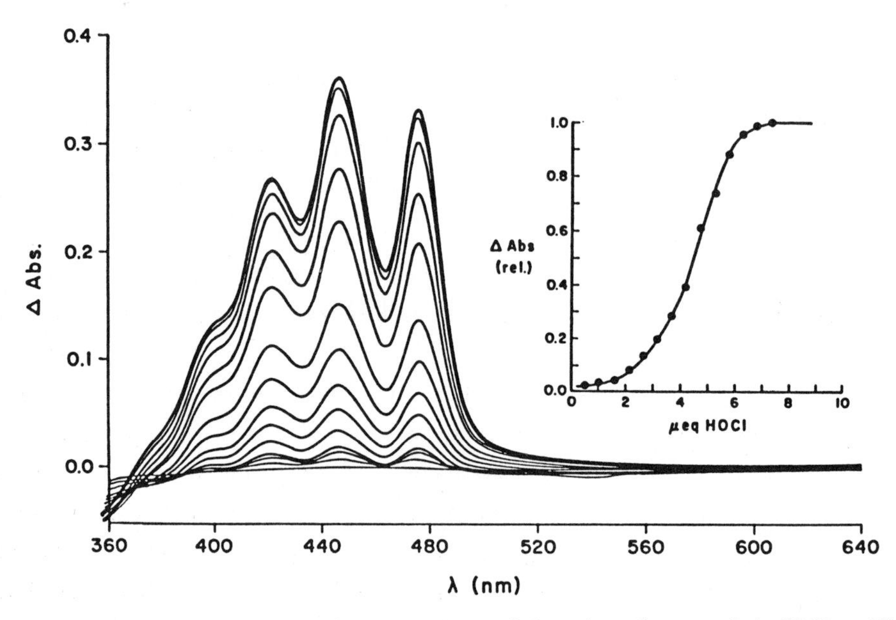

Fig. 12. Difference spectral titration of Sarcina lutea with HOCl. The difference curves obtained are identical to carotene absorption spectra. The inset gives the titrimetric change in absorbance at 447 nm corrected for dilution by titrant (from reference 52).

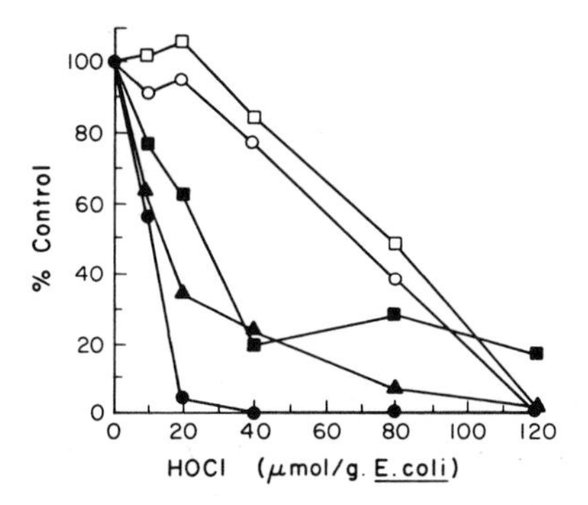

Fig. 13. Comparison of HOCl-promoted titrimetric loss of viability with respiratory function in E. coli. Closed circles: cell viability measured by quantitative pour-plate analysis; closed squares, O_2 respiratory rate; triangles, succinate dehydrogenase activity in membrane vesicles; open circles and squares, relative amplitudes of S_1 and S_3 FeS centers in succinate dehydrogenase in the membrane vesicles.

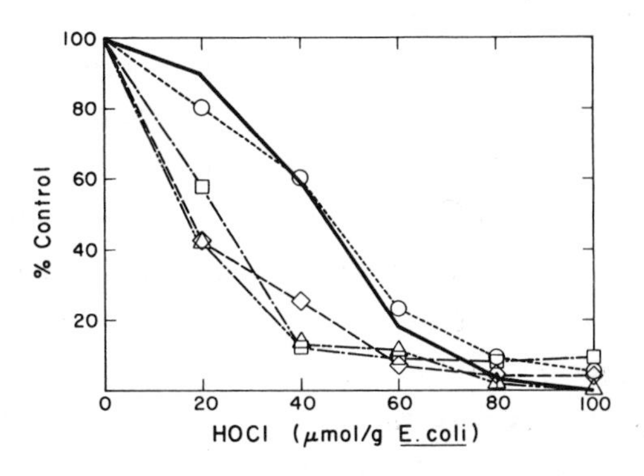

Fig. 14. ^{14}C-labeled metabolite uptake by HOCl-treated E. coli. Solid line, cell viability; circles, thiomethylgalactoside uptake; squares, leucine uptake; diamonds, glutamine uptake; triangles, proline uptake (from reference 58).

dehydrogenase activity and relative intensity of FeS EPR signals in subcellular plasma membrane particles prepared from the cells, are plotted as a function of concentration of added HOCl. From the data, one infers that destruction of the microbial respiratory chain is an early oxidative event and might be associated with dehydrogenase inhibition, but probably not as a consequence of reaction at the FeS redox centers. Because respiratory loss lags behind viability loss on the titrimetric scale, there are apparently other reactions that are the primary contributors to cellular inactivation. In pursuing a number of correlations of this type, we have found that HOCl-sensitive biomolecules within the bacterial cytosol, e.g., adenine nucleotides [59] and the sulfhydryl-dependent enzymes aldolase [52] and β-galactosidase [58], are not oxidatively damaged until HOCl addition exceeds by 3- to 4-fold the amount required for sterilization. These results implicate the plasma membrane as the locale of the lethal lesion(s). Furthermore, metabolite transport across the membrane is inhibited in a manner that parallels or precedes viability loss (Figure 14) [58,62]. Several distinct mechanisms for active transport of nutrients and ions exist in bacteria, involving either cotransport of protons or coupled hydrolysis of ATP or phosphate ester bonds (Figure 15) [63]. Transport loss might arise from direct oxidative inactivation of the transport proteins, loss of driving force for their accumulation by membrane depolarization and/or ATP hydrolysis, or loss of the ability to maintain any chemical gradients across the membrane as a consequence of nonspecific destruction of membrane integrity. The last possibility has been excluded by other experiments which have shown that heavily oxidized cells retain unimpaired proton conductances and glycerol impermeabilities [58], and normal chemiosmotic potentials [59]. Because the membrane remains polarized, the mechanism of inhibition of transport systems coupled to proton symport must involve direct oxidative inactivation of the membrane-localized transport proteins. Furthermore, massive hydrolysis of intracellular ATP attends exposure of _E. coli_ to lethal amounts of HOCl, so that the alternative possibility that the driving force for active transport is lost is sufficient to account for inactivation of ATP-dependent systems [59]. Inhibition of phosphoenolpyruvate-dependent glucose uptake (Figure 15, mechanism 3) also precedes titrimetrically cellular inactivation, although the molecular basis for this inhibition has not yet been ascertained.

The observation that HOCl-inactivated _E. coli_ are unable to maintain proper ATP levels has provided the first real clue to the microbicidal mechanism, since cells that are unable to store metabolic energy cannot

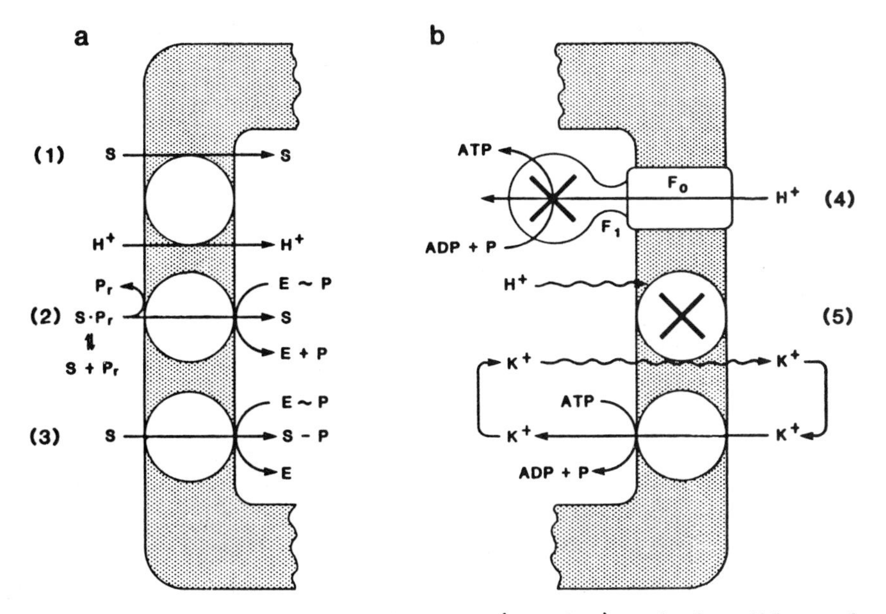

Fig. 15. Bacterial transport mechanisms (panel a) and plausible mechanisms for HOCl-induced phosphoanhydride bond hydrolysis (panel b). Panel a: substrate (S) uptake coupled to proton translocation (1); substrate uptake coupled to hydrolysis of a phosphoryl donor metabolically derived from ATP (2), Pr is a substrate-specific periplasmic binding protein; substrate uptake by group translocation driven by hydrolysis of phosphoenolpyruvate (3). Panel b: HOCl inhibition of the ATP-hydrolyzing F_1 subunit of ATP synthase blocks chemiosmotically-coupled ATP synthesis (4); futile cycle caused by HOCl-induced leak in a K^+-specific proton symporter with attempted compensation by a K^+-translocating ATPase (5).

undertake biosynthetic functions essential to repair and growth [64]. Net loss of ATP might be a consequence of impaired synthetic capabilities or enhanced utilization. An example of the first type is modification of the proton-translocating ATP synthase [63]. We have now determined that the hydrolytic capabilities of this enzyme decrease in parallel with loss of viability in E. coli [62]. The ATP-synthesizing capabilities are almost certainly also lost, although this remains to be established. A hypothetical example of the second type is loss of the gating mechanism in an ATP-independent ion transport system, e.g., K^+ (or PO_4^{3-}) such that uncontrolled efflux of ions occurs. Other ATP-dependent, K^+ (or PO_4^{3-})

transport systems could become engaged in a "futile cycle" to attempt to compensate for the leak [65] (Figure 15). Futile cycles appear to be less important mechanisms for net ATP hydrolysis in HOCl-oxidized E. coli than ATP synthetase inactivation because hydrolysis rates following inactivation are not markedly dependent upon oxidant dose levels [62]. The notion that inhibition of ATP synthesis and metabolite transport constitutes the microbicidal reaction mechanism is particularly appealing because loss of these metabolic capabilities is certainly lethal to all cells and therefore is capable of accounting for the universal character of HOCl toxicity.

MPO-Independent Mechanisms

Although there exists a considerable body of evidence supporting the primary role of MPO catalysis in cellular disinfection, it is clear that the neutrophil possesses other effective means of inactivating cells. Individuals with hereditary MPO-deficiency, characterized by neutrophils which lack peroxidase activity but exhibit normal phagocytosis and a stimulated respiratory response, generally do not suffer the life-threatening infections common to CGD patients [1]. Comparison of the response of MPO-deficient and CGD neutrophils to S. aureus, illustrated in Figure 3, suggests that at least part [67] of the remaining bactericidal activity is oxidative in character. This remaining oxygen-dependent toxicity has generally been described in terms of reactions involving the immediate products of the respiratory burst, O_2^- and H_2O_2, or secondary reactions which are best described as metal-catalyzed Fenton reactions, producing either hydroxyl radical or compounds capable of reacting similarly. The use of chemical and enzymatic probes has figured prominently in attempts to identify the actual toxins produced. The premises of this approach are that the reactive compounds are accessible to the trapping agents and that the ensuing chemistry is well understood. In hindsight, it appears that these conditions are seldom met in phagocytic systems.

Superoxide Ion and Hydrogen Peroxide. Although the putative role of superoxide in oxygen toxicity in general remains controversial [68], O_2^- is probably not directly involved in intraphagosomal bactericidal reactions. Stoichiometric studies [13-15] of the respiratory burst have shown that nearly all of the O_2 consumed is converted to O_2^-. Because these measurements rely upon reaction with chemical trapping agents located in the extracellular medium, the O_2^- formed by the NADPH oxidase must be efficiently released into the aqueous phase. Disproportionation to H_2O_2 and O_2 is also nearly quantitative, so there is no indication throughout the primary respiratory sequence of appreciable reaction of O_2^- with the neutrophil plasma membranes. These observations are consistent with expectations, since O_2^- exhibits very limited redox chemistry in water [69]. Similarly, H_2O_2 alone appears to be generally ineffective as an antimicrobial agent at low to moderate concentrations, which may not be surprising since H_2O_2 is a respiratory end product of many aerobic organisms.

Hydroxyl Radical. The action of various chemical and enzymatic probes in the presence of stimulated neutrophils has been taken to indicate that the toxic compound produced is OH radical or a chemically similar species. Typical results from an early, and relatively self-consistent, study [70], are given in Figure 16. Cophagocytosis of enzymes adsorbed onto latex spheres with bacteria provided a means to introduce the enzymes into the phagosomal reaction environment. Either active catalase or superoxide dismutase (SOD) partially protected bacteria from the phagosomal toxins, but the denatured enzymes did not. The results were interpreted to indicate that neither O_2^- nor H_2O_2 alone was the bactericidal agent, but that it was a product of their reaction, presumably hydroxyl radical. Consistent with this interpretation, hydrophilic compounds capable of scavenging •OH also afforded some protection. One disturbing feature of these results is the protection by SOD since, if MPO-mediated reactions are dominant, enhancing H_2O_2 formation is expected to potentiate, not inhibit, killing.

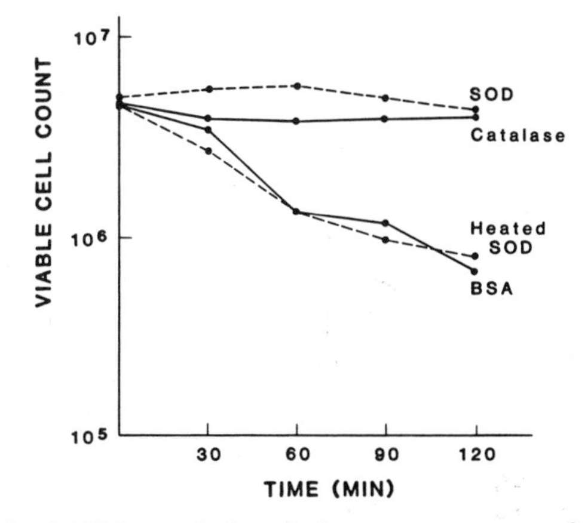

Fig. 16. Phagocytic killing of Staphylococcus aureus by human neutrophils The effect of cophagocytosis of latex particles containing bound enzymes is shown (Adapted from reference 70). SOD, superoxide dismutase; BSA, bovine serum albumin.

Other probe studies have not proven so unambiguous. Hydroxyl spin-trapped adducts have been observed [71-73] by EPR in solutions of stimulated neutrophils containing 5,5-dimethyl-1-pyrroline-1-oxide (DMPO), but they may be formed [72,73] by reduction of the DMPO-OOH perhydroxy adduct, rather than direct trapping of •OH. Ethylene was produced from 2-keto-4-thiomethylbutyric acid in the presence of stimulated neutrophils [74,75], a reaction originally thought to be diagnostic for the •OH radical. However, the reactions appeared to be predominantly MPO-catalyzed, a conclusion supported by demonstration of reaction of the probe with the cell-free MPO-H_2O_2-Cl^- model system [76]. Comparison of probe studies of other diverse reaction systems does not give recognizable patterns of behavior. Some reactions are preferentially inhibited by SOD or catalase, whereas others appear to be inhibited equally well by either enzyme, but show little response to addition of OH-trapping agents.

The notion that OH is a primary phagosomal toxin also presents a conceptual problem. Hydroxyl radical is a powerful oxidant with the capacity to react with virtually all organic molecules. To be effective, endogenous antimicrobial agents should be able to select for vulnerable cellular sites, thereby minimizing expenditure of leukocytic metabolic energy. This property is found, for example, in HOCl, but not in $^1\Delta O_2$, which is rapidly physically deactivated in aqueous solution [47]. Uncontrolled OH formation in the phagosomal milieu would likewise be expected to give inefficient killing because indiscriminant oxidation of cellular wall and membrane components should occur with little consequence to cell viability. However, the very appealing suggestion has been made that in cells OH formation is site-specific [77-80] (Figure 17). The reasoning is that, since the uncatalyzed Fenton reaction is too slow to be biologically significant [81], only metal-catalyzed reactions [82] will occur. The

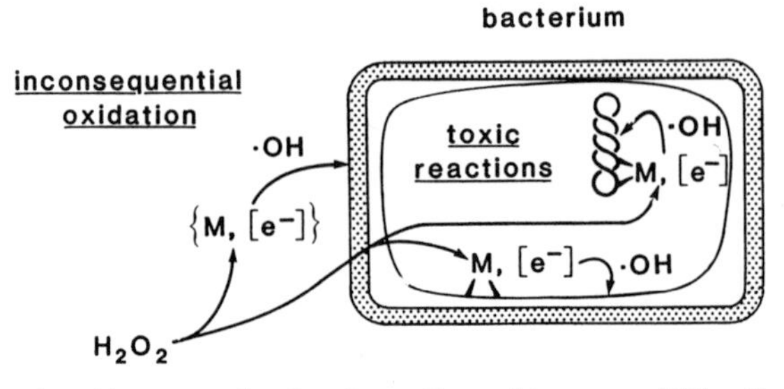

Fig. 17. Schematic diagram depicting the site-specific Fenton mechanism of H_2O_2 antimicrobial action.

reactants are therefore H_2O_2 and endogenous one-electron reducing agents, rather than free OH. Hydroxyl radical is then generated in situ at the biological metal binding site, which is likely to be involved in essential cellular function. The site is correspondingly lost because OH immediately attacks the surrounding biological material. Thus, it is envisioned that selectivity is conferred upon the system by the location of the catalyst. In support of this concept, redox metal ions (Fe(III), Cu(II)) are found to dramatically enhance the toxicity of reducing agents towards viruses [79] and possibly animals, as well as enhancing their ability to inactivate enzymes [77,78,80], in a manner consistent with the site-specific mechanism. The model is also consistent with observations that the toxicity of H_2O_2 towards bacteria increases with Fe(III) uptake but that exogenously added Fe(III) is protective [83]. Sequestration of OH generating sites also provides a rationalization for the varying response to exogenous chemical probes found in diverse reaction systems.

CONCLUDING COMMENTS

Major advances are presently being made in our understanding of leukocyte biochemistry, particularly aspects dealing with respiratory activation, the components of the respiratory chain, the ultimate oxidative toxins responsible for antimicrobial activity and their mechanisms of action. Inactivation by hypochlorous acid appears to involve disruption of energy-transducing cellular elements, whereas other oxidative microbicidal reactions may involve "site-specific" Fenton chemistry. Studies to identify the metabolic dysfunctions attending cellular death should allow better definition of the mechanisms of toxicity in the latter case. Other aspects of leukocyte biochemistry not addressed in this review, including nonoxidative antimicrobial mechanisms and reactions involving other phagocytic cells, are also developing rapidly.

Acknowledgements

JKH has benefited enormously from collaborative interactions, instructional advice, and stimulating discussions with colleagues and associates, most notably J. M. Albrich, W. C. Barrette, Jr., T. R. Green, S. J. Klebanoff and H. Rosen, and from research support from the National Institutes of Health (#AI-15834) and the Medical Research Foundation of Oregon.

REFERENCES

1. For a comprehensive review, see: S. J. Klebanoff and R. A. Clark, "The Neutrophil: Function and Clinical Disorders," North-Holland, Amsterdam (1978).
2. B. Dewald, M. Baggiolini, J. T. Curnutte, and B. M. Babior, Subcellular localization of the superoxide-forming enzyme in human neutrophils, J. Clin Invest. 63:21 (1979).
3. T. Yamaguchi, K. Sato, K. Shimada, and K. Kakinuma, Subcellular localization of O_2^- generating enzyme in guinea pig polymorphonuclear leukocytes; fractionation of subcellular particles by using a Percoll density gradient, J. Biochem. 91:31 (1982).
4. J. A. Badwey and M. L. Karnovsky, Production of superoxide by phagocytic leukocytes: a paradigm for stimulus-response phenomena, Curr. Top. Cell Regul. 28:183 (1986).
5. S. S. Sibbett and J. K. Hurst, Structural analysis of myeloperoxidase by resonance Raman spectroscopy, Biochemistry 23:3007 (1984).
6. J. Schultz and K. Kaminker, Myeloperoxidase of the leukocyte of normal human blood. I. Content and localization, Arch. Biochem. Biophys. 96:465 (1962).
7. P. Elsbach, On the interaction between phagocytes and micro-organisms, N. Engl. J. Med. 16:846 (1973).
8. A. W. Segal, M. Geisow, R. Garcia, A. Harper, and R. Miller, The respiratory burst of phagocytic cells is associated with a rise in vacuolar pH, Nature 290:406 (1981).
9. Z. A. Cohn, The fate of bacteria within phagocytic cells I. The degradation of isotopically labeled bacteria by polymorphonuclear leucocytes and macrophages, J. Exp. Med. 117:27 (1968).
10. R. J. Selvaraj, B. B. Paul, R. R. Strauss, A. A. Jacobs, and A. J. Sbarra, Oxidative peptide cleavage and decarboxylation by the myeloperoxidase-hydrogen peroxide-chloride ion antimicrobial system, Infect. Immun. 9:255 (1974).
11. E. M. Ayoub and J. G. White, Intraphagocytic degradation of Group A streptococci: Electron microscope studies, J. Bacteriol. 98:728 (1969).
12. G. L. Mandell, Bactericidal activity of aerobic and anaerobic polymorphonuclear leukocytes, Infect. Immun. 9:337 (1974).
13. R. K. Root and J. A. Metcalf, Hydrogen peroxide release from human granulocytes during phagocytosis. Relationship to superoxide anion formation and cellular catabolism of hydrogen peroxide: Studies with normal and cytochalasin B-treated cells, J. Clin. Invest. 60:1266 (1977).
14. T. R. Green and D. E. Wu, The NADPH:O_2 oxidoreductase of human neutrophils. Stoichiometry of univalent and divalent reduction of O_2, J. Biol. Chem. 261:6010 (1986).
15. T. R. Green and K. L. Pratt, A reassessment of product specificity of the NADPH:O_2 oxidoreductase of human neutrophils, Biochem. Biophys. Res. Commun. 142:213 (1987).
16. R. T. Briggs, D. B. Drath, M. L. Karnovsky, and M. J. Karnovsky, Localization of NADH oxidase on the surface of human polymorphonuclear leukocytes by a new cytochemical method, J. Cell Biol. 67:566 (1975).
17. R. T. Briggs, J. M. Robinson, M. L. Karnovsky, and M. J. Karnovsky, Superoxide production by polymorphonuclear leukocytes. A cytochemical approach, Histochemistry 84:371 (1986).
18. T. R. Green, R. E. Schaefer, and M. T. Makler, Orientation of the NADPH-dependent superoxide generating oxidoreductase on the outer membrane of human PMN's, Biochem. Biophys. Res. Commun. 94:262 (1980).

19. G. L. Babior, R. E. Rosin, B. J. McMurrich, W. A. Peters, and B. M. Babior, Arrangement of the respiratory burst oxidase in the plasma membrane of the neutrophil, J. Clin. Invest. 67:1724 (1981).
20. J. E. Harrison and J. Schultz, Studies on the chlorinating activity of myeloperoxidase, J. Biol. Chem. 251:1371 (1976).
21. J. K. Hurst, P. A. G. Carr, F. E. Hovis, and R. J. Richardson, Hydrogen peroxide oxidation by chlorine compounds. Reaction dynamics and singlet oxygen formation, Inorg. Chem. 20:2435 (1981).
22. A. M. Harper, M. F. Chaplin, and A. W. Segal, Cytochrome b_{-245} from human neutrophils is a glycoprotein, Biochem. J. 227:783 (1985).
23. A. R. Cross, F. K. Higson, O. T. G. Jones, A. M. Harper, and A. W. Segal, The enzymic reduction and kinetics of oxidation of cytochrome b_{-245} of neutrophils, Biochem. J. 204:479 (1982).
24. A. R. Cross, O. T. G. Jones, A. M. Harper, and A. W. Segal, Oxidation-reduction properties of the cytochrome b found in the plasma-membrane fraction of human neutrophils, Biochem. J. 194:599 (1981).
25. A. R. Cross, J. F. Parkinson, and O. T. G. Jones, Mechanism of the superoxide-producing oxidase of neutrophils. O_2 is necessary for the fast reduction of cytochrome b_{-245} by NADPH, Biochem. J. 226:881 (1985).
26. A. W. Segal and O. T. G. Jones, Novel cytochrome b system in phagocytic vacuoles of human granulocytes, Nature (London) 276:515 (1978).
27. A. W. Segal and O. T. G. Jones, Absence of cytochrome b reduction in stimulated neutrophils from both female and male patients with chronic granulomatous disease, FEBS Lett. 110:111 (1980).
28. D. R. Light , C. Walsh, A. M. O'Callaghan, E. J. Goetzl, and A. I. Tauber, Characteristics of cofactor requirement for the superoxide-generating NADPH oxidase of human polymorphonuclear leukocytes, Biochemistry 20:1468 (1981).
29. K. Kakinuma, M. Kaneda, T. Chiba, and T. Ohnishi, Electron spin resonance studies on a flavoprotein in neutrophil plasma membranes. Redox potentials of the flavin and its participation in NADPH oxidase, J. Biol. Chem. 261:9426 (1986).
30. T. A. Gabig and B. A. Lefker, Catalytic properties of the resolved flavoprotein and cytochrome b components of the NADPH dependent O_2^- generating oxidase from human neutrophils, Biochem. Biophys. Res. Commun. 118:430 (1984).
31. D. R. Crawford and D. L. Schneider, Identification of ubiquinone-50 in human neutrophils and its role in microbicidal events, J. Biol. Chem. 257:6662 (1982).
32. C. C. Cunningham, L. R. De Chatelet, P. I. Spach, J. W. Parce, M. J. Thomas, C. J. Lees, and P. S. Shirley, Identification and quantitation of electron transport components in human polymorphonuclear neutrophils, Biochim. Biophys. Acta 682:430 (1982).
33. T. G. Gabig and B. A. Lefker, Activation of human neutrophil NADPH oxidase results in coupling of electron carrier function between ubiquinone-10 and cytochrome b_{559}, J. Biol. Chem. 260:3991 (1985).
34. A. R. Cross, O. T. G. Jones, R. Garcia, and A. W. Segal, The subcellular localization of ubiquinone in human neutrophils, Biochem. J. 216:765 (1983).
35. R. Lutter, R. van Zwieten, R. S. Weening, M. N. Hamers, and D. Roos, Cytochrome b, flavins, and ubiquinone-50 in enucleated human neutrophils (polymorphonuclear leukocyte cytoplasts), J. Biol. Chem. 259:9603 (1984).
36. G. A. Glass, D. M. DeLisle, P. de Togni, T. G. Gabig, B. H. Magee, M. Markert, and B. M. Babior, The respiratory burst oxidase of human neutrophils. Further studies of the purified enzyme, J. Biol. Chem. 261:13247 (1986).

37. P. F. Urban and M. Klingenberg, Redox potentials of ubiquinone and cytochrome in the respiratory chain, Eur. J. Biochem. 9:510 (1969).
38. P. Bellavite, O. T. G. Jones, A. R. Cross, E. Papini, and P. Rossi, Composition of partially purified NADPH oxidase from pig neutrophils, Biochem. J. 223:639 (1984).
39. A. G. Segal, P. G. Heyworth, S. Cockcroft, and M. M. Barrowman, Stimulated neutrophils from patients with autosomal recessive chronic granulomatous disease fail to phosphorylate a M_r-44,000 protein, Nature (London) 316:547 (1985).
40. P. G. Heyworth and A. W. Segal, Further evidence for the involvement of a phosphoprotein in the respiratory burst oxidase from human neutrophils, Biochem. J., 239:723 (1986).
41. T. Hayakawa, K. Suzuki, S. Suzuki, P. C. Andrews, and B. M. Babior, A possible role for protein phosphorylation in the activation of the respiratory burst in human neutrophils. Evidence from studies with cells from patients with chronic granulomatous disease, J. Biol. Chem. 261:9109 (1986).
42. J. K. Hurst, J. M. Albrich, T. R. Green, H. Rosen and S. Klebanoff, Myeloperoxidase-dependent fluorescein chlorination by stimulated neutrophils, J. Biol. Chem. 259:4812 (1984).
43. M. B. Grisham, M. M. Jefferson, D. F. Melton, and E. L. Thomas, Chlorination of endogenous amines by isolated neutrophils: Ammonia-dependent bactericidal, cytotoxic, and cytolytic activities of the chloramines, J. Biol. Chem. 259:10404 (1984).
44. C. S. Foote, T. E. Goyne, and R. I. Lehrer, Assessment of chlorination by human neutrophils, Nature 301:715 (1983).
45. S. J. Weiss, R. Klein, A. Slivka, and M. Wei, Chlorination of taurine by human neutrophils: Evidence for hypochlorous acid generation, J. Clin. Invest. 70:598 (1982).
46. J. M. Zglinczynski and T. Stelmaszynska, Chlorinating ability of human phagocytizing leukocytes, Eur. J. Biochem. 56:157 (1975).
47. C. S. Foote, Mechanisms of photosensitized oxidation, Science 162:963 (1968).
48. R. C. Allen, Halide dependence of the myeloperoxidase-mediated antimicrobial system of the polymorphonuclear leukocyte in the phenomenon of electronic excitation, Biochem. Biophys. Res. Commun. 63:675 (1975).
49. H. Rosen and S. J. Klebanoff, Formation of singlet oxygen by the myeloperoxidase-mediated antimicrobial system, J. Biol. Chem. 252:4803 (1977).
50. A. M. Held, D. J. Halko, and J. K. Hurst, Mechanisms of chlorine oxidation of hydrogen peroxide, J. Am. Chem. Soc. 100:5732 (1978).
51. E. Sander and W. P. Jencks, General acid and base catalysis of the reversible addition of hydrogen peroxide to aldehydes, J. Am. Chem. Soc. 90:3817 (1968).
52. J. M. Albrich, C. A. McCarthy, and J. K. Hurst, Biological reactivity of hypochlorous acid: Implications for microbicidal mechanisms of leukocyte myeloperoxidase, Proc. Natl. Acad. Sci. USA 78:210 (1981).
53. A. M. Held and J. K. Hurst, Ambiguity associated with use of singlet oxygen trapping agents in myeloperoxidase-catalyzed reactions, Biochem. Biophys. Res. Commun. 81:878 (1978).
54. B. D. Cheson, R. L. Christensen, R. Sperline, B. E. Kohler, and B. M. Babior, The origin of the chemiluminescence of phagocytosing granulocytes, J. Clin. Invest. 58:789 (1976).
55. J. C. Morris, Kinetics of reactions between aqueous chlorine and nitrogen compounds, in: "Principles and Applications of Water Chemistry," S. D. Faust and J. V. Hunter, eds., Wiley, New York (1967).

56. B. B. Paul, A. A. Jacobs, R. R. Strauss and A. J. Sbarra, Role of the phagocyte in host-parasite interactions. XXIV. Aldehyde generation by the myeloperoxidase-hydrogen peroxide antimicrobial system: A possible in vivo mechanism of action, Infect. Immun. 2:414 (1970).
57. E. L. Thomas, Myeloperoxidase-hydrogen peroxide-chloride antimicrobial system: Effect of exogenous amines on antibacterial action against Escherichia coli, Infect. Immun. 25:110 (1979).
58. J. M. Albrich, J. H. Gilbaugh III, K. B. Callahan, and J. K. Hurst, Effects of the putative neutrophil-generated toxin, hypochlorous acid, on membrane permeability and transport systems of Escherichia coli, J. Clin. Invest. 78:177 (1986).
59. W. C. Barrette, Jr., J. M. Albrich, and J. K. Hurst, Hypochlorous acid-promoted loss of metabolic energy in Escherichia coli, manuscript submitted.
60. E. L. Thomas, Myeloperoxidase, hydrogen peroxide, chloride antimicrobial system: Nitrogen-chlorine derivatives of bacterial components in bactericidal action against Escherichia coli, Infect. Immun. 23:522 (1979).
61. J. M. Albrich and J. K. Hurst, Oxidative inactivation of Escherichia coli by hypochlorous acid. Rates and differentiation of respiratory from other reaction sites, FEBS Lett. 144:157 (1982).
62. W. C. Barrette, Jr., and J. K. Hurst, unpublished observations.
63. F. M. Harold, "The Vital Force: A Study of Bioenergetics," W. H. Freeman, New York (1986).
64. C. J. Knowles, Microbial metabolic regulation by adenine nucleotide pools, Symp. Soc. Gen. Microbiol. 27:241 (1977).
65. W. Epstein and L. Laimins, Potassium transport in Escherichia coli: Diverse systems with common control by osmotic forces, Trends Biochem. Sci. 5:21 (1980).
66. S. J. Klebanoff and C. B. Hamon, Role of myeloperoxidase-mediated anti-microbial systems in intact leukocytes, J. Reticuloendothel. Soc. 12:170 (1972).
67. P. Elsbach and J. Weiss, A reevaluation of the roles of the oxygen-dependent and oxygen-independent microbicidal systems of phagocytes, Rev. Infect. Dis. 5:843 (1983).
68. J. A. Fee, Is superoxide important in oxygen poisoning? Trends Biochem. Sci. 7:84 (1982); B. Halliwell, Superoxide and superoxide-dependent formation of hydroxyl radicals are important in oxygen toxicity, Trends Biochem. Sci. 7:271 (1982).
69. D. T. Sawyer and J. S. Valentine, How super is superoxide? Acc. Chem. Res. 14:393 (1981).
70. R. B Johnston, Jr., B. B. Keele, Jr., H. P. Misra, J. E. Lehmeyer, L. S. Webb, R. L. Baehner, and K. V. Rajagopalan, The role of superoxide anion generation in phagocytic bactericidal activity. Studies with normal and chronic granulomatous disease leukocytes, J. Clin. Invest. 55:1357 (1975).
71. M. R. Green, H. A. O. Hill, M. J. Okolow-Zubkowska, and A. W. Segal, The production of hydroxyl and superoxide radicals by stimulated human neutrophils--measurements by epr spectroscopy, FEBS Lett. 100:23 (1979).
72. H. Rosen and S. J. Klebanoff, Hydroxyl radical generation by polymorphonuclear leukocytes measured by electron spin resonance spectroscopy, J. Clin. Invest. 64:1725 (1979).
73. B. E. Britigan, G. M. Rosen, Y. Chai, and M. S. Cohen, Do human neutrophils make hydroxyl radical? Determination of free radicals generated by human neutrophils activated with a soluble or particulate stimulus using electron paramagnetic resonance spectrometry, J. Biol. Chem. 261:4426 (1986).

74. S. J. Weiss, P. K. Rustagi, and A. F. LoBuglio, Human granulocyte generation of hydroxyl radical, J. Exp. Med. 147:316 (1978).
75. S. J. Klebanoff and H. Rosen, The role of myeloperoxidase in the microbicidal activity of polymorphonuclear leukocytes, Ciba Found. Symp. 65:263 (1979).
76. S. J. Klebanoff and H. Rosen, Ethylene formation by polymorphonuclear leukocytes. Role of myeloperoxidase, J. Exp. Med. 148:490 (1978).
77. T. Navok and M. Chevion, Transition metals mediate enzymatic inactivation by favism-inducing agents, Biochem. Biophys. Res. Commun. 122:297 (1984).
78. E. Shinar, T. Navok, and M. Chevion, The analogous mechanism of enzymatic inactivation induced by ascorbate and superoxide in the presence of copper, J. Biol. Chem. 258:14778 (1983).
79. A. Samuni, J. Aronovitch, D. Godinger, M. Chevion, and G. Czapski, On the cytotoxicity of vitamin C and metal ions. A site-specific Fenton mechanism, Eur. J. Biochem. 137:119 (1983).
80. A. Samuni, M. Chevion, and G. Czapski, Unusual copper-induced sensitization of the biological damage due to superoxide radicals, J. Biol. Chem. 256:12632 (1981).
81. G. J. McClune and J. A. Fee, Stopped flow spectrophotometric observation of superoxide dismutation in aqueous solution, FEBS Lett. 67:294 (1976).
82. D. A. Rowley and B. Halliwell, Superoxide-dependent and ascorbate-dependent formation of hydroxyl radicals in the presence of copper salts: A physiologically significant reaction? Arch. Biochem. Biophys. 225:279 (1983), and references therein.
83. J. E. Repine, R. B. Fox, and E. M. Berger, Hydrogen peroxide kills Staphylococcus aureus by reacting with iron to form hydroxyl radical, J. Biol. Chem. 256:7094 (1981).

MECHANISMS OF DIOXYGEN ACTIVATION IN METAL-CONTAINING MONOOXYGENASES: ENZYMES AND MODEL SYSTEMS

Joan Selverstone Valentine, Judith N. Burstyn, and Lawrence D. Margerum

Department of Chemistry and Biochemistry
University of California, Los Angeles
Los Angeles, California 90024

INTRODUCTION

Monooxygenase enzymes catalyze reactions in which one atom of oxygen, derived from dioxygen, is incorporated into an organic substrate while the other atom of oxygen is reduced by two electrons to form water.[1,2]

$$R\text{-}H + O_2 + 2e^- + 2H^+ \longrightarrow R\text{-}O\text{-}H + H_2O$$

The enzymes of this type that have been characterized contain some type of redox-active cofactor, such as a flavin,[3] or a metal ion, or both.[4,5] The metalloenzymes, to which we are restricting our present discussion, have been found to contain heme, non-heme iron, or copper at their active sites.[6-8]

The reaction mechanisms of these enzymes have not been fully elucidated. Our knowledge is most advanced in the case of cytochrome P450, but even in this case, where we have available a crystal structure of an enzyme-substrate complex[9] and extensive information about related reactions of low molecular weight analogues of the enzymes,[10] the details of the process by which dioxygen is activated continue to elude us. Some of the fundamental questions regarding the reaction mechanisms of this class of enzymes that remain to be answered are: (a) how are O_2 and substrate bound in each active site, (b) what is the nature of the "active" oxidant, and (c) what is the mechanism of reaction of the "active" oxidant with the substrate?

Studies of metal-dioxygen complexes (including metal-superoxide and metal-peroxide complexes) and the reactions of dioxygen, superoxide and peroxides with metal complexes[11] suggest possibilities for the mode in which dioxygen may be reacting with the metal center in monooxygenase enzymes. If we consider the reaction of a single metal ion or complex with dioxygen, the likely intermediates in this reaction are those shown in Scheme I. Among the species depicted, it is unknown which one represents the active oxidant that transfers an oxygen atom to substrate in the enzyme systems . Moreover, it is not known if there is one type of reaction mechanism that operates in all or most of the monooxygenase enzymes or if each type of enzyme follows a different mechanism.

The purpose of this paper is to describe briefly the properties of some of the better characterized metal-containing monooxygenase enzymes, to com-

Scheme I

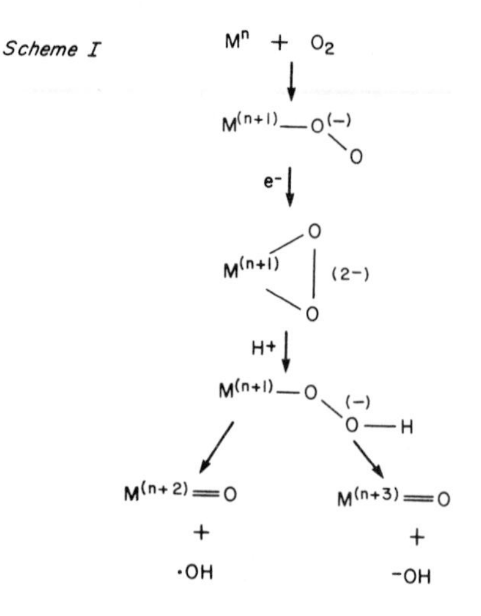

pare and contrast theories concerning the detailed steps in the dioxygen activation process for each enzyme, and to describe our recent findings concerning the reactivity of dioxygen complexes of metalloporphyrin complexes that may be analogous to intermediates formed in reactions of monooxygenases. We have limited ourselves to discussing systems in which the process of dioxygen activation appears to involve only one metal ion coordinated to the dioxygen moiety. Thus we do not discuss here, for example, tyrosinase, which contains a two-copper dioxygen binding site,[12-21] and phenylalanine hydroxylase, which contains a reduced pterin as a bound cofactor in addition to iron[4,22-30,33] or copper.[31] These latter systems will be discussed in a subsequent paper.

MONOOXYGENASE ENZYMES

Cytochrome P450

The mechanism of reaction of cytochrome P450 has been extensively studied in many laboratories using either the enzyme itself or synthetic analogues, and a relatively detailed understanding of the steps of the reaction mechanism has been achieved.[6] These are summarized in Figure 1. The reaction appears to proceed in the following sequence: 1) the substrate binds to the ferric form of the enzyme and the enzyme is subsequently reduced to the ferrous state, 2) the ferrous state binds dioxygen to form an oxy complex similar to that in oxymyoglobin, 3) the oxy complex is then reduced leading to the formation of the active oxidant bound to the iron center, and 4) oxygen is transferred to the bound substrate regenerating the ferric enzyme.

The X-ray crystal structure of cytochrome $P450_{cam}$ provides valuable information concerning the mode of substrate binding to the enzyme, but, at the same time, gives us few clues to the nature of the dioxygen activation steps subsequent to dioxygen binding.[9] The substrate, camphor, binds 4 Å above a pyrrole ring directly adjacent to the iron atom, which is the dioxygen binding site. Camphor is hydrogen bonded to a tyrosine residue and its binding seems to be designed to direct the 5-exo position toward the iron atom, where the active oxidant is undoubtedly generated. It is clear from analysis of this crystal structure that the specificity of this enzyme comes not from selectivity of the active oxidant but from the enzyme-substrate interaction.

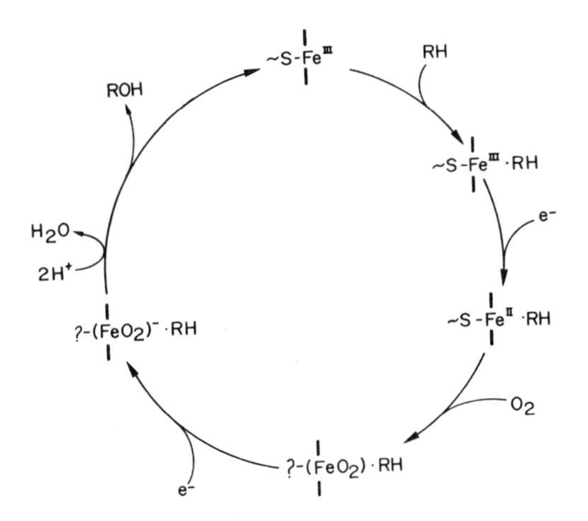

Figure 1. The enzymatic cycle of Cytochrome P450.

A hypothetical model for the mechanism of generation and the nature of the active oxidant in cytochrome P450 reactions has been proposed by Groves on the basis of studies of the reactions of iron porphyrin complexes.[10,32] In this model (Scheme II), it is proposed that an iron hydroperoxo complex undergoes heterolytic O-O bond cleavage to form an Fe(IV)-oxo complex of a one-electron oxidized porphyrin and that this latter species, **1**, is the active oxidant. Species **1** then reacts with substrate, abstracting a hydrogen atom to form an Fe(IV)-hydroxo complex (**2**). The resulting radical center on the substrate then recombines with OH from the iron center to form hydroxylated substrate regenerating the ferric form of the enzyme.

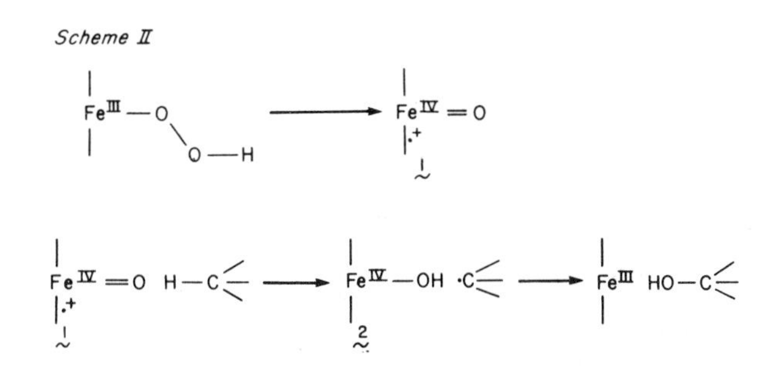

Hydrocarbon Monooxygenase

Specific substrate binding plays an important role in the specificity of several non-heme monooxygenase systems. Coon and coworkers[7,34,35] were the first to show that the monooxygenase system from Pseudonomas oleovorans (POM), which catalyzes terminal methyl group ω-hydroxylation, consists of three protein components: rubredoxin, a flavoprotein reductase, and a non-heme iron monooxygenase. The physiological substrate of POM is n-octane, generating 1-octanol as product.

$$CH_3CH_2CH_2CH_2CH_2CH_2CH_2CH_3 + O_2 + 2e^- + 2H^+ \longrightarrow CH_3CH_2CH_2CH_2CH_2CH_2CH_2CH_2OH + H_2O$$

May and coworkers have shown that this monooxygenase system can also be used in vitro to catalyze stereospecific olefin epoxidation[36-41] and aldehyde formation from terminal alcohols;[42,43] plus, sulphoxidation and S-dealkylation[43] and O-demethylation.[41] The propensity of this enzyme system to catalyze reactions at the methyl termini of linear alkanes suggests strongly that, as in the case of cytochrome P450, the specificity of these reactions is due to binding of substrate to the enzyme in a specific orientation rather than to selectivity of the activated metal-oxygen species.

Information regarding the chemical or electronic nature of the Fe-active site of POM is unknown, so that little can be said about the composition of the "activated Fe-oxygen" species. Information gleaned about the mechanism comes from substrate inhibition studies and has lead to the following postulates:[41] (1) Initial oxygen attack (by the "activated Fe-oxygen") occurs exclusively at the terminal carbon resulting in the formation of terminal epoxides and alcohols, but not ketones, from terminal olefins. (2) Stereochemical and configurational studies suggest that both epoxidation and hydroxylation occur through reaction of enzyme with the substrate to generate intermediates with cationic or radical character. (3) The "activated Fe-oxygen" species can attack either face of the olefin leading to loss of configuration during epoxidation, a process not seen in P450 chemistry. Hydrophobic binding of substrate apparently forces the closure to epoxide to occur preferentially giving R-(+) epoxides.

Dopamine β-monooxygenase

Dopamine β-monooxygenase (DBM) catalyzes the hydroxylation of dopamine to the neurotransmitter norepinephrine in vivo.[8] It is a tetrameric enzyme containing four active sites per tetramer with two copper atoms per active site. There is no evidence that the copper atoms are bridged and both are classified as Type II copper;[8] the process of dioxygen binding and activation appears to involve single copper atoms.[44] In vitro, the enzyme is not limited to hydroxylation and can be used to catalyze the oxygenation of a variety of substrates such as aryl-substituted phenylethylamines[45-47] and to catalyze benzylic oxidations of sulfides,[46,48] olefins[49-51] and aldehydes.[52]

Oxidation by dopamine β-monooxygenase has been proposed by Klinman and coworkers[53,54] to proceed through the following steps (Figure 2). First, the catalytic Cu(II) center is reduced by ascorbate, which is followed by binding of dopamine and dioxygen. General acid catalysis of dioxygen reduction is required and may be provided by the presence of a protonated base. Partial O-O bond homolysis generates electrophilic character on an oxygen atom which, in turn, initiates C-H bond homolysis. This process leads to a transition state, in which both C-H bond breaking (H atom abstraction) and O-H bond making are important, consistent with the observation of significant primary and secondary isotope effects.[53,55] Finally, the transition state yields a radical dopamine intermediate which rapidly recombines with an oxygen atom bound to copper [$Cu(II)-O^- \longleftrightarrow Cu(III)-O^{2-}$]. The proposed alkoxide-Cu(II) complex then undergoes slow dissociation to yield the hydroxylated product.

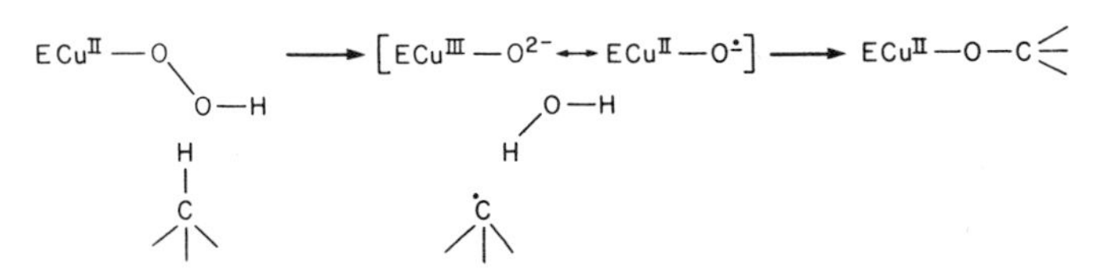

Figure 2. Hydroperoxide Mechanism proposed for Dopamine β-monooxygenase by Klinman and coworkers. Homolytic O-O bond cleavage is concerted with H atom abstraction, followed by oxygen rebound. E=Enzyme.

Mechanisms of Dioxygen Activation in Monooxygenase Enzymes

A key unanswered question in our current picture of the mechanism of dioxygen activation in monooxygenase enzymes has to do with the sequence of events involved in O-O bond cleavage and attack on the enzyme-bound substrate. There is very strong evidence indicating that the catalytic sequence of dioxygen activation commences with binding of dioxygen to the reduced metal center,[11,56] presumably forming a peroxide or, more likely, a hydroperoxide complex. In each case where selectivity of attack on the substrate is observed, it is believed that the site to be oxygenated is directed toward the metal center by specific substrate binding to the enzyme. At issue is the nature of the oxygen-containing species that reacts with the enzyme-bound substrate.

In the case of cytochrome P450, it has been proposed that O-O bond cleavage precedes attack on substrate.[57] It is generally believed that this cleavage is heterolytic, although homolytic cleavage of peroxide derivatives by cytochrome P450 has also been observed.[58] The thermodynamic barrier for homolytic bond cleavage of free H_2O_2 is 51 kcal/mol,[59] whereas that for heterolytic reductive cleavage of H_2O_2 has been estimated to be 20-34 kcal/mol.[53,59]

Homolytic: $H_2O_2 \longrightarrow 2\ OH\cdot$

Heterolytic: $H_2O_2 + e^- \longrightarrow \cdot OH + OH^-$

In the case of heme-containing systems, it is believed that the activation barrier for O-O bond cleavage can be lowered by the complexation of the resulting oxygen atom by the iron porphyrin center [$Fe^{n+}(P)$], i.e.:

Homolytic: $Fe^{III}(P)\text{-}OOH \longrightarrow Fe(IV)(P){=}O + \cdot OH$

Heterolytic: $Fe^{III}(P)\text{-}OOH \longrightarrow [Fe(IV)(P^+){=}O]^+ + OH^-$

The estimated activation barrier for homolytic cleavage in DBM is lowered to 30 kcal/mol due to Cu binding of hydroperoxide.[53] The activation barrier for heterolytic cleavage is apparently lowered in peroxidases[60] by interaction of the hydroperoxide ligand with conserved residues in the active site which help to stabilize the charge-separated transition state. The crystal structure of cytochrome $P450_{cam}$ provides no comparable clues to the mechanism of O-O bond cleavage; there are no amino acid residues situated in such a way that they might protonate the dioxygen ligand, or stabilize hydroxide or water as it is formed by heterolytic cleavage of the hydroperoxide ligand.[9]

The role of the metal center in either heterolytic or homolytic O-O bond cleavage is that of a reducing agent.

$$M^{n+}\text{-}OOH^{-} \longrightarrow M^{(n+1)+}\text{-}O^{2-} + \cdot OH$$

In the case of the heme systems, the accessibility of high oxidation states of the iron plus porphyrin ligand is well-documented by model studies.[10] In the case of the non-heme systems, however, comparable high-valent metal oxo species have not been characterized and therefore the analogous mechanism in such systems is not as appealing. While high oxidation states of iron and copper are known to exist in certain complexes, they generally are supported by ligands such as oxide and fluoride that are particularly resistant to oxidation[61] and not by ligands typically found in metalloenzyme active sites, such as imidazole or phenol. One possible explanation is that in the non-heme systems (and possibly in the heme systems as well[62]), the oxygen is not present as an oxide ligand but has been inserted in a nitrogen metal bond, forming an N-oxide complex, i.e.,

$$HOO\text{-}M^{n+}\text{-}N(\text{ligand}) \longrightarrow HO\text{-}M^{n+}\text{-}O\text{-}N(\text{ligand})$$

$$HO\text{-}M^{n+}\text{-}O\text{-}N(\text{ligand}) + \text{substrate} \longrightarrow HO\text{-}M^{n+}\text{-}N(\text{ligand}) + \text{substrate(O)}$$

Such a species cannot be ruled out in reactions of iron-EDTA complexes with hydroperoxides recently described by Bruice and coworkers.[63] Another possibility is that it is the hydroperoxide complex that reacts with the substrate and that bond formation from O to substrate is concerted with O-O bond breaking, as proposed by Klinman for dopamine β-monooxygenase,[53] thus providing compensation for the cost of O-O bond cleavage in the transition state (Figure 3). In fact, it is interesting to speculate that for each of these enzymes the mechanism by which the substrate is oxidized could be dependent on the reactivity of the substrate. One could envision certain substrates that could react with the metal-bound hydroperoxide ligand prior to or concerted with O-O bond cleavage. This is a possibility that is difficult to assess because we have so little information concerning the reactivity of HO_2^- when complexed to different metals.

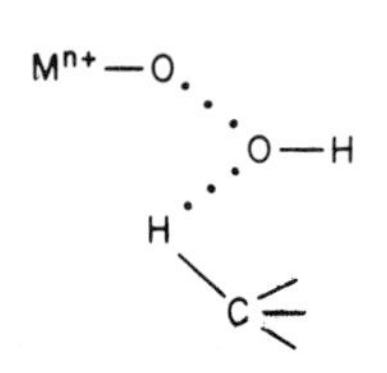

Figure 3. Hypothetical transition state in the hydroperoxide mechanism of Klinman and coworkers. Coupling of O-O bond breaking with formation of O-H gives a lower energy pathway.

REACTIONS OF METALLOPORPHYRIN PEROXO COMPLEXES

In order to probe the mechanism of O-O bond cleavage and dioxygen activation, we have studied the reactivity of two metalloporphyrin peroxide complexes, $(MnTPPO_2)^-$ and $(FeTPPO_2)^-$ under a variety of reaction conditions.

These complexes are at the same oxidation level as the ferric peroxide or hydroperoxide intermediate in the proposed mechanism for the reaction of cytochrome P450 (Figure 1). These complexes contain an intact oxygen-oxygen bond and therefore offer an opportunity to examine the conditions, if any, under which this bond may be cleaved to generate a high valent oxo species. We had previously isolated and characterized the anionic manganese porphyrin peroxide[64] and iron porphyrin peroxide[65] complexes, $(MnTPPO_2)^-$ and $(FeTPPO_2)^-$. The iron complex was originally characterized by observation of its oxygen-oxygen stretching frequency by IR and its rhombic EPR spectrum.[65] The elemental analysis of the crystalline solid, prepared as the tetramethylammonium salt using octaethylporphyrin as the ligand, was consistent with the formulation $[Me_4N][FeOEPO_2]$. This solid was further studied by Mössbauer and ESR spectroscopy and by determination of its magnetic susceptibility.[66] The iron in this complex is clearly high spin and highly rhombic, which is unusual for a ferric heme. A crystal structure of the manganese complex, $[K(K222)][MnTPPO_2]$,[64] as the potassium-cryptate salt using tetraphenylporphyrin as the ligand, revealed a triangularly bound peroxide ligand with a O-O bond distance of 1.45 Å (Figure 4). The manganese was displaced from the plane of the porphyrin by 0.76 Å, an unusually large out-of-plane displacement. This geometry results in a reordering of the d orbital energy levels so that the complex is predicted to be high spin d^4 with the energy of the d_{yz} orbital exceeding that of the $d_{x^2-y^2}$ orbital. A similar geometry could account for the rhombic symmetry observed for the iron complex.

We wished to study these complexes in order to determine the conditions under which the oxygen-oxygen bond could be activated to promote oxygenation of substrates. The first type of study was a comparison of the reactivity of the peroxide ligand complexed to a metalloporphyrin as compared with that of other metalloperoxide complexes. It had been previously observed that the metalloperoxides which epoxidized olefins were also able to oxidize butyl anion to butoxide.[67] These metalloperoxides were classified as "electrophilic" while those that did not react with butyl anion were classed

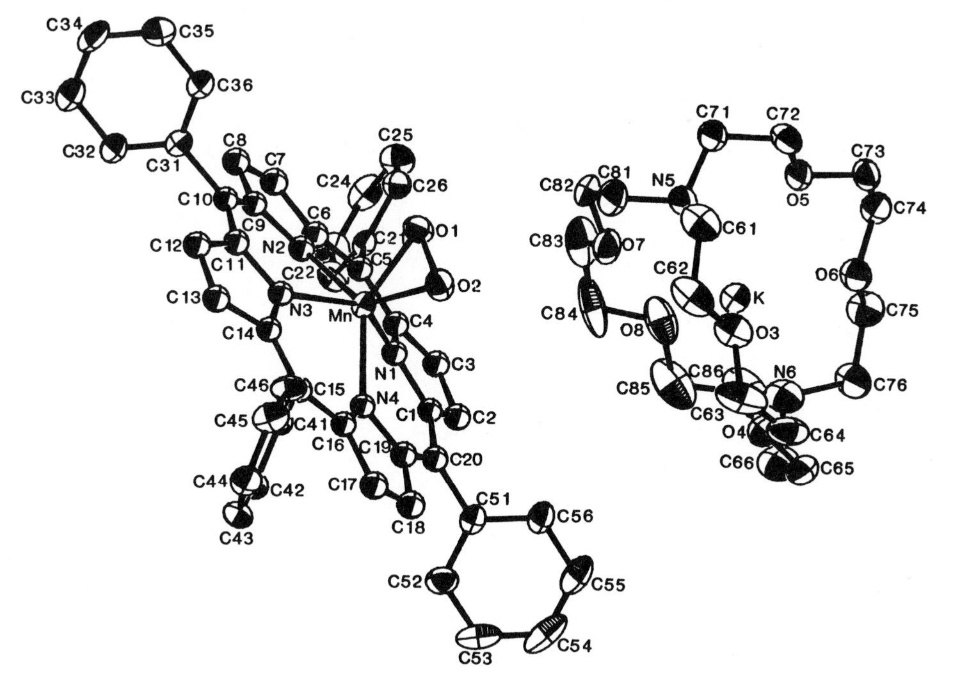

Figure 4. ORTEP plot of $[K(K222)][MnTPPO_2]$ showing both the porphyrin anion and potassium cryptate cation. The hydrogen atoms have been omitted for clarity.

as "nucleophilic" peroxides.[67] In general, the electrophilic peroxides are those containing high valent metal atoms from the left side of the periodic table, i.e., Ti^{IV}, V^{V}, Cr^{VI}, Mo^{VI}. The group VIII metalloperoxides, i.e., Pt^{II}, Pd^{II}, Ir^{III}, Rh^{III}, Ru^{II}, are generally nucleophilic. The iron and manganese porphyrin peroxo complexes were found not to react with butyl anion and therefore appear to be of the nucleophilic type.

Two other characteristic reactions of metalloperoxides were studied: the reaction with SO_2 to form sulfate and the oxidation of triphenylphosphine to form triphenylphosphine oxide.[68] The iron complex, $(FeTPPO_2)^-$ was observed to react with SO_2 in THF to give free sulfate anion.[69] The same complex gave relatively low yields of phosphine oxide (16%). The manganese complex gave similarly low yields of phosphine oxide. Both of these reactions with triphenylphosphine occurred quite slowly, at rates comparable to the rate of decomposition of the peroxo complex itself, suggesting that the small amount of oxygenation is occurring through reaction with a decomposition product of the peroxo complex. These results indicate that the metalloporphyrin peroxides are not themselves highly reactive oxygenating species.

We concluded from the reactions described above that releasing the reactivity of the peroxide ligand in the metalloporphyrin peroxide complexes would require that the ligand be converted to a different peroxide species and/or that the O-O bond be cleaved. This problem was addressed by studying the effect of potential activating agents on the ability of these peroxide complexes to oxidize triphenylphosphine or olefins. In particular, we began by examining the possibility that the metalloperoxides might be activated by Lewis acids. This type of mechanism has been proposed for cytochrome P450,[70] in which addition of either protons or an acylating group promotes the cleavage of the oxygen-oxygen bond to generate the high valent oxo species by heterolytic cleavage (Figure 5).

The absence of potential acylating groups in the crystal structure of the soluble cytochrome P450 from Pseudomonas putida[9] has reduced the probability of a facilitating acyl group, but the potential role for protons remains. Groves and coworkers have previously demonstrated that the addition of acyl halide to the manganese porphyrin peroxide in the presence of an olefin results in the production of epoxides. As shown in Figure 6, he proposed that a acylperoxide complex was formed which spontaneously decomposed to give the Mn(V)oxo species which was responsible for the oxidation.[71-72] We repeated the same type of reactivity experiments with the

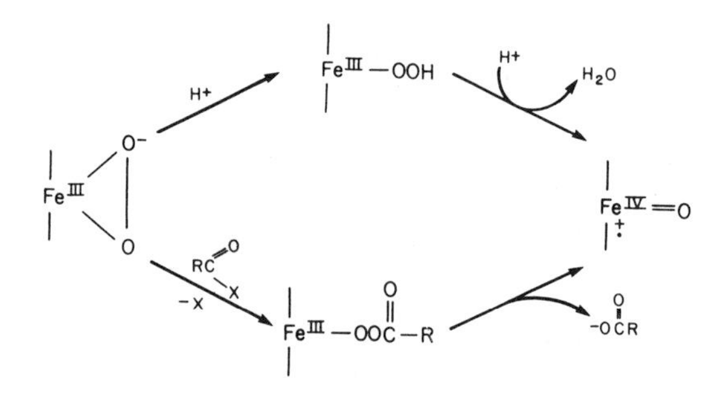

Figure 5. Activation of Cytochrome P450 by addition of protons or acylating groups to promote heterolytic O-O bond cleavage.

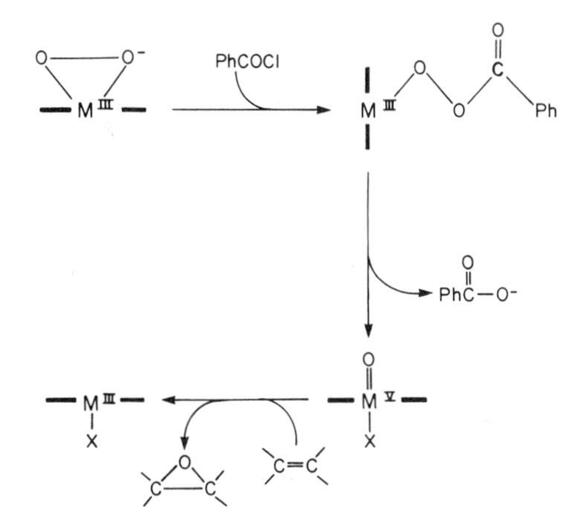

Figure 6. The mechanism of reaction of a metalloporphyrin with acyl chloride (M=Mn).

analogous iron-peroxide complex. We observed that when the iron porphyrin peroxide was reacted with olefin and acyl halide, in the manner described by Groves, oxidation did occur but the products were more characteristic of a radical reaction than of the oxygen insertion seen in the iron and iodosylbenzene system. Table I compares the product distributions for these two reactions and for the Fenton reaction, an iron-catalyzed radical reaction.[73] The predominance of the allylic oxidation products in the reaction of the iron peroxo complex with the acyl halide suggest that a radical mechanism is involved. This argues against the spontaneous heterolytic cleavage of the oxygen-oxygen bond since such a reaction pathway would have generated the same species as was generated in the iodosylbenzene reaction. Table I also shows a similar comparison for the manganese complex. The high yield of epoxide in this latter case is consistent with the conclusion of Groves and coworkers[71-72] that the manganese reaction does proceed via the Mn(V)-oxo species, although the presence of a fair amount of the ketone suggests that other pathways may also be occurring. When either of the metalloporphyrin peroxides were reacted with protons under similar conditions, it was not possible to trap any reactive intermediates with either olefin or triphenylphosphine. Spectroscopic evidence suggests that the protons do react but that the resulting species are highly unstable.

Since the peroxide ligand in the iron and manganese porphyrin peroxo complexes was not activated to give P450-type products by these Lewis acids, other activation mechanisms were considered. One possibility that has been suggested[1] is that the axial cysteine ligand bound to iron in the enzyme may play a critical role in facilitating O-O bond cleavage (Figure 7).

In order to explore the possible effects that thiols or thiolates might have on the reactivity of this system, we investigated the effect of the addition of benzenethiol to the iron peroxide complex in the presence of triphenylphosphine. We found that addition of this reducing agent to this system resulted in a rapid reaction which yielded a 50-80% conversion to the phosphine oxide. When thiol was used instead of acyl halide in the reactions with cyclohexene as added substrate, oxidation was observed although, again, the products appeared to be the result of a radical process. The observation of any oxidation promoted by the addition of a thiol reducing agent is novel and may be important in understanding the role of the unique thiolate ligand in the cytochromes P450. The radical paths observed in the model system may be promoted by the loss of RS radical, forming the diphenyldisulfide. This cannot happen in the enzyme due to the constraints of the protein backbone.

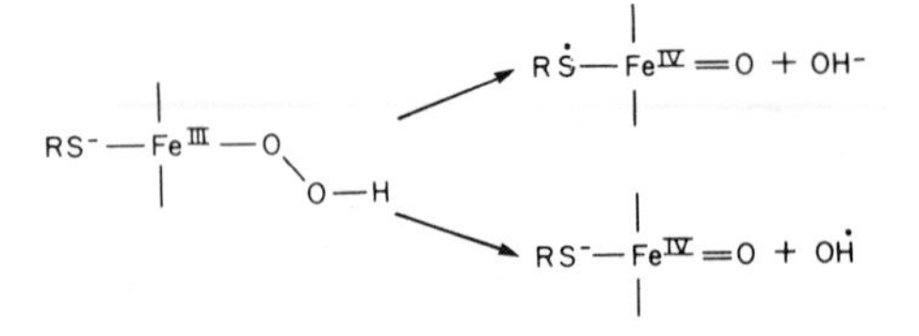

Figure 7. Possible role of axial cysteine (RS) ligand bound to iron in Cytochrome P450.

Table I. Comparison of the cyclohexene oxidation product ratios formed with various oxidants.

Reactants	Products		
M^a + Oxidant + cyclohexene	cyclohexene -oxide	cyclohexene -1-ol	cyclohexene -1-one
$(MnTMPO_2)^-$ + m-Cl-C_6H_5COCl	1.0	0	0.79
MnTMPCl + C_6H_5IO	1.0	0.33	0.01
$(FeTMPO_2)^-$ + m-Cl-C_6H_5COCl	1.0	2.79	5.21
FeTMPCl + C_6H_5IO	1.0	0.09	0.01
Fe^{2+} + H_2O_2	1.0	5	3

[a]The peroxo complexes were prepared *in situ* by the reaction of the MTMPCl (TMP is tetramesitylporphyrin), 1.5 mM in CH_3CN, with two equivalents of 18-crown-6 and five equivalents of KO_2. After removal of excess KO_2 this solution was combined under inert atmosphere with a solution of 1.5 equivalents of the activating agent under study and a hundred-fold excess of cyclohexene, also in CH_3CN. The formation of cyclohexene oxidation products was assayed by GC/MS.

CONCLUSIONS

While the model systems have been successful in mimicking the product distribution in cytochrome P450, the involvement of the "iron(V)oxo" species in the enzyme systems has yet to be conclusively demonstrated. It seems unlikely that non-heme monooxygenases react via a high-valent oxo species in the absence of the stabilizing porphyrin ligand. There exists the distinct possibility that multiple reaction pathways exist, some of which may not involve prior cleavage of the O-O bond, and that the course of the reaction may be dictated by the substrate. The presence of the cysteine ligand in cytochrome P450 may play an important role in the oxygen activation process in this enzyme, serving either to facilitate O-O bond cleavage or oxygen atom transfer. As more becomes known about the non-heme monooxygenases, we will be able to get a clearer idea of the requirements for oxygen activation and gain an appreciation for the variety of potential mechanisms involved.

ACKNOWLEDGEMENTS

Support of this work by the National Science Foundation is gratefully acknowledged.

REFERENCES

1. White, R. E.; Coon, M. J. Ann. Rev. Biochem, 49, 315-56 (1980).
2. Hayaishi, O., In "Molecular Mechanisms of Oxygen Activation"; Hayaishi, O.; Editor; (Academic: New York, N. Y.), (1974); p 7.
3. Guenferich, F. P.; Macdonald, T. L. Acc. Chem. Res., 17, 9 (1984).
4. Kaufman, S.; Fisher, D. B., In "Molecular Mechanisms of Oxygen Activation"; Hayaishi, O.; Editor; (Academic: New York, N. Y.), (1974); p 285.
5. Hamilton, G. A., In "Metal Ions in Biology, Vol. 3: Copper Proteins"; Spiro, T. G.; Editor; (John Wiley and Sons: New York, N. Y.), (1981); p. 205.
6. "Cytochrome P-450: Structure, Mechanism, and Biochemistry" Ortiz de Montellano, P. R.; Editor, (Plenum Press: New York, N. Y.), (1986).
7. Ruettinger, R. T.; Griffith, G. R.; Coon, M. J. Arch. Biochem. Biophys., 183, 528 (1977).
8. Villafranca, J. J., In "Metal Ions in Biology, Vol. 3: Copper Proteins"; Spiro, T. G.; Editor; (John Wiley and Sons: New York, N. Y.), (1981); p 263.
9. Poulos, T. L.; Finzel, B. C.; Gunsalus, I. C.; Wagner, G. C.; Kraut, J. J. Biol. Chem., 260(30), 16122-30 (1985).
10. McMurry, T. J.; Groves, J. T., In "Cytochrome P-450: Structure, Mechanism, and Biochemistry"; Ortiz de Montellano, P. R.; Editor; (Plenum Press: New York, N. Y.), (1986); p 1-28.
11. "Metal Ions in Biology, Vol. 2: Metal Ion Activation of Dioxygen" Spiro, T. G.; Editors, (Wiley-Interscience: New York, N. Y.), (1980).
12. Mason, H. S. Adv. Exp. Med. Biol., 74(Iron Copper Proteins), 464-9 (1976).
13. Strothkamp, K. G.; Jolley, R. L.; Mason, H. S. Biochem. Biophys. Res. Commun., 70(2), 519-24 (1976).
14. Strothkamp, K. G.; Mason, H. S. Biochem. Biophys. Res. Commun., 61(3), 827-32 (1974).
15. Makino, N.; McMahill, P.; Mason, H. S.; Moss, T. H. J. Biol. Chem., 249(19), 6062-6 (1974).
16. Jolley, R. L. J.; Evans, L. H.; Makino, N.; Mason, H. S. J. Biol. Chem., 249(2), 335-45 (1974).
17. Makino, N.; Mason, H. S. J. Biol. Chem., 248(16), 5731-5 (1973).
18. Jolley, R. L. J.; Evans, L. H.; Mason, H. S. Biochem. Biophys. Res. Commun., 46(2), 878-84 (1972).
19. Huber, M.; Hintermann, G.; Lerch, K. Biochemistry, 24(22), 6038-44 (1985).
20. Wilcox, D. E.; Porras, A. G.; Hwang, Y. T.; Lerch, K.; Winkler, M. E.; Solomon, E. I. J. Am. Chem. Soc., 107(13), 4015-27 (1985).
21. Lerch, K. Mol. Cell. Biochem., 52(2), 125-38 (1983).
22. Fisher, D. B.; Kirkwood, R.; Kaufman, S. J. Biol. Chem., 247(16), 5161-7 (1972).
23. Bloom, L. M.; Gaffney, B. J.; Benkovic, S. J. Biochemistry, 25(15), 4204-10 (1986).
24. Dix, T. A.; Benkovic, S. J. Biochemistry, 24(21), 5839-46 (1985).
25. Dix, T. A.; Bollag, G. E.; Domanico, P.; Benkovic, S. J., Biochemistry, 24(12), 2955-8 (1985).
26. Lazarus, R. A.; Benkovic, S. J.; Kaufman, S. J. Biol. Chem., 258(18), 10960-2 (1983).
27. Lazarus, R. A.; DeBrosse, C. W.; Benkovic, S. J. J. Am. Chem. Soc., 104(24), 6869-71 (1982).
28. Gottschall, D. W.; Dietrich, R. F.; Benkovic, S. J.; Shiman, R. J. Biol. Chem., 257(2), 845-9 (1982).
29. Lazarus, R. A.; Dietrich, R. F.; Wallick, D. E.; Benkovic, S. J. Biochemistry, 20(24), 6834-41 (1981).

30. Moad, G.; Luthy, C. L.; Benkovic, P. A.; Benkovic, S. J. J. Am. Chem. Soc., 101(20), 6068-76 (1979).
31. Pember, S. O.; Villafranca, J. J.; Benkovic, S. J. Biochemistry, 25(21), 6611-19 (1986).
32. Groves, J. T. J. Chem. Educ., 62(11), 928-31 (1985).
33. Wallick, D. E.; Bloom, L. M.; Gaffney, B. J.; Benkovic, S. J. Biochemistry, 23(6), 1295-302 (1984).
34. Peterson, J. A.; Basu, D.; Coon, M. J. J. Biol. Chem., 241, 5162-5164 (1966).
35. Peterson, J. A.; Kusunose, M.; Kusunose, E.; Coon, M. J. J. Biol. Chem., 242, 4334-4340 (1967).
36. May, S. W.; Abbott, B. J. Biochem. Biophys. Res. Commun., 48(5), 1230-4 (1972).
37. May, S. W.; Abbott, B. J. J. Biol. Chem., 248(5), 1725-30 (1973).
38. May, S. W.; Schwartz, R. D. J. Amer. Chem. Soc., 96(12), 4031-2 (1974).
39. May, S. W.; Steltenkamp, M. S.; Schwartz, R. D.; McCoy, C. J. J. Am. Chem. Soc., 98(24), 7856-8 (1976).
40. May, S. W.; Gordon, S. L.; Steltenkamp, M. S. J. Am. Chem. Soc., 99(7), 2017-24 (1977).
41. Katopodis, A. G.; Wimalasena, K.; Lee, J.; May, S. W. J. Am. Chem. Soc., 106(25), 7928-35 (1984).
42. May, S. W.; Padgette, S. R. Bio/Technology, 1(8), 677-86 (1983).
43. May, S. W.; Katopodis, A. G. Enzyme Microb. Technol., 8(1), 17-21 (1986).
44. Klinman, J. P.; Krueger, M.; Brenner, M.; Edmondson, D. E. J. Biol. Chem., 259(6), 3399-402 (1984).
45. Kaufman, S.; Friedman, S. Pharmacol. Rev., 17, 71 (1965).
46. May, S. W.; Phillips, R. S.; Mueller, P. W.; Herman, H. H. J. Biol. Chem., 256(16), 8470-5 (1981).
47. Klinman, J. P.; Krueger, M. Biochemistry, 21(1), 67-75 (1982).
48. May, S. W.; Phillips. R. S., J. Am. Chem. Soc., 102(18), 5981-3 (1980).
49. May, S. W.; Mueller, P. W.; Padgette, S. R.; Herman, H. H.; Phillips, R. S. Biochem. Biophys. Res. Commun., 110(1), 161-8 (1983).
50. Colombo, G.; Rajashekhar, B.; Giedroc, D. P.; Villafranca, J. J. Biochemistry, 23(16), 3590-8 (1984).
51. Padgette, S. R.; Wimalasena, K.; Herman, H. H.; Sirimanne, S. R.; May, S. W. Biochemistry, 24, 5826-5839 (1985).
52. Bossard, M. J.; Klinman, J. P. J. Biol. Chem., 261(35), 16421-7 (1986).
53. Miller, S. M.; Klinman, J. P. Biochemistry, 24(9), 2114-27 (1985).
54. Ahn, N.; Klinman, J. P. Biochemistry, 22(13), 3096-106 (1983).
55. Miller, S. M.; Klinman, J. P. Biochemistry, 22(13), 3091-6 (1983).
56. Ochiai, E. I., "Bioinorgganic Chemistry: An Introduction" (Allyn and Bacon: Boston), (1977).
57. White, R. E.; Sligar, S. G.; Coon, M. J. J. Biol. Chem., 255, 11108 (1980).
58. Ortiz de Montellano, P. R., In "Cytochrome P-450: Structure, Mechanism, and Biochemistry"; Ortiz de Montellano, P. R.; Editor; (Plenum Press: New York, N. Y.), (1986); p 218-271.
59. see ref. 56, p. 266.
60. Poulos, T. L.; Kraut, J. J. Biol. Chem., 255(17), 8199-8205 (1980).
61. Cotton, F. A.; Wilkinson, G., "Advanced Inorganic Chemistry: A Comprehensive Text", 4th Ed, (Wiley: New York, N. Y.), (1980).; p 765 and 818.
62. Groves, J. T.; Watanabe, Y. J. Am. Chem. Soc., 108(24), 7836-7837 (1986).
63. Balasubramanian, P. N.; Bruice, T. C. J. Am. Chem. Soc., 108(18), 5495-5503 (1986).
64. VanAtta, R. B.; Strouse, C. E.; Hanson, L. K.; Valentine, J. S. J. Am. Chem. Soc., 109(5), 1425-34 (1987).

65. McCandlish, E.; Miksztal, A. R.; Nappa, M.; Sprenger, A. Q.; Valentine, J. S.; Stong, J. D.; Spiro, T. G. J. Am. Chem. Soc., 102(12), 4268-71 (1980).
66. Burstyn, J. N.; Roe, J. A.; Miksztal, A. R.; Shaevitz, G. L.; Valentine, J. S., submitted for publication.
67. Regen, S. L.; Whitesides, G. M. J. Organomet. Chem., 59, 293-297 (1973).
68. Mimoun, H.; Postel, M.; Casabianca, F.; Fischer, J.; Mitschler, A. Inorg. Chem., 21, 1303-06 (1982).
69. Miksztal, A. R.; Valentine, J. S. Inorg. Chem., 23(22), 3548-52 (1984).
70. Sligar, S. G.; Murray, R. I., In "Cytochrome P-450: Structure, Mechanism, and Biochemistry"; Ortiz de Montellano, P. R.; Editor; (Plenum Press: New York, N. Y.), (1986); p 429-504.
71. Groves, J. T.; Watanabe, Y.; McMurry, T. J. J. Am. Chem. Soc., 105(17), 4489-90 (1983).
72. Groves, J. T.; Watanabe, Y. Inorg. Chem., 25(27), 4808-10 (1986).
73. Hamilton, G. A. J. Am. Chem. Soc., 86, 3390-91 (1964).

RADICAL CATION PATHWAYS FOR SELECTIVE CATALYTIC OXIDATION BY MOLECULAR OXYGEN

Dennis P. Riley and Milton R. Smith

Monsanto Company

800 N. Lindbergh Boulevard
St. Louis, Missouri 63167

INTRODUCTION

The selective oxidative conversion of a particular molecule to a desired product utilizing the abundant and inexpensive oxidant oxygen often represents a desirable method for upgrading the value of a raw material. All too often, of course, this type of selective chemistry does not exist. It has been a goal of our research to discover new catalytic pathways which will permit us to utilize oxygen as a selective oxidant. During our research into better methods of oxidizing waste thioethers and in converting tertiary amines to their amine oxides, we discovered that these substrates are subject to a novel autoxidation process which under high oxygen concentrations, elevated temperatures, and polar solvents yields almost exclusively the sulfoxide product.[1] The mechanism of this unusual autoxidation appears to involve an initial unfavorable electron transfer step (eq. 1), followed by triplet oxygen (in high concentration) trapping the resultant radical cation (eq. 2).[2]

$$(1) \quad R_2S + {}^3O_2 \rightleftharpoons R_2\overset{+}{S}\cdot + O_2^{\overline{\cdot}}$$

$$(2) \quad R_2\overset{+}{S}\cdot + {}^3O_2 \rightleftharpoons R_2\overset{+}{S}O\text{-}O\cdot$$

Back donation of an electron from superoxide to the oxygenated radical cation yields a zwitterionic species (eq. 3) whose chemistry is known to yield sulfoxide upon exposure to additional thioether (eq. 4).[3]

$$(3) \quad R_2\overset{+}{S}O\text{-}O\cdot + O_2^{\overline{\cdot}} \longrightarrow R_2\overset{+}{S}\text{-}OO^- + O_2$$

$$(4) \quad R_2\overset{+}{S}O\text{-}O^- + R_2S \xrightarrow{k_4} 2\ R_2S{\rightarrow}O$$

We also extended these high pressure autoxidation studies to other substrates, such as olefins and alkynes.[2] In these systems we also observe products whose appearance are not predicted based on known autoxidation pathways under ambient conditions where allylic hydroperoxides and their decomposition products, ketones and alcohols, are the major products of olefin oxidation,[4-6] and propargylic hydroperoxides and their corresponding α-alcohol and α-ketone decomposition products are the major products of alkyne oxidations.[4,7]

Since the initial electron-transfer step is unfavorable and is rate-determining in this slow autoxidation reaction, we believed that with the use of a suitable one-electron oxidants it should be possible to catalyze or initiate a selective oxygen oxidation of an electron-rich substrate. In this report we document our success in catalyzing the selective oxygen oxidation of thioethers to sulfoxides[8] and alkynes to yield carboxylic acids derived from site specific cleavage of the triple bond. After extensive screening of oxidants only one system has been found which catalyzes this chemistry and that utilizes Ce(IV) salts as the catalyst.

EXPERIMENTAL

All of the thioethers and alkynes used in these studies were purchased from Aldrich Chemical Co. and distilled before use. Sulfoxide standards were prepared by standard procedures using H_2O_2,[9], and $(NH_4)_2Ce(NO_3)_6$ and $Ce(NO_3)_3 \cdot 6H_2O$ were purchased from Alfa-Ventron. HPLC grade acetonitrile was distilled before use and distilled, de-ionized water was used in all cases.

Electronic spectra were monitored using matched quartz cells in a Hitachi 110A UV-VIS spectrophotometer over the range of 200-500 nm. All high-pressure catalytic runs used an apparatus analogous to that reported previously.[10] Gas uptake measurements were made by utilizing a pressurized external calibrated stell tube connected directly to the reactor. Pressure drop in this calibrated external tube could be correlated to moles of O_2 consumed during the reaction. Reactions were also monitored by gas chromatography using a Varian Model 3400 GC with a flame ionization detector and analyzed using a 15m OV101 capillary column. Yields were determined by utilizing dodecane as an internal standard and by comparison to calibrated solutions. Electrochemical studies were performed using a Bioanalytical Systems CV-1B cyclic voltammograph and voltammograms were recorded on a Houston Instruments 100 XY recorder. All cyclics were recorded in dry methylene chloride using 0.5M tetra-n-butylammonium tetrafluoroborate as a supporting electrolyte. A single two-electrode cell was used which contained a glassy-carbon working electrode and a Pt reference utilizing the Fe(II,III) couple of ferrocene as an internal standard (all potentials are corrected to SHE).

RESULTS AND DISCUSSION

Sulfoxide Oxidations

In our attempts to catalyze the oxygen oxidation of thioethers to sulfoxides via a one-electron scheme described in eq. 1-4, a variety of one-electron oxidants were utilized including, NO^+, $Fe(bipy)_3^{3+}$, $Ru(bipy)_3^{3+}$, $Ag^+/K_2S_2O_8$, $Mo(CN)_8^{3-}$, $KBrO_3$, electrochemical cell, etc. In no case was any catalytic (non-stoichiometric) chemistry observed. In contrast we observe that Ce(IV) salts in catalytic amounts, especially $(NH_4)_2Ce(NO_3)_6$, give a very large rate enhancement to the selective O_2 oxidation of thioethers. In Table 1 are listed several examples of unoptimized Ce(IV) promoted oxygen oxidations. For comparison under similar conditions (100°C, 1000 psi O_2 pressure) thioethers require several days to autoxidize completely to sulfoxides. Cerium(IV) has a profound effect on the rates of O_2 oxidation and Ce(IV) is to date unique in promoting this chemistry.

In general with all the thioethers which we have investigated in detail (thioanisole, decylmethyl sulfide, tetrahydrothiophene, pentamethylene sulfide, and hexamethylene sulfide), we observe first-order thioether substrate kinetics at O_2 pressures above 30 psi (the lowest O_2 pressures we have studied). In Figure 1 is shown a typical reaction profile in which the uptake of oxygen is monitored as a function of time. The reaction proceeds rapidly to completion with a ½-life of 7 min. under the conditions noted.

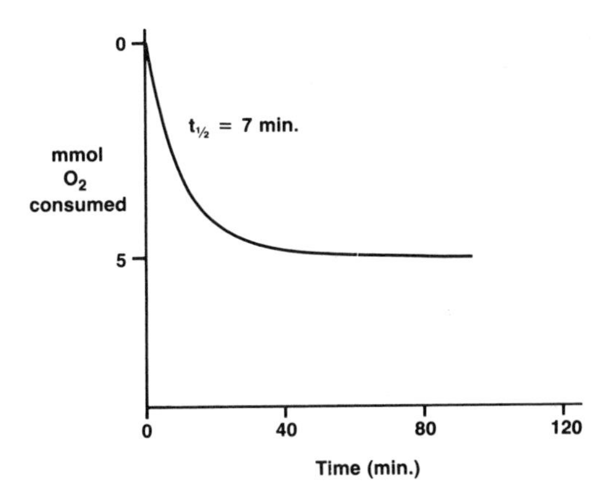

Fig. 1. The $(NH_4)_2Ce(NO_3)_6$ catalyzed molecular oxygen oxidation of tetrahydrothiophene at 75°C, 125 psi O_2 pressure, in 9:1 CH_3CN/H_2O with $[SR_2]$ = 1.0 M and sub/cat = 60.

Table 1. Examples of Both Catalyzed and Uncatalyzed High O_2 Pressure Autoxidation of Thioethers

Substrate	Temp (°C)	Time (hr)	Sulfoxide Product % Conv., % Selectivity
Thioanisole[a]	100	1	97, 95
Thioanisole[b]	100	60	84, 88
Diphenyl Sulfide[a]	100	3.5	37, 90
Diphenyl Sulfide[b]	100	60	5, 85
Tetrahydrothiophene[a]	60	0.5	89, 95
Tetrahydrothiophene[b]	100	24	90, 90
Pentamethylene Sulfide[a]	60	5.0	98, 95
Pentamethylene Sulfide[b]	100	24	76, 91

[a] Reactions run in CH_3CN containing $Ce(NH_4)_2(NO_3)_6$ at a substrate/cat. ratio of 20, all runs unoptimized and performed under 14 bar O_2 pressure, $[SR_2]$ = 0.16 M. [b] Reactions are run without catalyst in CH_3CN under 60 bar O_2 pressure, $[SR_2]$ = 0.16 M.

Analyses of reaction samples demonstrate two important points: i) the stoichiometry is such that for each mole of O_2 consumed two moles of sulfoxide are generated, and ii) the reaction is very clean; i.e., the reaction proceeds to > 90% thioether conversion with > 95% selectivity to the sulfoxide. The dependence of this reaction on the $[O_2]$ (or pressure) has been investigated over the range 30-1000 psi and found to be zero-order. The cerium dependence on the reaction rates (both the initial rates and the observed rate constants for thioether loss) is first-order in [Ce(IV)] added. In Figure 2 is shown a plot of one such study for thioanisole. Not only are the reactions first-order in total [Ce], but we observe a non-zero intercept consistent with the presence of the slow catalyst-free autoxidation reaction discussed above. Added counterion such as nitrate also inhibits the reaction.

The kinetic results in the Ce(IV) system are consistent with a rate-determining step in which the thioether is oxidized to its radical cation by Ce(IV) (eq. 5).

$$(5)\quad R_2S + Ce(IV) \xrightarrow{k_1} R_2\overset{+\cdot}{S} + Ce(III)$$

The reaction of Ce(IV) with 100-fold excesses of various thioethers was monitored at 25°C under N_2 and this oxidation was found to be slow, with ½-lives on the order of 1-2 hrs typically. Thus, at elevated temperatures the observed rates are consistent with this assignment of the rate-determining step.

In order to gain additional mechanistic insight, a number of other studies were carried out. For example, it is known that Ce(IV) will promote the stoichiometric oxidation of thioethers to sulfoxides by a hydration mechanism.[11]

$$(6)\quad R_2\overset{+\cdot}{S} + H_2O \longrightarrow R_2\overset{+\cdot}{S}\text{—}OH_2 \xrightarrow[-H^+]{Ce(IV)} R_2S\rightarrow O$$

Labeling studies in our system were attempted using 99.9% O-18 labeled H_2O. Unfortunately, in the Ce(IV) promoted system no conclusive incorporation information could be obtained by GC-mass spectra. This resulted from the fact

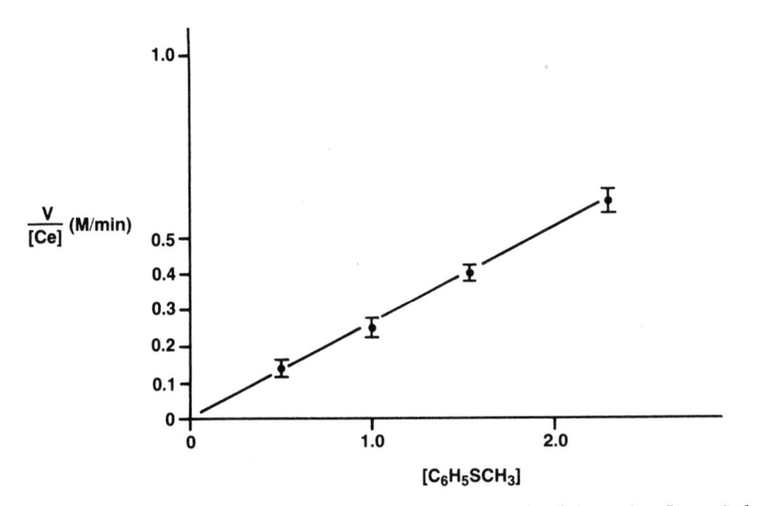

Fig. 2. Plot of initial rate of O_2 oxidation of thioanisole with O_2 (125 psi) catalyzed by $Ce(NH_4)_2(NO_3)_6$([Ce]=0.0258 M) in dry acetonitrile at 70°C as a function of $[SR_2]$.

that Ce(IV) and Ce(III) were shown to rapidly promote the sulfoxide oxygen/water oxygen exchange reaction at 70°C. To circumvent this problem and to demonstrate that O_2 is a suitable trapping agent of a thioether radical cation reactions were carried out under 200 psi O_2 pressure using thioanisole and tetrahydrothiophene at 70°C in an acetonitrile solvent mixture containing 10% $^{18}OH_2$ and a stoichiometric amount of $S_2O_8^{=}/Ag^{+}$ to generate the radical cation, $R_2S^{\dot{+}}$.[12] The reaction proceeds rapidly to yield the desired sulfoxide, which by GC-mass spectroscopy contains less than 1% ^{18}O incorporation for tetrahydrothiophene sulfoxide and 18% ^{18}O incorporation for thioanisole. Thus, $^3O_2$ can be a competent trapping agent for the proposed sulfur radical cation intermediate (eq. 2).

In separate studies using other oxidants, including constant current coulometry under O_2 pressures (up to 1000 psi), we have never observed any indication of radical cation chain chemistry,[13] only stoichiometric thioether oxidation. Clearly the Ce(IV,III) system is unique, but Ce(IV) must provide more than an initial potent oxidant to generate the sulfur radical cation. In fact, these reactions are truly catalytic in Ce. To regenerate Ce(IV) in these systems a potent oxidizing intermediate must be generated. Indeed, we believe that the oxygenated sulfur radical cation is likely to be very strongly oxidizing, in analogy to carbon systems.[13] Thus the catalytic cycle could be completed with the reoxidation of Ce(III) by the oxygenated sulfur radical cation (eq. 7):

$$(7) \quad R_2\overset{+}{S}OO\cdot + Ce(III) \underset{}{\overset{k_3}{\rightleftharpoons}} R_2\overset{+}{S}OO^- + Ce(IV)$$

The generation of sulfoxide is then completed by the known rapid reaction of thioether with the zwitterionic (of eq. 7), as described above (eq. 4).

The temperature dependence of this reaction (60-100°C) was studied for decyl methyl sulfide oxidations in 9:1 CH_3CN/H_2O under 200 psi O_2. The activation energy for this Ce catalyzed reaction was 10.6 kcal. The uncatalyzed autoxidation under the same conditions exhibits a much larger activation energy, 24 kcal.

An integrated rate expression based on the mechanistic scheme contained in equations 5,6,7, and 4 has been derived assuming a steady-state treatment for the concentration of the $R_2\overset{+}{S}\cdot$, $R_2\overset{+}{S}OO\cdot$, and $R_2\overset{+}{S}OO^-$ intermediates (eq. 8):

$$(8) \quad V = \frac{2k_1k_2k_3[SR_2][Ce(IV)][O_2]}{k_{-1}k_{-2} + k_{-1}k_3\,[Ce(III)] + k_2k_3[O_2]}$$

If the term $k_2k_3[O_2] >> k_{-1}k_{-2} + k_{-1}k_3[Ce(III)]$, then equation 8 reduces to equation (9):

$$(9) \quad V = 2k_2[SR_2][Ce(IV)], \text{ where } [Ce(IV)] \sim [Ce]_t$$

The assumption that $k_2k_3[O_2]$ is larger than the other terms of the denominator of eq. 8 is realistic since these reactions are carried out at $[O_2] \geq 0.1$ $\underline{M}$ and $[Ce]_t \leq 0.02$ $\underline{M}$ and O_2 trapping of radical cations could be fast.[14] This mechanistic picture is also consistent with the O_2 uptake data which confirms the overall stoichiometry.

To further test the validity of equation (9) to this catalytic system, aliquots of reaction samples were taken from the reactor and diluted into cold acetonitrile and their electronic spectra were run over the region 200-500 nm. In Figure 3 are shown examples of Ce(IV), Ce(III), and an actual experimental run aliquot diluted to 1.0×10^{-3} M in Ce. The spectrum of the experimental sample clearly appears to be dominated by the lower energy extinction absorbance with the extinction coefficient ∿ 2500 characteristic of Ce(IV). In separate additions spectra of both Ce(III) and Ce(IV) (in short times) were found to be relatively insensitive to the presence of other donors, such as SR_2 or $R_2S{\longrightarrow}O$. Thus, in these systems the spectral fingerprint is consistent with $[Ce(IV)] \cong [Ce]_t$.

Another important prediction that the mechanistic scheme proposed here allows us to make is that owing to the background radical cation autoxidation these reactions should be autocatalytic in Ce(III) with no added Ce(IV). Since the same oxygenated radical cation intermediate $R_2\overset{+}{S}\text{-O-O}^-$ is present in both the non-catalytic and Ce catalyzed regimes, this intermediate should be formed in the absence of Ce(IV) and slowly convert Ce(III) to Ce(IV), leading to autocatalysis. In Figure 4 is shown an example of such an experiment. We observe the slow conversion of thioether to sulfoxide. As the reaction approaches completion (monitored by O_2 uptake) an additional 10 mmol of thioether was added to the reactor. At this point the reaction proceeded rapidly at the same rate as if Ce(IV) had initially been used. These results demonstrate a principle we have shown in separate experiments; namely, the active catalyst species is generated and maintained in these systems allowing recycle of the active catalyst.

One intriguing aspect of these Ce(IV) catalyzed oxidations was the observed effect of ring size in a series of cyclic thioethers on the observed rates of reaction. In Table 2 are listed some examples of cyclic thioether oxidations. The five-membered ring thioether oxidizes in these systems much more rapidly than the 6- or 7-membered ring system. While this may be due at least in part to the fact that tetrahydrothiophene is easier to oxidize than the other thioethers, it is not clear at this point if other factors (e.g., steric) affecting the binding to Ce(IV) may be operative. Nevertheless, these results suggest that if a high energy intermediate, the persulfoxide, $R_2\overset{+}{S}OO^-$, is present, it would not be able to discriminate between different ring size thioethers when it reacts with thioether to generate two molecules of sulfoxide (eq. 4). Thus, a mixed thioether reaction should reveal the validity of this step in the reaction sequence. In Figure 5 we show an example of such a Ce(IV) catalyzed mixed thioether oxygen oxidation. Using a 1:1 ratio of tetrahydrothiophene and pentamethylene sulfide, we find that the initial rate is very fast (comparable to tetrahydrothiophene alone, but that the rate decreases faster than a first-order decay). In addition at the ∿ 25% conversion point an aliquot of the reaction revealed by GC that a significant amount of pentamethylene sulfoxide is present. The ratio of $\overline{(CH_2)_4-S}{\longrightarrow}O$ to $\overline{(CH_2)_5-S}{\longrightarrow}O$ equaled 2.8. This number is actually very close to the value of three which the mechanism proposed here predicts for early in the reaction. In the early stages of the reaction most of the $R_2S\cdot$ generated will arise from the 5-membered ring thioether, assuming oxidation of thioether by Ce(IV) is rate-determining. The sequence of reactions is shown here and this sequence predicts three moles of 5-membered ring sulfoxide and one 6-membered ring sulfoxide per every initiation by Ce(IV) (eqs. 10-13, where m.r. = membered ring):

$$\text{(10)}\quad R_2S\ \text{(5-m.r.)} + Ce(IV) \longrightarrow R_2\overset{+\cdot}{S}\ \text{(5-m.r.)} + Ce(III)$$

$$\text{(11)}\quad R_2\overset{+\cdot}{S}\ \text{(5-m.r.)} + {}^3O_2 \longrightarrow R_2\overset{+}{S}OO\cdot\ \text{(5-m.r.)}$$

$$\text{(12)}\quad R_2\overset{+}{S}OO\cdot\ \text{(5-m.r.)} + Ce(III) \longrightarrow R_2\overset{+}{S}OO^-\ \text{(5-m.r.)} + Ce(IV)$$

$$\text{(13a)}\quad \tfrac{1}{2}\ R_2\overset{+}{S}OO^-\ \text{(5-m.r.)} + \tfrac{1}{2}\ R_2S \longrightarrow R_2S{\longrightarrow}O\ \text{(5-m.r.)}$$

$$\text{(13b)}\quad \tfrac{1}{2}\ R_2\overset{+}{S}OO^-\ \text{(5-m.r.)} + \tfrac{1}{2}\ R_2S\ \text{(6-m.r.)} \longrightarrow \tfrac{1}{2}\ R_2S{\longrightarrow}O\ \text{(5-m.r.)} + \tfrac{1}{2}\ R_2S{\longrightarrow}O\ \text{(6-m.r.)}$$

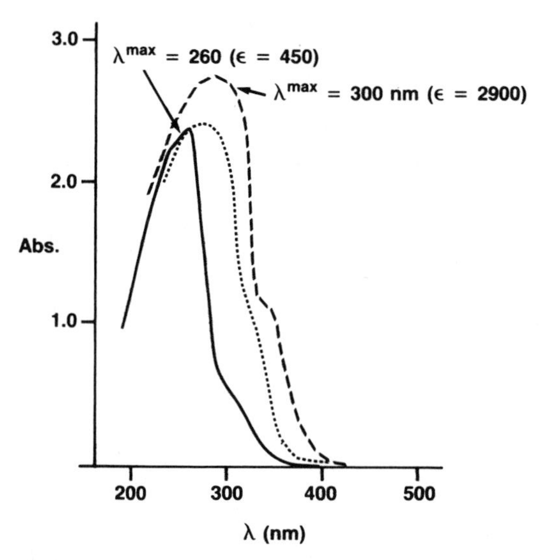

Fig. 3. The electronic spectra of various cerium species in CH_3CN (—) $Ce(NO_3)_3 \cdot 6H_2O$, 5.6×10^{-3} $\underline{M}$; (---) $(NH_4)_2Ce(NO_3)_6$, 9.0×10^{-4} M; and (···) an actual reaction aliquot (thioanisole) diluted to 1.0×10^{-3} M in Ce.

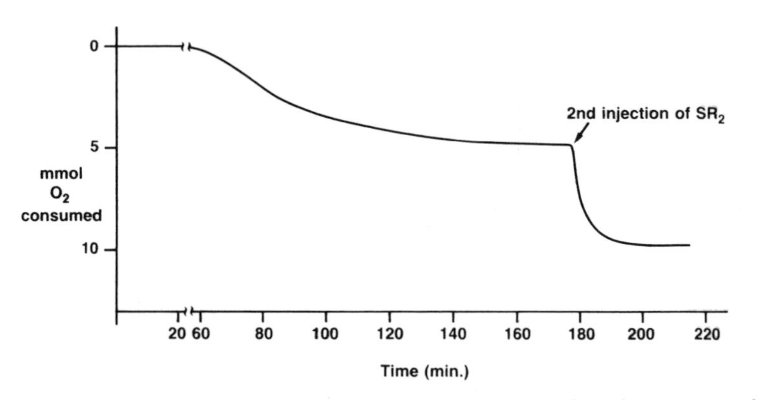

Fig. 4. The autoxidation of tetrahydrothiophene ($[SR_2]_i = 1.0$ $\underline{M}$) in the presence of $Ce(NO_3)_3 \cdot 6H_2O$ ($[Ce(III)] = 0.05$ M) in 9:1 CH_3CN/H_2O at 70°C under 125 psi O_2.

Since equations 13a and 13b are equally probable, the ratio of 5-membered ring to 6-membered ring sulfoxides produced is 3. Our observation of a 2.8 to 1 ratio is in excellent agreement with this and supports the intermediacy of such a high-energy intermediate as $R_2\overset{+}{S}OO^-$.

Table 2. The Ce(IV) Catalyzed Oxygen Oxidation of Cyclic Thioethers at 60°C in CH_3CN under 200 psi O_2 Pressure.

$\overline{(CH_2)_n S}$ n[a,b]	k_{obs} (cat.) ($hr.^{-1}$)	$E_{1/2}$ (V)[c]	Selectivity at 90% conv.	k_{obs}(uncatalyzed)[d] (hr^{-1})
4	8.9	1.68	93	0.27
5	0.11	1.76	89	0.11
6	0.12	1.80	88	0.18

[a] $[SR_2]$ = 0.2 M. [b] $[(NH_4)_2Ce(NO_3)_6] = 1.0 \times 10^{-3}$ M. [c] Irreversible.

[d] T = 115°C and P_{O_2} = 1000 psi.

Alkyne Autoxidations

The autoxidation of alkynes under high O_2 pressures, in polar solvents and at elevated temperatures proceeds slowly in the absence of any catalyst but yields some unexpected products (Table 3). For example a symmetrical

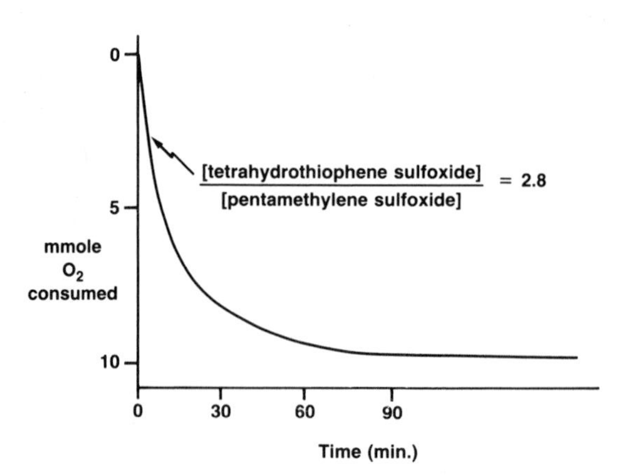

Fig. 5. The Ce catalyzed molecular oxygen oxidation of tetrahydrothiophene (1.0 M) plus pentamethylene sulfide (1.0 M) in 9:1 CH_3CN/H_2O under 125 psi O_2 with Sub_t/cat = 100.

internal alkyne yields cleavage products; e.g., carboxylic acids as in 6-dodecyne yields the hexanoic acid and shorter chain congeners, but no longer chain acids. This is inconsistent with the normal propargylic autoxidation pathway since if acids were derived by the α-CH activation, longer chain acids, heptanoic acid for example, should be present. Phenyl substituted alkynes are much less reactive, but do autoxidize. The major product observed for 1-phenyl-1-octyne is the 1-phenyl-1,2-octanedione. Again this indicates that under these conditions an α-CH autoxidation pathway is not operative. In analogy to the thioether case and to olefin systems which are known to exhibit radical cation chain autoxidation pathways,[13] we propose that a similar pathway exists for alkynes in polar media, at elevated temperatures, and high O_2 concentrations (eqs. 14-16).

$$(14)\quad R_1C{\equiv}CR_2 + O_2 \rightleftharpoons [R_1C{=}CR_2]^{+\cdot} + O_2^{\cdot -}$$

(15)

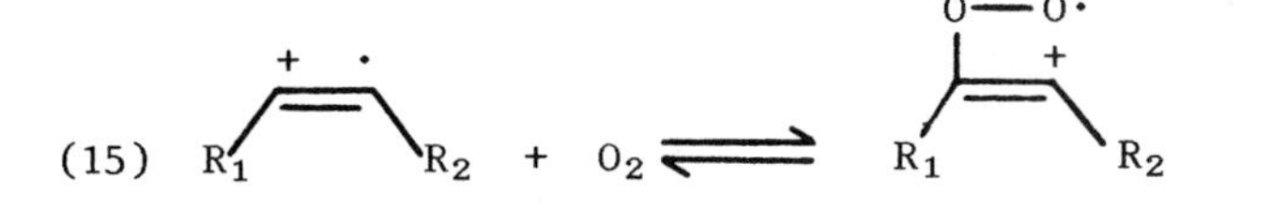

(16)

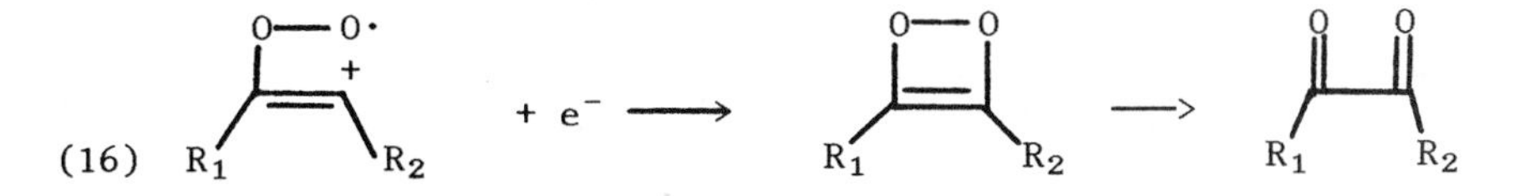

Table 3. Alkyne Autoxidations in CH_3CN

Alkyne	P_{O_2} (bar)	Temp. (°C)	Sub/cat	k_{obs} (hr^{-1})	% Conv. (time, hrs.)	Yields as % Conversion
6-dodecyne	15	110	No Cat.	0.42	66 (6)	Acids: 18% C_6, 7.5% C_5 3% C_4; C_6/C_5 = 2.4
6-dodecyne	15	110	20	0.68	93 (3)	Acids: 84% C_6, 4% C_5 C_6/C_5 > 20
PhC≡CH	70	115	No Cat.			No Rxn
PhC≡CH	70	115	20	fast	100 (0.1)	65% Benzoic Acid
PhC≡CPh	70	115	No Cat.			No Rxn
PhC≡CPh	70	115	20	0.6	90 (3)	61% Benzoic acid, 23% benzil
$PhC{\equiv}CC_6H_{13}$	70	90	No Cat.	slow	15 (24)	85% 1-phenyl-1,2-octanedione
$PhC{\equiv}CC_6H_{13}$	70	90	20	0.8	95 (3)	65% benzoic acid 34% heptanoic 3% hexanoic 12% 1-phenyl-1,2-octanedione

The dioxatene structure produced as in eq. 16 would be expected to rearrange to the α-diketone, as observed in these systems and in other alkyne systems.[15] Control reactions reveal under the reaction conditions the α-diketone will undergo oxidative site-specific cleavage yielding two moles of acid at rates comparable to the overall observed alkyne autoxidation rates demonstrating the chemical competence of the diketone intermediate.

In the thioether system it was noted that the high pressure constant current coulometry (HPCPC) oxidation of thioether in the presence of a high concentration of O_2 does not yield any evidence of chain chemistry, only stoichiometric oxidation. Even though electrochemical initiation studies with alkynes were inconclusive with regard to chain chemistry, due to the severe problems with filming of the electrodes in these systems an important discovery, especially in view of the endoenergic nature of the initial electron-transfer step (eq. 4), is that certain alkynic substrates exhibit unusual electrochemical behavior at higher oxygen pressures. Indeed we observe an oxygen concentration dependence on the formation of a new easier-to-oxidize species in the HPCV (Figure 6). That this is a new species and not due to surface absorption phenomena was confirmed by the linear nature of the plots of Po_2 versus integrated current and the integrated current versus $(\text{scan rate})^{1/2}$ which intercepts the current axis at the origin. These preliminary studies and results suggest that we are actually observing the oxidation of an oxygen adduct, the nature of which has been the subject of much discussions.[16 19] If we are indeed observing the direct oxidation of an oxygen adduct, the implications for oxidation catalysis are significant. The apparent reduction in the alkyne oxidation potential when these substrates are complexed to O_2 is striking and may account for the type of chemistry we observe; i.e., initial electron transfer occurs because the oxidation potential is nearly one volt lower at high O_2 pressure. A number of electron-rich compounds are known to form such O_2 adducts, including olefins and alkynes, with measured equilibrium formation constants on the order 5-15 M^{-1}.[17 19] In our own system the linear O_2 dependence on the increase in current in the O_2-dependent oxidation wave or the decrease in current with the O_2 (free) substrate oxidation wave allows us to calculate equilibrium constants in this range, 5-15 M^{-1}. These correlations support the view that an alkyne oxygen adduct is oxidized by oxygen.

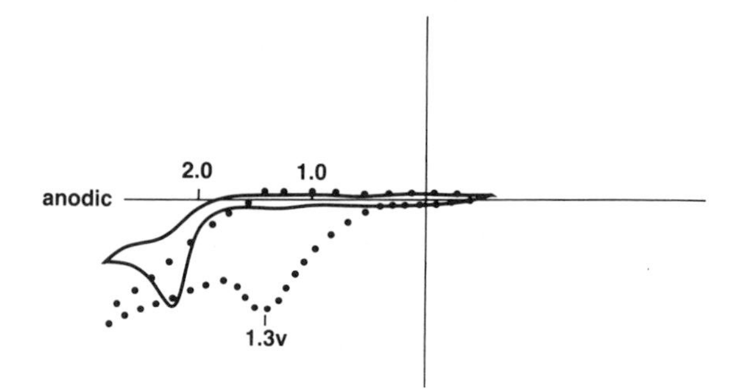

Fig. 6. Cyclic voltammogram for 6-dodecyne under N_2(—) and under 1000 psig O_2 pressure (····) in CH_3CN.

When various one-electron oxidants were tried as catalysts for this chemistry under high O_2 pressures (> 100 psi), only the $Ce(NH_4)_2(NO_3)_6$ salt gave any catalysis, as in analogy to the thioether system. As can be noted in Table 3, not only does Ce(IV) have a profound catalytic rate-enhancing effect, it also dramatically increases the selectivity for site specific cleavage. In the 1-phenyl-1-octyne system the diketone is an observed product, which appears to be a competent intermediate. In separate studies using added 1-phenyl-1,2-octanedione in the autoxidation, we found that it is oxidized further to yield selectively benzoic and heptanoic acids at rates comparable to the observed overall rates of alkyne oxidation. Further ^{18}O-labeling studies in CH_3CN at 90° under 200 psi O_2 with 1% $^{18}OH_2$ present in the solvent reveal by GC-mass spectra analyses that the product diketone contains only ^{16}O-labeled oxygens. This indicates that the O_2 and not water is the source of oxygen in this reaction.

Preliminary kinetic studies in the 6-dodecyne system have revealed that the reactions are first-order both in substrate (both initial rate and log [alkyne] versus time slots are linear), and on $[Ce(IV)]_{added}$ (Figure 7) with a non-zero intercept corresponding to the background catalyst-free autoxidation. The rate dependence on the oxygen pressure is more complex in that saturation kinetics are observed (Figure 8). Finally the Ce(IV) catalyzed 6-dodecyne autoxidation exhibits a linear arrhenius plot with an activation energy of 16.0 kcal/mole over the temperature range 60-120°C.

A radical cation mechanistic picture again best fits our experimental results. The strong oxidant Ce(IV) is involved in the rate-determining step, as is apparently the alkyne. The HPCV reveals that a new species, which we believe is the spin-orbit coupled oxygen adduct of 6-dodecyne is much easier to oxidize and it is this species which Ce(IV) oxidizes. From a thermodynamic standpoint this is much more satisfactory since Ce(IV) reduces at ∿ +1.3V in our system, the "bare" alkyne oxidation at +2.4V

suggests a high endoenergic barrier. The O_2-adduct oxidizes at about + 1.4V, very near the Ce(IV) potential. We propose that Ce(IV) oxidizes the alkyne-O_2 adduct formed in a rapid pre-equilibrium (eq. 17 and 18).

$$\text{(17)} \quad \text{alkyne} + O_2 \underset{}{\overset{K_2}{\rightleftharpoons}} [\text{alkyne-}O_2]$$

$$\text{(18)} \quad [\text{alkyne-}O_2] + \text{Ce(IV)} \xrightarrow{k_2} [\text{alkyne-}O_2]^{+\cdot} + \text{Ce(III)}$$

The alkyne-O_2 radical cation could then rearrange to the dioxatene radical cation (eq. 19), a potent oxidant in analogy to olefin radical cations.[13,16] The dioxatene radical cation could then reoxidize Ce(III) to Ce(IV) and produce the dioxatene which thermally rearranges rapidly to the α-diketone.

$$\text{(19)} \quad [\text{alkyne-}O_2]^{+\cdot} \xrightarrow{k_3}$$

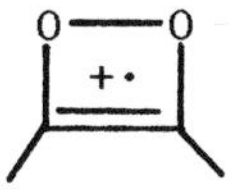

O—O
+·

$$\text{(20)} \quad + \text{Ce(III)} \xrightarrow{\text{fast}} \quad \longrightarrow$$

O—O +· O—O O O

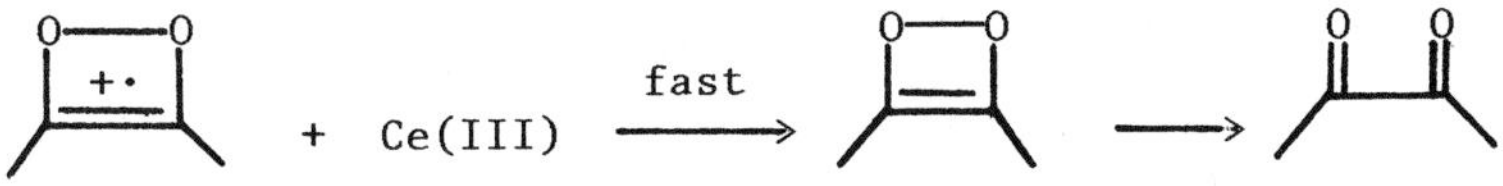

By noting that $[\text{alkyne}]_t = [\text{alkyne}] + [\text{alkyne-}O_2]$ an integrated rate expression can be derived (eq. 21):

$$\text{(21)} \quad V = \frac{-d[\text{alkyne}]_t}{dt} = \frac{k_2K_1[\text{alkyne}][O_2][\text{Ce(IV)}]}{1 + K_1[O_2]}$$

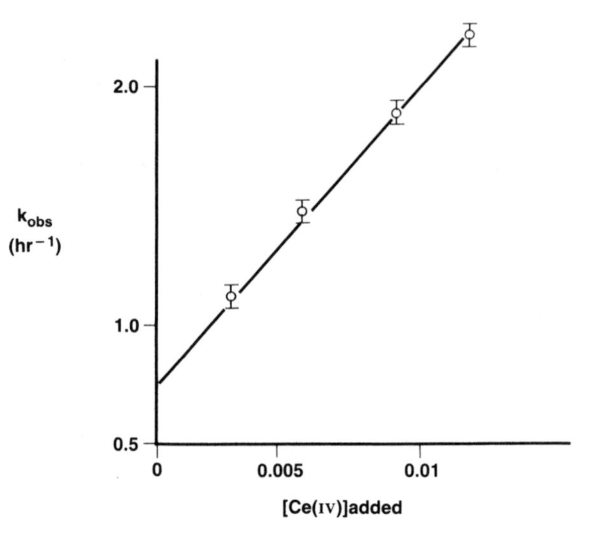

Fig. 7. 6-Dodecyne Autoxidation Catalyzed by [Ce(IV)] in CH_3CN(P_{O_2} = 15 bar and T = 110°C).

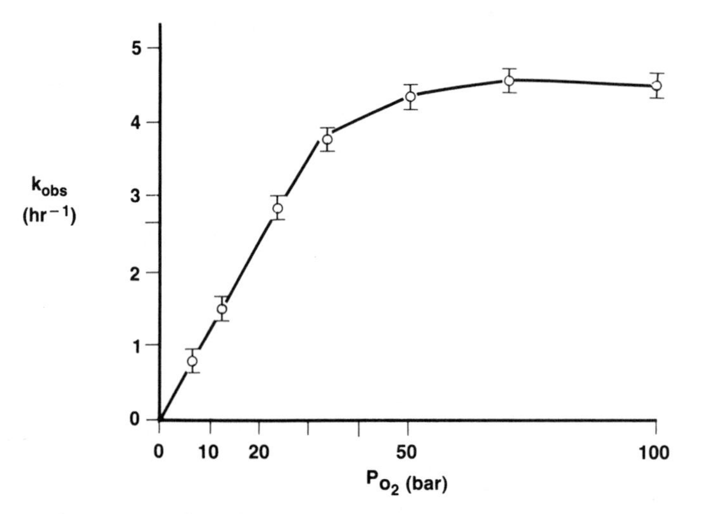

Fig. 8. Dependence of the Observed Rate of Autoxidation of 6-Dodecyne on P_{O_2}(T = 120°C, Sub/cat = 20).

This rate expression is consistent both with the observed 1st-order $[alkyne]_t$ and $[Ce(IV)]_t$ dependences (where $[Ce(IV)] \cong [Ce]_t$) and the observed O_2 saturation kinetics.

CONCLUSIONS

We have reported that the Ce(IV) ion, a potent oxidant, is a very effective catalyst for the selective molecular oxygen oxidation of thioethers and alkynes, promoting the formation of sulfoxides and site selective cleavage, respectively. In these systems Ce(IV) promotes O_2 dioxygenase-like reactivity in analogy to biological systems. This means that both atoms of the O_2 molecule are utilized in the product and expensive and cumbersome co-reductant systems are not required to activate oxygen. This reaction also is important because it appears to proceed via a radical cation pathway.

REFERENCES

1. P. E. Correa and D. P. Riley, J. Org. Chem., 50 (10), 1787 (1985).
2. a) D. P. Riley, P. E. Correa, and G. Hardy, submitted to J. Am. Chem. Soc.; b) Portions of this work were described at the California Institute of Technology Industrial Affiliates Oxidation Symposium, Feb. (1986).
3. C. S. Foote and J. W. Peters, J. Am. Chem. Soc., 93, 3795 (1971).
4. M. G. Simic, J. Chem. Educ., 58(2), 125 (1981).
5. K. U. Ingold, Acc. Chem. Res., 2, 1 (1969).
6. a) D. E. Van Sickle, F. R. Mayo, R. M. Arluck, and M. G. Zyz, J. Am. Chem. Soc., 89(4), 967 (1967); b) D. F. Van Sickle, F. R. Mayo, F. S. Gould, and R. M. Arluck, J. Am. Chem. Soc., 89(4), 977 (1967).
7. a) W. Pritzkow, T. S. S. Rao, J. Prakt. Chem., 327 (6), 887 (1985); b) L. T. Dao, B. Karla, W. Pritzkow, W. Schmidt-Renner, V. Voerckel, and L. Willecke, J. Prakt. Chem., 326 (1), 73 (1984).
8. D. P. Riley and P. E. Correa, J. Chem. Soc., Chem. Commun., 1097 (1986).
9. L. Fieser and M. Fieser, Reagents for Organic Synthesis, Wiley, New York, Vol 1, p. 471 (1967).
10. D. P. Riley and R. E. Shumate, J. Chem. Educ., 61, 923 (1984).
11. a) T. -L. Ho, Syn. Commun., 9(4), 237 (1979); and b) T. -L. Ho, Synthesis, 561 (1972).
12. F. Minisci, A. Atterio, and C. Giordano, Acc. Chem. Res., 16, 27 (1983).
13. a) S. F. Nelsen and M. F. Teasley, J. Org. Chem., 51, 3221 (1986); b) S. F. Nelsen and K. J. Akaba, J. Am. Chem. Soc., 103, 2096 (1981); c) S. F. Nelsen, M. F. Teasley, and D. L. Kapp, J. Am. Chem. Soc., 108, 5503 (1986).
14. M. Simic and E. Hayon, J. Phys. Chem., 15, 1677 (1971).
15. N. J. Turro, V. Ramamurthy, K. -C. Liu, A. Krebs, and R. Kemper, J. Am. Chem. Soc., 98(21), 6758 (1976).
16. S. F. Nelsen, D. L. Kapp, F. Gerson, and J. Lopez, J. Am. Chem. Soc., 108, 1027 (1986).
17. A. L. Buchachenko, Russ. Chem. Rev., 54(2), 117 (1985).
18. H. Tsubomura and R. S. Mulliken, J. Am. Chem. Soc., 82, 5966 (1960).
19. D. F. Evans, J. Chem. Soc., 1981 (1961).

VANADIUM CATALYZED AUTOXIDATION OF HYDROGEN SULFIDE

H. W. Gowdy, D. D. Delaney, and D. M. Fenton

UNOCAL Science & Technology Division
376 South Valencia Ave.
Brea, California 92621, USA

ABSTRACT

The Unisulf process utilizes a vanadium-based solution for absorbing H_2S from gas streams and oxidizing it to elemental sulfur. Unisulf solution offers the same high H_2S absorption efficiency as Stretford solution, with better recovery during overloading episodes. In addition, the vanadium complexing agents used in Unisulf suppress the formation of thiosulfate and sulfate. Consequently, Unisulf does not have the severe solution disposal problems associated with the operation of Stretford plants.

In December 1985, a Stretford plant in the United States was converted to Unisulf. Unisulf chemicals were added to the existing Stretford solution while the plant remained onstream. No operating problems occurred during this transition. Since that time the plant has continued to run smoothly and the H_2S emissions have remained nil. The thiosulfate initially present in the Stretford solution decomposed to sulfate and elemental sulfur. After this thiosulfate had disappeared, no further increase in sulfate has been observed. Furthermore, no thiosulfate has been detected since the conversion to Unisulf.

Another Unisulf plant was started up in December 1986, at Unocal's Santa Maria Refinery in California. The design sulfur production of the plant is 5.6 tonnes per day. This is a BSR/Unisulf application treating Claus tail gas. In the front end of the plant the sulfur species are catalytically converted to H_2S in a BSR reactor. In the tail end of the plant the H_2S is converted to sulfur in the Unisulf solution.

The Unisulf Process was developed at the Unocal Science & Technology Division in Brea, California.

BACKGROUND

The Unisulf Process is one of several processes jointly licensed by the Unocal Science & Technology Division and the Ralph M. Parsons Company in the field of sulfur technology. The others include the Selectox Process and the Beavon Sulfur Removal Process (BSRP)[1]. This latter process consists of two main operating sections: I. the hydrogenation/hydrolysis reactor where all forms of sulfur in the Claus tail gas are converted to hydrogen sulfide by reaction with hydrogen or water; and II. the Stretford section, where the hydrogen sulfide is oxidized to sulfur. The Stretford process was developed by the Northwest Gas Board in England (now part of the British Gas Corporation)[2].

There are over 50 BSRP plants operating, under construction or in design in the United States, Europe and Japan. The process has been a success with respect to the removal of sulfur compounds in Claus tail gas. However, in the Stretford process, vanadium is chelated by anthraquinone disulfonic acid (ADA) isomers and by carbonate and bicarbonate. The resulting homogeneous complex permits thiosulfate and sulfate to be continuously formed.

With fresh Stretford solution which contains no thiosulfate and which operates below 38°C (100°F), the initial thiosulfate production rate can be as low as 0.3 to 0.5 g/l/day. (All thiosulfate values in this paper are calculated as $Na_2S_2O_3 \cdot 5H_2O$.) As the thiosulfate concentration increases, so does its rate of production, reaching about 1 g/l/day at 75 g/l and 2 g/l/day at 250 g/l, at which point the solution disposal is usually begun. At one BSRP unit, the thiosulfate level was allowed to increase to 400 g/l. Before this time the thiosulfate rate had been 1 g/l/day. However, at 400 g/l, it increased to over 10 g/l/day, meaning that over 70 percent of all the H_2S in the feed gas was being converted to thiosulfate rather than to sulfur.

Stretford solution is very temperature sensitive. At a gas treating plant the thiosulfate production rate increased from 1 to 3 g/l/day when the solution temperature increased from 38 to 52°C (100 to 125°F). Consequently, plants using direct autoclaves (which operate at 130-145°C [260-290°F]) produce considerably more thiosulfate than those using filters or centrifuges.

In general then, in the Stretford Process, substantial quantities of hydrogen sulfide are continuously oxidized to thiosulfate and sulfate. As these by-products build up in solution, the corrosion rate increases. Also, the thiosulfate production rate increases markedly as the thiosulfate concentration increases. Base is consumed in the formation of thiosulfate and sulfate. In order to limit corrosion and thiosulfate production, and ultimately to prevent component precipitation, solution must be purged. The effective life of the solution can be anywhere from six to eighteen months, depending on the operating conditions. Thus, there is chemical consumption due to direct decomposition, as in the case of base and ADA, and due to replacement of purged solution.

Other operating problems encountered in Stretford plants include foaming due to bacteria buildup, and maintenance and operating problems with the autoclave system.

It was expected that the disposal of used Stretford solution would become more difficult in the future due to increasingly strict environmental regulations. Therefore work was begun at the Unocal Science & Technology Division (Brea, California) to develop a solution in which thiosulfate and sulfate production is reduced sufficiently to permit a solution life of 7 to 10 years without the need for solution disposal. The Unisulf process has met this objective.

THE UNISULF PROCESS

General

The Unisulf process utilizes a family of vanadium-based solutions with the following general constituents:

1. Vanadium
2. Sodium carbonate/bicarbonate buffer
3. Thiocyanate
4. Carboxylate (citrate is usually used)
5. Aromatic sulfonate complexing agent

In the absorber section of the process, hydrogen sulfide is oxidized to sulfur as the vanadium (V) is reduced to vanadium (IV).

$$(2V^{+5})_n + HS^- + CO_3^{=} \text{ ----> } (2V^{+4})_n + S + HCO_3^- \qquad (1)$$

In the oxidizer section of the process, vanadium (IV) is reoxidized to vanadium (V).

$$(2V^{+4})_n + 2HCO_3^- + 1/2\ O_2 \text{ ----> } (2V^{+5})_n + 2CO_3^{=} + H_2O \qquad (2)$$

The chelating agents used to complex vanadium determine whether or not by-product thiosulfate or sulfate is formed. The chelating agents used in Unisulf solution minimize by-product salt formation[3,4,5]. The actual components and their concentrations depend on the specific processing application.

Two experimental units have been used in the Unisulf development work. A bench-scale unit was used to screen new solutions. It was operated for over 4 years. A pilot plant has been operated almost continuously for several years. In addition data are now being collected from several commercial plants. This work has shown that no measurable sulfur solubilization occurs over a wide range of component concentrations.

Characteristics of Unisulf Solution

In Unisulf solution, no measurable thiosulfate is produced. With a Unisulf solution containing no added thiosulfate, the thiosulfate concentration remains below the analytical detection limit. Furthermore, when thiosulfate is added to these solutions it is decomposed to sulfur and sulfate, ultimately decreasing below the minimum analytical detection limit. Consequently any thiosulfate added to a commercial plant because of Claus plant upsets resulting in SO_2 breakthrough would not remain. Likewise, the thiosulfate present in the commercial Stretford solution when the Unisulf chemicals were added in December 1985 decomposed to sulfur and sulfate within two months.

In a new solution there is an initial sodium sulfate production of about 0.03 g/l/day. At this rate, 120 g/l sodium sulfate would be reached in 11 years. Additional experiments have been done with sodium sulfate added to the Unisulf solutions. This work showed that the measured rate of sulfate production decreases to zero as the sulfate concentration increases. At the converted Stretford plant, no change in the sulfate concentration has been observed following the decomposition of existing thiosulfate.

Except for citrate, there is no measurable decomposition of Unisulf chemicals. The required citric acid makeup rate is about 0.3 g/l/day.

Absence of Bacterial Growth

No bacterial growth has been detected in the pilot plant or commercial Unisulf solutions. This is due mainly to the substantial concentrations of thiocyanate present.

Many of the Stretford plants processing non-coke oven gas have been found to contain substantial quantities of bacteria. Large bacterial populations cause several problems, including conversion of thiosulfate to sulfate (leading to higher ADA losses), higher base demand and severe foaming. Biocides have recently been used to control bacteria in Stretford plants, but the biocides decompose, thus adding to the plant operating costs[6]. Furthermore, there is some indication that their effectiveness is not sustained.

Coke-oven gas contains cyanide, which is converted to thiocyanate in Stretford solution. Some coke-oven gas Stretford plants have been in operation for over 20 years, and no bacteria has been found as long as the solution contains around 50 g/l or more of sodium thiocyanate[6]. Therefore, it is the demonstrated capability of thiocyanate which gives the Unisulf solution its strong resistance to bacterial growth.

UNISULF PROCESS COMMERCIALIZATION

General

Several Unisulf plants are now operating. This paper will describe the first one, a Stretford plant (part of a BSRP plant) converted to Unisulf in December 1985; and the most recent one, a newly constructed BSR/Unisulf plant at Unocal's Santa Maria Refinery.

First Commercial BSR/Unisulf Plant

The design sulfur production rate of this plant is 2.5 LT/D. It was designed to use Stretford solution. A rotary drum filter is used to recover the sulfur from the froth. The first Stretford solution lasted about 15 months. In December 1985, the plant had run with the second charge of Stretford solution for about 2 months. The solution contained about 10 g/l sodium thiosulfate pentahydrate and 22 g/l sodium sulfate.

Unisulf chemicals were added in mid-December to convert the plant to the Unisulf Process. These chemicals were added while the plant was onstream. No operating problems occurred during this transition. Since that time the plant has continued to run smoothly and the H_2S emissions have remained nil. The thiosulfate present at the changeover has since been decomposed to sulfur and sulfate. No increase in sulfate has been observed since the thiosulfate disappeared.

The thiosulfate and sulfate concentrations for January through March of 1986 are shown in Figures 1 and 2*. These figures show that

---

* These thiosulfate concentrations have been obtained by a polarographic method. The test loses accuracy below 1 g/l due to interference from the thiocyanate present in the solution. All of these thiosulfate values were confirmed by HPLC.

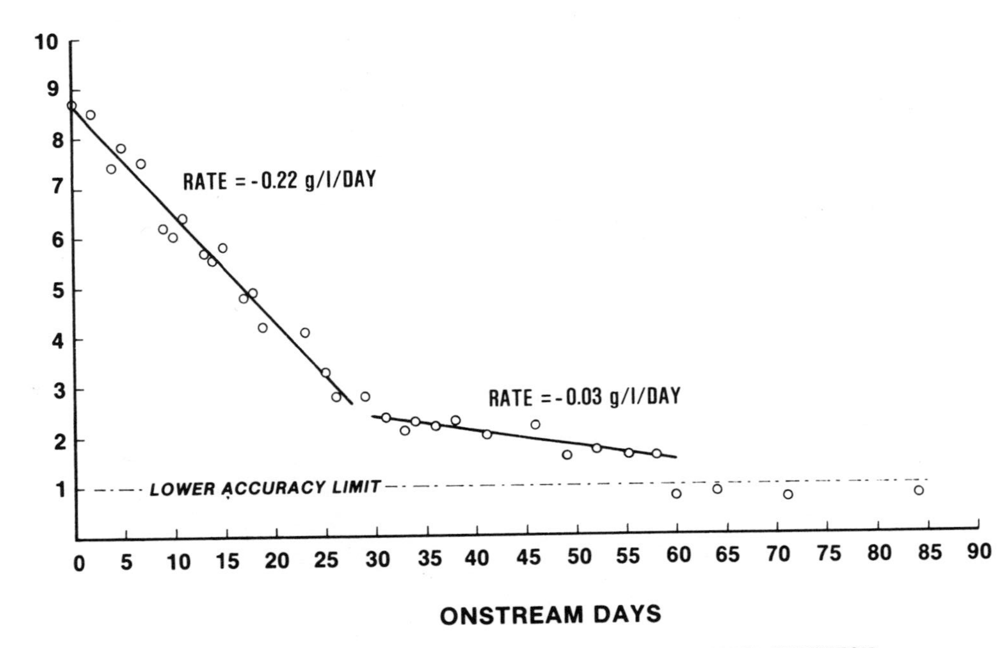

Fig. 1. THIOSULFATE DECOMPOSITION IN COMMERCIAL UNSULF SOLUTION

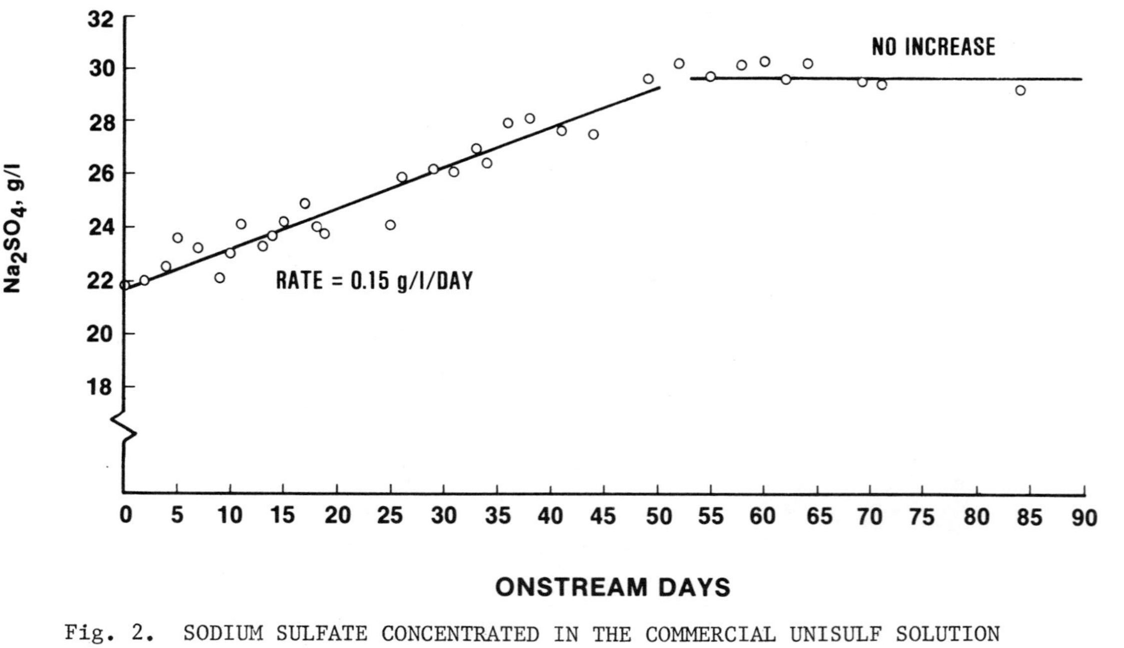

Fig. 2. SODIUM SULFATE CONCENTRATED IN THE COMMERCIAL UNISULF SOLUTION

during January, the thiosulfate decomposed at a rate of 0.22 g/l/day. During this same period, the sulfate increased at a rate of 0.15 g/l/day. Since a rate of 0.03 g/l/day sulfate is normal initially, the net sulfate production resulting from the thiosulfate conversion is approximately 0.12 g/l/day.

Oxidation of all the sulfur in the decomposing thiosulfate would have produced 0.25 g/l/day sulfate by the following stoichiometry:

$$2O_2 + Na_2S_2O_3 \cdot 5H_2O + Na_2CO_3 \text{ ----> } 2Na_2SO_4 + CO_2 + 5H_2O \qquad (3)$$

Thus this reaction would have produced about twice the sulfate actually seen.

One of the sulfur atoms in thiosulfate is in the zero valence state. If this atom was converted to elemental sulfur and the remaining sulfite oxidized to sulfate, then the stoichiometry would be:

$$\tfrac{1}{2}O_2 + Na_2S_2O_3 \cdot 5H_2O \text{ ----> } Na_2SO_4 + S + 5H_2O \qquad (4)$$

This reaction would produce 0.13 g/l/day sulfate from a loss of 0.22 g/l/day thiosulfate, and thus represents the actual data. An 8% drop in the calculated total soluble sulfur during this period confirms that some of the sulfur in the thiosulfate ended up as elemental sulfur, as predicted from equation 4.

Once the thiosulfate concentration had decreased below 3 g/l, its decomposition rate slowed appreciably and it took almost another month before its concentration dropped below 1 g/l. Since this time, no sulfate production has been measured.

Thiosulfate Decomposition

No thiosulfate is formed in Unisulf solution. Furthermore, any thiosulfate added to the solution due to sulfur dioxide dissolving in Unisulf solution during Claus plant upsets is unstable throughout the entire operating pH range of Unisulf plants. The thiosulfate decomposes to sulfur and sulfate regardless of solution pH. However, the split between sulfur and sulfate does depend on pH range. The thiosulfate decomposition rate is also dependent on solution pH. In the usual pH range of 8.5 to 9.0 this rate is 3 times as fast as that measured in the commercial plant discussed in the previous section.

BSR/Unisulf Plant at Unocal's Santa Maria Refinery, Arroyo Grande, California

General. This new BSR/Unisulf plant was started up in October 1986. Its design sulfur production rate is 5.6 tonnes per day. It processes tail gas from two Claus units.

BSR Section. This is of standard design with the gas flowing sequentially to the following equipment:

1. Reducing gas generator.
2. Hydrogenation/hydrolysis reactor.
3. Process gas cooler.
4. Contact condenser/desuperheater.

The reactor contains a cobalt-molybdenum catalyst.

Unisulf Section (Figure 3). The design feed gas to the Unisulf section is 7,854 Nm^3/h (7.05 MM scfd), containing 2.0 mol% H_2S. This H_2S is removed in an absorber containing a bottom spray section and 3 sections of splash plate packing above. This absorber was designed to reduce the H_2S to below 10 ppmw. All actual measurements have been below 1 ppmw.

The combustor on top of the absorber will automatically oxidize the H_2S to SO_2 should an upset occur which results in H_2S levels above 10 ppmw in the gas leaving the absorber.

The reaction tank is provided by retaining the Unisulf solution for a few minutes in the bottom of the absorber. The reaction tank liquid level is controlled by an external weir box with removable dividers.

The 3-stage oxidizer consists of one cylindrical tank divided into three equal 120° sectors. Solution and sulfur enters the first stage through a standpipe near the tank bottom. The solution then overflows into the second stage and underflows into the third stage. The sulfur froth overflows from each stage to the next, ultimately flowing down a chute to the froth tank. The oxidized solution leaves the third stage by passing beneath an underflow weir and then flows by gravity back to the balance tank.

Oxidation air is provided to each oxidizer stage by means of turbine aerators.

Sulphur Recovery Section (Figure 4). The froth contains about 8 wt% sulfur. The froth is first processed in a rotary drum filter in order to remove most of the Unisulf solution, and to concentrate the sulfur slurry in order to increase the capacity of the vertical-basket centrifuge which follows.

The final cake contains just 20 wt% moisture and is powdery rather than paste-like. Consequently the cake can be dropped through a chute without plugging and readily conveyed using a double-screw conveyor directly to the liquified sulfur pit where it is dropped through a chute onto the top of the liquid sulfur. Maintaining sufficient agitation on the surface of the liquid sulfur allows the cake to be quickly melted. Due to the low moisture content, little steam is evolved.

Analyses of the final sulfur cake yield about 200 ppmw ash and less than 5 ppmw vanadium. Hence it is more pure than sulfur typically produced in autoclaves. The final liquid sulfur product is used by Unocal for sulfuric acid manufacture.

Initial Plant Performance. This plant has now been onstream for several months and has operated without significant problems.

The gas leaving the absorber contains nil H_2S. No thiosulfate has been produced. As discussed in a previous section, in Unisulf solution the sulfate concentration builds up slowly and then levels out. It is still well below the 30 g/l maximum at which the first Unisulf plant has remained for over one year.

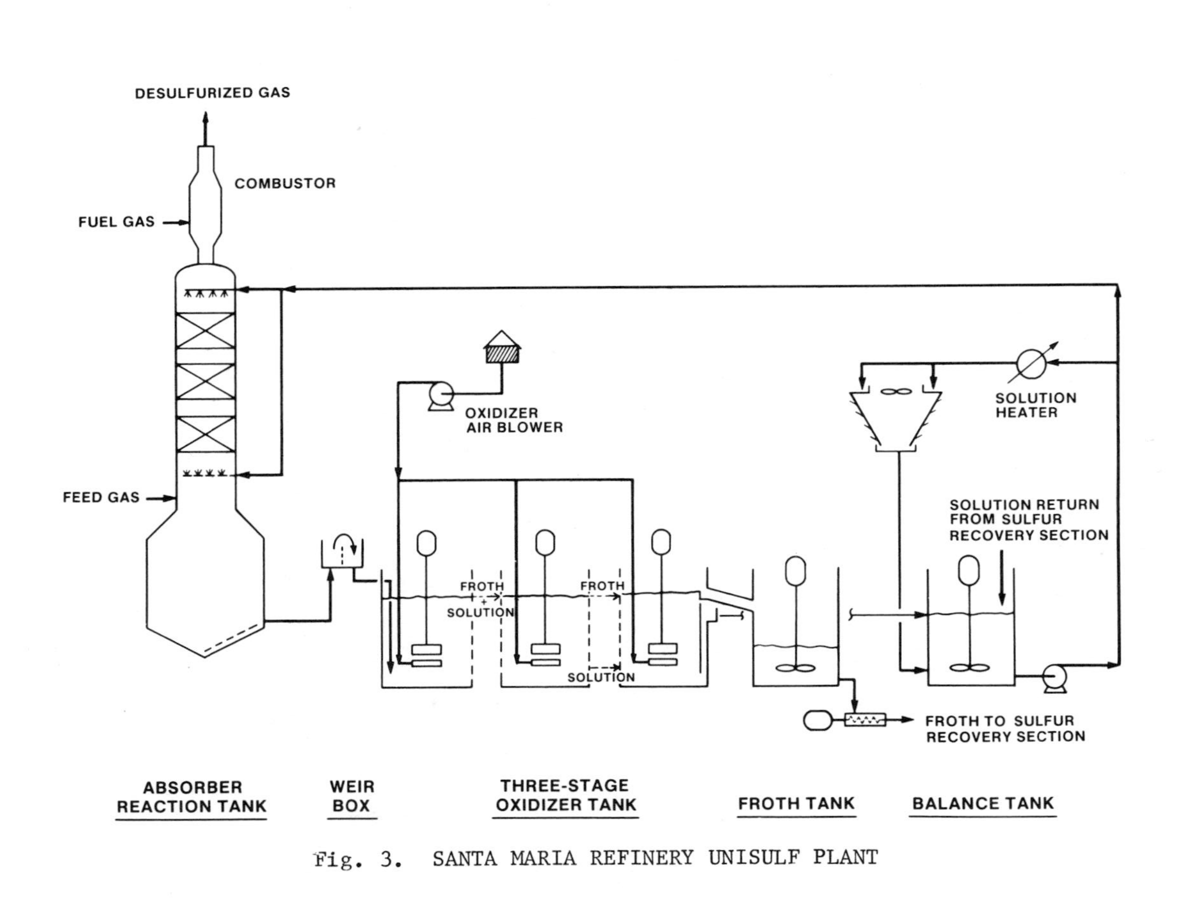

Fig. 3. SANTA MARIA REFINERY UNISULF PLANT

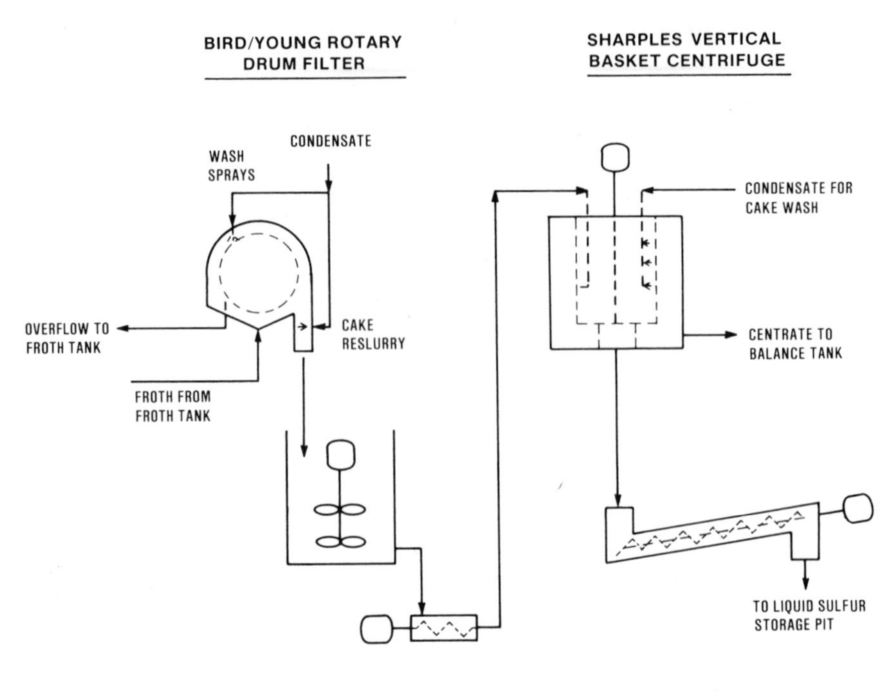

Fig. 4. SANTA MARIA REFINERY UNISULF PLANT SULFUR RECOVERY SECTION

Corrosion probes have been placed in the weir box to measure the rate for reduced solution (which typically has a higher corrosion rate than oxidized solution) and in the discharge of the circulation pump, a highly turbulent area where any erosion/corrosion will be detected. The corrosion rates at both locations are continually less than 0.025 mm/yr (less than 1 mil/yr).

CONCLUSION

The Unisulf Process presents a significant benefit to users of the Stretford Process. Unisulf solution is very stable. In over one year of operation of the first commercial Unisulf plant, no thiosulfate or sulfate has been produced. Operating problems historically associated with the Stretford Process have been resolved.

REFERENCES

1. R. H. Hass, J. W. Ward, H. W. Gowdy, "Recent Developments in Selectox and Unisulf Sulfur Recovery Technologies", Sulphur - 84 International Conference, Calgary, Alberta, Canada (1984).

2. A. L. Kohl, F. C. Riesenfeld, "Gas Purification", Gulf Publishing Company, Houston, Third Edition, 476-487 (1979).

3. D. M. Fenton, H. W. Gowdy, (Unocal Corporation), "Method for Removing Hydrogen Sulfide from Gas Streams", U. S. Patent 4,283,379 (1981).

4. H. W. Gowdy, D. M. Fenton, (Unocal Corporation), "Method for Removing Hydrogen Sulfide from Gas Streams", U. S. Patent 4,325,936 (1982).

5. H. W. Gowdy, D. M. Fenton, (Unocal Corporation), "Method for Removing Hydrogen Sulfide from Gas Streams", U. S. Patent 4,432,962 (1984).

6. B. M. Wilson, R. D. Newell, "H_2S Removal by the Stretford Process - Further Development by the British Gas Corporation", AIChE National Meeting, Atlanta, Georgia, 1984.

COPPER CATALYZED OXIDATIVE CARBONYLATION OF METHANOL TO DIMETHYL CARBONATE

G. Lee Curnutt and A. Dale Harley

Central Research, Inorganic Materials and Catalysis

Laboratory, The Dow Chemical Company, Midland, MI 48641

ABSTRACT

The vapor-phase oxidative carbonylation of methanol to produce dimethyl carbonate (DMC) over heterogeneous supported copper catalysts is reported. The products are DMC, carbon dioxide, methyl formate, and methyl acetate. Under acidic conditions, dimethyl ether is also produced. The effects of the support, reaction conditions, and various promoters on the rate, selectivity, and deactivation profile are discussed.

The deactivation process has been found to proceed via sintering and the replacement of chloride ligands by hydroxide. Regeneration may be accomplished by treatment with hydrochloric acid which restores the chloride content of the copper.

INTRODUCTION

The controlled oxidation of carbon monoxide in the presence of an alcohol is a well known route to the dialkyl esters of carbonic acid. The oxidative carbonylation of methanol to dimethyl carbonate (DMC) has been thoroughly investigated using homogeneous catalysts and was commercialized in Italy in 1983 by EniChem Sintesi SPA. Mauri et al.[1] have recently reviewed the uses for DMC in the production of a wide variety of organic chemicals. The commercial process is run in a CSTR as a slurry reaction[2,3] catalyzed by cuprous chloride. Numerous patents[4-10] have been issued claiming improved catalytic systems for the production of DMC. These processes are carried out in the liquid phase and suffer from problems associated with halide corrosion. More recent studies[11] have focused attention on homogeneous copper catalysts which form non-corrosive solutions with methanol and on nitrogen-containing promoters.

In contrast, methanol oxidative carbonylation over heterogeneous catalysts has received very little attention. Cipriani and Perrotti[12] prepared a heterogeneous catalyst by complexing cuprous chloride with poly-4-vinylpyridine and produced DMC by passing a mixture of liquid methanol, CO, and O_2 through a reactor charged with this catalyst. However, the rate at which DMC was produced was low and the leachable chloride produced corrosive solutions.

The literature contains no reports of the oxidative carbonylation of methanol having been carried out in the vapor phase. This paper describes copper-containing heterogeneous catalysts which were developed to produce DMC via a vapor-phase oxidative carbonylation. The best results have been obtained with an activated carbon supported cupric chloride catalyst promoted with potassium chloride. The pathway by which the heterogeneous catalyst deactivates has been established, and a regeneration procedure has been developed.

EXPERIMENTAL SECTION

Catalyst Synthesis

Preparation of the pyridine complex of copper methoxychloride, $C_5H_5NCu(OCH_3)Cl$, 1. Complex 1 was prepared by the method of Finkbeiner et al.[13] The composition of the green precipitate was confirmed by elemental analysis and infrared spectroscopy.

Catalyst preparation. Catalysts[14] derived from complex 1 were prepared by dissolving 1 in pyridine and using the solution to impregnate various types of supports by the incipient wetness technique. All supports used in catalyst preparation were commercially available products. A typical catalyst preparation consisted of dissolving 0.8 g of 1 in 20.0 ml of pyridine and impregnating 12.0 g of activated carbon (12-20 mesh). The pyridine was removed from the catalyst in a stream of dry helium. The dried catalyst analyzed as 1.47 wt % copper. A typical oxidative carbonylation experiment employed 1.0-2.0 g of catalyst.

A representative procedure for the preparation of catalysts containing cupric chloride is as follows. Anhydrous $CuCl_2$ (5.5 g) was dissolved in 40.0 ml of anhydrous ethanol and the solution used to impregnate 10.0 g of activated carbon. The catalyst was allowed to air dry overnight at ambient temperature. The partially dried catalyst was loaded into a Pyrex tube and heated to 140°C for two hours under a N_2 purge of 125 cc(STP)/min. A mixture of HCl (12 cc(STP)/min) diluted in air to a total flow rate of 138 cc(STP)/min was passed over the catalyst for one hour to replenish any chloride hydrolyzed during drying.

Adsorbed HCl was removed from the carbon support by purging with N_2 (125 cc(STP)/min) for three hours at 140°C. The dried catalysts contained copper at concentrations ranging from 4.9 to 21.3 wt %. Carbon-supported cupric chloride catalysts promoted with KCl, $MgCl_2$, or $LaCl_3$ were prepared by co-impregnation from aqueous solutions. The catalyst was then air dried and treated with HCl as described above. All salts and solvents were commercial products having the highest purity that was available. These products were used as received.

Oxidative Carbonylation Studies

Vapor-phase experiments. Oxidative carbonylation studies were carried out in a 1.3 cm o.d. x 30.5 cm plugged-flow, tubular reactor constructed of Hastelloy C276 and fitted with a concentric thermal well. The reactor was mounted inside a high temperature oven (Blue M, Model No. POM-206F-1). Temperatures were measured with a Hastelloy clad Type J thermocouple which extended axially into the catalyst bed. The middle section of the reactor was packed with 4 cc (ca. 1.0-2.0 g) of catalyst diluted with 8 cc of high purity SiC (8-12 mesh). Glass wool and SiC were used above and below the catalyst bed.

Methanol was contained in a N_2 padded feed cylinder attached to a weigh cell (Interface, Model MB-5). Liquid methanol was fed with a LC pump (Gilson Medical Electronics, Model 302) at a flow rate of ca. 0.06 ml/min to a vaporizer operated at 185°C. The vaporizer produced a steady flow (30 cc(STP)/min) of methanol vapor which was introduced into the CO/O_2 feed stream prior to flowing upwards through the reactor. The CO and O_2 were metered through a preheater prior to entering the reactor with thermal mass flow controllers (Brooks Instruments, Model 5810/5835) which had been calibrated at the pressure used in the experiments. Pressure was maintained in the system by a research control valve, constructed of Hastelloy C276 and fitted with a P-12 Stellite trim, which was connected to a Taylor 440D pneumatic controller. The feed stream consisting of 80.0 cc(STP)/min of CO, 30.0 cc(STP)/min of methanol, and 13.0 cc(STP)/min of O_2 was allowed to flow through the reactor under a pressure of 20.68 bars. Feed gases were of high purity grade (99.99%). Iron pentacarbonyl was removed from the CO by adsorption in a 13X molecular sieve trap. The O_2 contained 2.0-3.0% N_2 which served as the internal standard for the on-line gas chromatographic analyses.

The offgas was analyzed at 3-4 hour intervals for run times that typically lasted 100-200 hours. The offgas was sampled through a Valco GC valve and a microliter sampling loop. The valve and all product lines up to a cold trap were heat traced at 125°C. Analyses were carried out using a Hewlett-Packard HP5710-A gas chromatograph with thermal conductivity detectors. A temperature program of 60°C to 140°C at 8°C per minute with a 15 min hold at 60°C and a 30 min hold at 140°C was employed. The light gases (O_2, N_2, and CO) were separated on a 1.8 m 5A molecular sieve column. CO_2, H_2O, and the organics (methanol, methyl acetate, methyl formate, dimethyl ether, and DMC) were separated on a 1.8 m Poropak N column. These two columns were connected in series through a switching valve which prevented CO_2 contamination of the molecular sieve column. Standard mixtures were made up of the components found in the offgas in order to determine response factors. Integrations were performed using a Hewlett-Packard 3353 computer. The various parameters of the reactor were continuously monitored during an experiment with a microcomputer.

Condensable products were removed from the offgas in a dry ice cooled trap and a scrubber containing propylene glycol. These product solutions were analyzed on a Tenex column. The flow rate of the noncondensable products in the offgas was monitored with a wet test meter. Mass balance calculations were performed for a few of the experiments. The mass balances were found to range from 97 to 112%.

Catalyst regeneration. Catalysts were regenerated in the reactor by the following sequence of steps: 1) discontinuing the feed stream and purging the reactor with N_2 while cooling to room temperature; 2) drying in a nitrogen purge, 125 cc(STP)/min for two hours; and, 3) chlorination in a 10% HCl/90% N_2 stream for three hours and cooling to ambient temperature while maintaining the HCl/N_2 flow.

The reactivated catalysts were subjected to the same activity tests as described for the oxidative carbonylation reaction. The presence of oxygen during the regeneration had no adverse effect on the activity. Regenerated catalysts freed of adsorbed HCl by purging with an inert gas showed initially higher activity for a few hours but otherwise demonstrated comparable performance.

Catalyst characterization. The copper and chloride contents of the fresh and used catalysts were analyzed by neutron activation. Copper contents were also measured on selected catalysts by plasma emission spectroscopy (Spectrametrics, Model IV). Transmission electron microscopy (TEM) was performed using a modified JOEL 100C with a Kevex energy dispersive x-ray (EDX) detector. The samples were dispersed in n-hexane with an ultrasonicator and applied to a gold grid. Beam intensities were reduced to minimize the degradation of the sample. X-ray diffraction powder patterns were recorded on a Phillips Diffractometer with CuKα radiation as the source. BET surface areas were measured on selected catalysts using a Micromeritics DigiSorb 2500 with nitrogen as the adsorbate. The ash contents of the activated-carbon supports were determined from the results of weight measurements which were made on samples before and after they had been oxidized in air at 500°C.

RESULTS

Temperature

The non-catalytic thermal oxidation of CO was studied to establish the temperature at which CO_2 formation from this side reaction became important. The reactor was charged with SiC and purged with the CO/O_2 feed gas. GC analysis of the offgas showed that thermal oxidation was significant above 130°C. Experiments were carried out on a DARCO carbon-supported $CuCl_2$ catalyst to establish the lower temperature limit for the process. Below 100°C the reaction rate was insignificant. The oxidative carbonylation rate was found to increase over the initial 24 hours of operation and then began to decrease slowly with time. The oxidative carbonylation rates were measured in the temperature range of 110-130°C after at least 24 hours of continuous operation.

Pressure and Space Velocity

The effect of pressure on the rate of oxidative carbonylation was studied for $CuCl_2$ supported on carbon. The best results were obtained at 20.68 bars. Decreasing the pressure to 11.72 bars resulted in a 50% decrease in the rate. At a pressure of 41.36 bars, (T = 160°C), total oxygen consumption to deep oxidation of CO and methanol to form CO_2 was observed.

The gas hourly space velocity (GHSV) was varied over the catalyst derived from the pyridine complex of copper methoxychloride. At GHSV of 1800 and 726 h^{-1}, essentially the same reaction rates were observed. However, the selectivity to DMC based on CO consumption was substantially higher, (74% vs. 53%), at the higher GHSV. The oxidative carbonylation rates were measured at 20.68 bars with a GHSV in the range of 1800-2000 h^{-1}.

Support Effects

Table I contains performance data on catalysts derived from complex 1 supported on various types of carbons and inorganic oxides. The feed was converted to DMC at rates ranging from 1.8 to 4.8 lb DMC cu^{-1} ft^{-1} cat^{-1} h^{-1} for the activated-carbon-supported catalysts. The selectivity based on the amount of CO consumed varied from 41 to 76%. The major byproduct

Table 1. PERFORMANCE DATA ON PYRIDINE COPPER METHOXYCHLORINE SUPPORTED ON VARIOUS TYPES OF CARBONS AND IRNOGANIC OXIDES

Support	Support Precursor	Ash Content wt %	S_{BET} m^2g^{-1}	Rate of Oxidative Carbonylation[B]	DMC Selectivity % on CO
Darco* 12 x 20	Lignite	14.2	502	3.0	65
Nuchar HG-40	Bituminous Coal	6.8	902	2.7	76
Strem 06-0050	Bituminous Coal	5.3	724	4.8	45
Norit 8 x 20	Peat	5.0	—	1.8	41
Alfa 88765	Petroleum Coke	—	—	1.8	74
Witcarb* Lck	Petroleum Coke	1.9	—	2.0	63
Witcarb 965	Petroleum Coke	0.6	1037	3.5	48
SN—5701	Carbon Black Composite	—	219	1.1	42
MgO	—	—	23.8	0.4	61
ZnO	—	—	6.3	0.4	41
SiO_2 (Cabosil* M-5)	—	—	198	0.3	37
TiO_2 (Anatase)	—	—	11.9	0.1	34
Al_2O_3 (SA3232)	—	—	139	0.2	60

[A]Temp = 100°C; Press = 20.68 bars; gas hourly space velocity = 1800 h^{-1}; copper loading = 2.0 wt %
*Registered Trademark
[B]Units: lb DMC $cu^{-1}ft^{-1}cat^{-1}h^{-1}$

was CO_2 in all cases. The best overall performance was exhibited by the lignite based activated carbon DARCO 12 x 20. With this catalyst, the yields of methyl formate and methyl acetate were small. The catalytic activities and selectivities showed no correlation with neither the ash content of the carbon nor the total surface area of the catalyst.

Carbon black or the inorganic oxides were poorer supports for the copper complex than the activated carbons. Although the range of the selectivities were similar for the various types of supports, the carbonylation activity was markedly lower for the inorganic oxide carriers than the activated carbons.

Cupric Chloride Catalysts

Screening showed that DMC could be prepared on a commercial catalyst consisting of $CuCl_2$ intercalated into the layered structure of graphite. Furthermore, Hallgren[6] showed that $CuCl_2$ catalyzed the oxidative carbonylation of methanol to DMC in the liquid phase at temperatures above 170°C. Preliminary experiments showed that $CuCl_2$ impregnated onto DARCO 12 x 20 carbon was active, and studies were initiated to optimize the

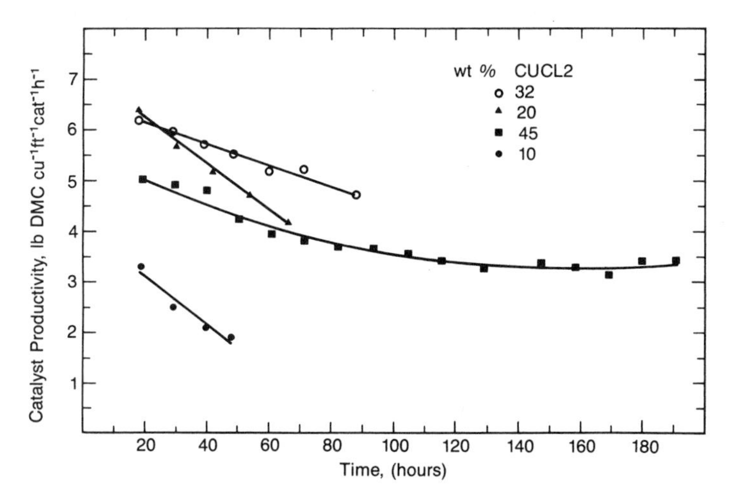

Figure 1. Activity-Time Curves for DARCO-Supported CUCL2 Catalysts.

results. Figure 1 shows the effect of loading on the productivity of $CuCl_2$/DARCO 12 x 20. The productivity increased with loading up to 20 - 30 wt %; higher loadings resulted in a decline in activity. Catalyst deactivation was somewhat dependent on loading. Figure 1 shows that the loss of activity with time was slightly lower as the concentration of $CuCl_2$ was increased.

Figure 2 shows how the selectivity was affected by loading. The products consisted mainly of DMC and CO_2 at each loading. The DMC selectivity based on CO consumption decreased linearly as the weight percent of $CuCl_2$ increased from 10 to 45. Other oxidation products formed on $CuCl_2$ besides DMC were methyl formate and methyl acetate. The amounts of these low molecular weight esters were nearly independent of the $CuCl_2$ loading, and their yields also varied little with run time. These catalysts exhibit optimum performance when $CuCl_2$ is supported on the DARCO 12 x 20 carbon at a loading of approximately 20 wt %.

Several metal chlorides and oxides were investigated as promoters for the reaction utilizing DARCO-supported catalysts containing either the pyridine complex of $Cu(OCH_3)Cl$, 1, or $CuCl_2$ as the active component. The addition of Ag_2O or the rare earth chlorides, $CeCl_3$ or $LaCl_3$, resulted in catalysts with lower activity. These additives were tested on catalysts derived from complex 1 and were added in a second impregnation step.

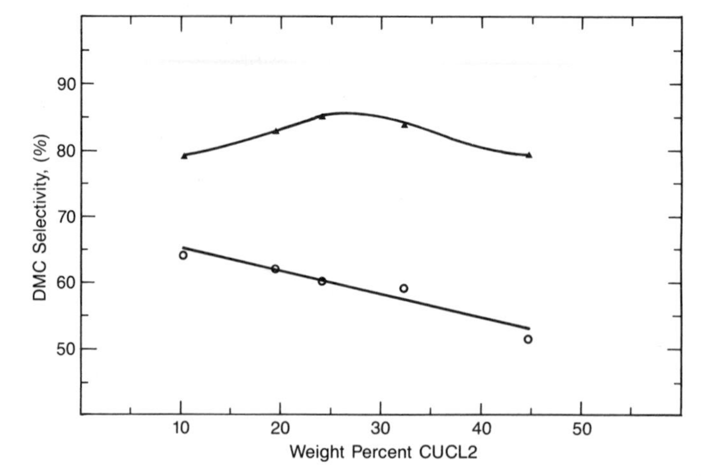

Figure 2. The Effect of CUCL2 Loading on DMC Selectivity.
(▲) Selectivity Based on CH_3OH
(O) Selectivity Based on CO

The $KCl/CuCl_2$ and $MgCl_2/CuCl_2$ combinations offered increased catalyst stability relative to unpromoted $CuCl_2$. A comparison of the activities is shown in Figure 3. These catalysts were prepared by co-impregnation from aqueous solutions of salts and contained the optimum amount (20%) of $CuCl_2$ and 5 wt % additive. While $LaCl_3$ was found to be harmful, $MgCl_2$ and particularly KCl were effective in retarding deactivation. A 22% increase in catalyst productivity was noted for the $KCl/CuCl_2$ catalyst at 60 hours on stream. Beyond this point, the catalyst productivity gradually declines and reaches a steady-state value of 4 lb DMC cu^{-1} ft^{-1} cat^{-1} h^{-1} after 100 hours. We made no attempt to optimize the loading of these promoters.

Catalyst Deactivation

The activity-time curves presented in Figure 1 show the gradual deactivation of the $CuCl_2$ catalysts with time on stream at 115°C. In a 200 hour test with a 45 wt % $CuCl_2$/DARCO catalyst at this temperature, the DMC productivity dropped from 5.0 to 3.5 lb DMC cu^{-1} ft^{-1} cat^{-1} h^{-1} in the initial 100 hours, but thereafter remained nearly constant. XRD was used to identify the copper phases present on fresh and used catalysts.

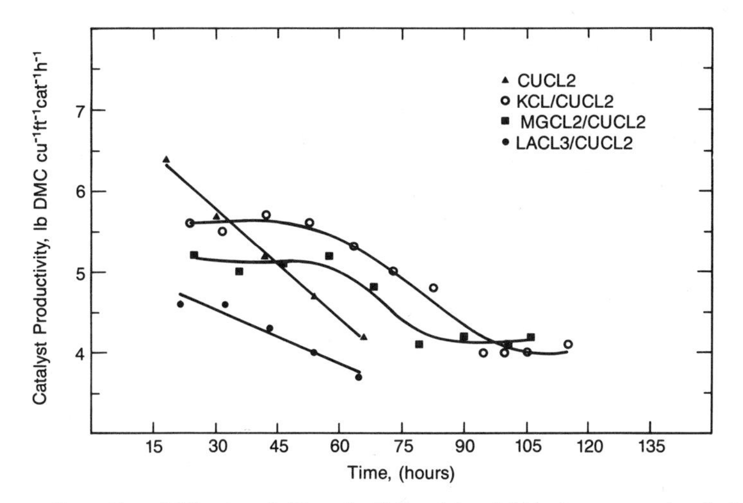

Figure 3. The Effect of Metal Chloride Additives on Activity.
Catalyst Loading: 20 wt% Additive

Figure 4 shows x-ray diffraction patterns on a fresh DARCO-supported cupric chloride catalyst containing 32.3 wt % $CuCl_2$ and on the deactivated catalyst which had been used for 90 hours. The fresh catalyst showed only crystalline α-quartz which is present in the support and trace amounts of $CuCl_2 * 2H_2O$. No crystalline $CuCl_2$ phase was detected by XRD on the fresh catalyst containing less than 45 wt % $CuCl_2$.

The used catalyst showed paratacamite, $Cu_2(OH)_3Cl$, as the only crystalline copper phase. The XRD results indicate that during deactivation the chloride ligand is replaced by hydroxide. Table II shows the results of XRD measurements on samples of fresh and used catalysts at 10.3 wt % $CuCl_2$. After using this catalyst for 90 hours, the XRD showed that the crystalline copper phases present were Cu(OH)Cl and $Cu_2(OH)_3Cl$. Elemental analysis indicated that copper was not lost from the catalyst during use. The results in Table II show that the weight ratio of chloride to copper agrees quite well with the theoretical value ($CuCl_2$ = 1.2) for the fresh catalyst, but this ratio decreases nearly five fold for the deactivated catalyst. This observation agrees with the XRD data identifying the phases of lower degrees of chlorination, Cu(OH)Cl and $Cu_2(OH)_3Cl$, as major components in the deactivated catalyst. The chloride

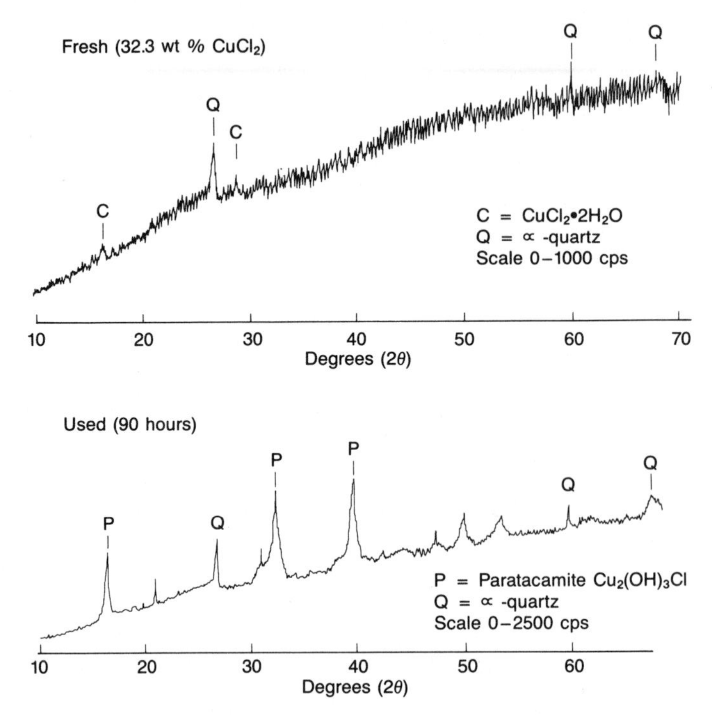

Figure 4. X-ray Diffraction Patterns on DARCO-Supported Cupric Chloride Catalysts.

abstracted from the copper formed methyl chloride as evidenced from the results of headspace analyses performed on the product solutions.

Crystalline $CuCl_2$ was not observed by XRD on fresh catalysts containing less than 45 wt % $CuCl_2$. This suggests that the $CuCl_2$ phase is present as an amorphous film or monolayer or as particles less than 5 nm in diameter. TEM analyses were used to further characterize the copper dispersion and particle sizes before and after use to determine if sintering had occurred. The micrograph in Figure 5 for the fresh catalyst shows one region above center which was high in copper (EDX), but most regions do not show discernible particles. Prolonged exposure of the specimen to the electron beam caused the copper species to agglomerate and to form spherical particles. These spheres exhibited only copper X-rays

Table 2. Chemical Analysis by Neutron Activation and X-Ray Diffraction Measurement on Selected Darco-Supported $CuCl_2$ Catalysts

Catalyst Sample	Copper Crystalline Phases	Copper Loading wt %	wt % Ratio Chloride / Copper
Fresh (10.3 wt % $CuCl_2$)	—	5.1 ± 0.2	1.10 (1.12)[A]
Used (90 Hours)	Cu(OH)Cl $Cu_2(OH)_3Cl$	5.1 ± 0.2	0.24
HCl Treated[B]	$Cu_2(OH)_3Cl$ (trace)	5.2 ± 0.3	1.12

[A]Theory
[B]Treatment: One hour in flowing 10% HCl/N_2 at 125°C followed by three hour N_2 purge

which suggests that the chloride was removed as a result of copper reduction by the electron beam. The nature and size of the particles prior to exposure to the beam can only be extrapolated from the micrograph to be an amorphous film or discrete particles less than 2 nm in diameter. The micrograph in Figure 6 for the used catalyst shows very large agglomerates ranging in size from 200-1000 nm as indicated by the dark irregularly shaped regions. These agglomerates exhibited a strong copper signal (EDX) and contained only minor amounts of chloride. The used catalyst was not as sensitive to beam damage as the fresh catalyst. These agglomerates were not present in the fresh catalyst.

We summarize the main points regarding the deactivation data below:

1) The copper species on the fresh catalyst below loadings of 45 wt % $CuCl_2$ is highly dispersed.

2) The presence of large agglomerates of reduced copper species on the used catalyst suggests that migration and sintering occurs during use.

3) $CuCl_2$ is unstable under the reaction conditions and reacts with water and methanol resulting in chloride loss and incorporation of hydroxyl groups. The abstracted chloride is ultimately incorporated into methyl chloride.

Regeneration Procedure

A number of chloride-containing compounds were tested as feed additives which might decompose under reaction conditions and prevent catalyst deactivation through replenishment of abstracted chloride.

Fig. 5. Electron Micrograph of $CuCl_2$/DARCO 12 x 20. Fresh Catalyst containing 19.5 wt% $CuCl_2$. The scale bar represents 200 nm.

Fig. 6. Electron Micrograph of $CuCl_2$/DARCO 12 x 20. Used catalyst containing 19.5% $CuCl_2$. The scale bar represents 200nm.

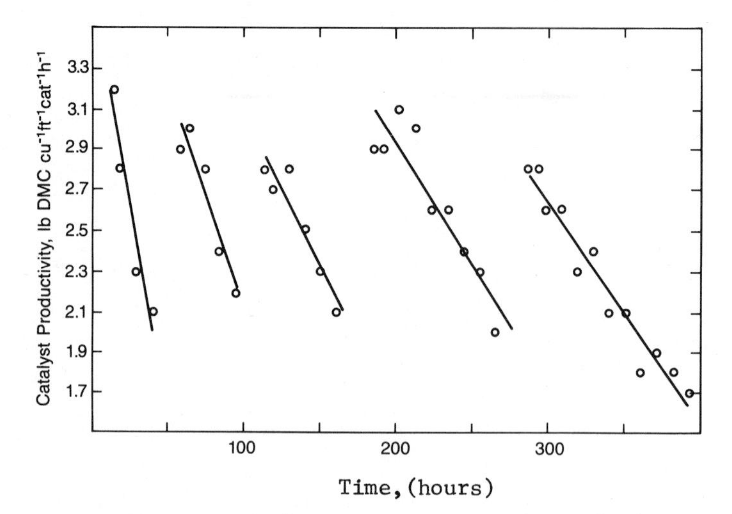

Fig. 7. Activity-Time Curves for DARCO-Supported CUCL2 Regenerated with HCL. Conditions: Temperature = 115°C; Pressure = 208.68 bars; Wt% CUCL2 =

Incorporation of 100-200 ppm of methyl chloride, ethylene dichloride, or hydrochloric acid into the feed stream resulted in a partial loss of activity and failed to prevent catalyst deactivation. Substantial amounts of dimethyl ether were produced upon addition of hydrochloric acid to the feed. Table II shows the analysis of an aged 10.3 wt % $CuCl_2$/DARCO catalyst that had been regenerated by treatment with 10% HCl/N_2 for one hour at 125°C. The weight ratio of chloride to copper increased from 0.24 for the used catalyst to the theoretical value of 1.12 after treatment with HCl. The XRD data presented in Table II indicate that the regeneration procedure converted most of the copper hydroxychloride phases to dichloride.

Activity data for multiple regenerations obtained during a 400 hour test are plotted in Figure 7. The activity profile of the regenerated catalyst was similar to that of the fresh catalyst over the four cycles. The regenerated catalyst displayed DMC/CO_2 selectivities and trace byproduct levels comparable to that of the fresh catalyst. The rate of deactivation was found to be somewhat dependent on the number of regenerations. Figure 7 shows that the loss of activity with time was slightly lower after each of the four regenerations. The HCl treatment might lower the rate of deactivation by redispersing the copper. TEM analyses of regenerated catalysts are recommended for further study to establish if HCl affects the redispersion of the copper.

DISCUSSION

Reaction Pathway

The initial step in the proposed mechanism for the vapor-phase methanol oxidative carbonylation is coordination of methoxy groups to divalent copper atoms. This proceeds under process conditions via a reaction between methanol and cupric chloride according to equation (1).

$$CuCl_2 + 2\ CH_3OH \text{ ----------> } Cu(OCH_3)Cl + CH_3Cl + H_2O \qquad (1)$$

Methoxylation of the divalent copper atoms is followed by CO insertion into a Cu-O bond to form a labile carbomethoxide species which couples with a neighboring methoxy group to form DMC according to equation (2).

$$Cu(COOCH_3)Cl + Cu(OCH_3)Cl \text{ ----------> } (OCH_3)_2CO + 2\ CuCl \qquad (2)$$

This reductive elimination step produces two Cu(I)Cl species which are reoxidized to Cu(II) by molecular oxygen. The Cu(II) atoms are remethoxylated with methanol. The two protons lost from the two methanol molecules combine with atomic oxygen to form the byproduct water as shown in equation (3).

$$2\ Cu(I)Cl + 1/2\ O_2 + 2\ CH_3OH \text{ ---------> } 2\ Cu(OCH_3)Cl + H_2O \qquad (3)$$

The CO insertion mechanism is analogous to that proposed by Koch et al.[15] and Romano et al.[2] for the cuprous chloride catalyzed formation of DMC in a liquid phase oxidative carbonylation. The insertion of CO into copper-oxygen bonds has also been proposed by Saegusa et al.[16] to account for the DMC formed during the carbonylation of cupric alkoxide compounds. Although the copper carbomethoxide intermediate is unknown, the insertion of CO into $Pt-OCH_3$[17] and $Hg-OCH_3$[18] bonds to form the corresponding metal carbomethoxides is well documented. Furthermore, DMC has been recently observed as one of the products formed in the carbonylation of methanol using a catalyst derived from $PtCl_2$.[17] Further support for the proposed mechanism is provided by the fact that the copper catalysts which show significant activity are ones which have linear chainlike structures. Willett and Breneman[19] showed that the structure of $C_5H_5NCu(OCH_3)Cl$ consisted of methoxy-bridged dimers connected to each other by asymmetric chlorine bridges, forming a one dimensional linear chain. Wells[20] reported for cupric chloride that each copper atom was bonded to four neighboring chlorine atoms, forming planar $CuCl_4$ groups which share opposite edges. As in the case of $C_5H_5NCu(OCH_3)Cl$, stacked, linear polynuclear copper chains are formed in the solid state. These polynuclear structures would be expected to be important in the coupling step where two copper atoms are needed to accept the two electrons released during reductive elimination of DMC.

The aforementioned byproducts observed on carbon-supported cupric chloride are methyl formate, methyl acetate, and CO_2. Under acidic conditions, dimethyl ether is also formed from the dehydration of methanol. In the absence of oxygen, CO_2 and neither ester were observed suggesting that these byproducts are formed from oxidative reactions rather than from the carbonylation of methanol or from the decomposition of DMC. Scheme 1 accounts for all of the products observed during the vapor-phase reactions between CO, O_2, and methanol on carbon-supported cupric chloride.

The Role of Promoters and Activated Carbon

The results shown in Figure 7 indicate that the addition of $MgCl_2$, and particularly KCl, to $CuCl_2$ lowers the rate of deactivation. The reactions of KCl and $MgCl_2$ with $CuCl_2$ to form the chlorocuprates, $KCuCl_3$ and $MgCuCl_4$ * 6H2O, respectively, are well known.[21] These chlorocuprates have a higher ratio of chloride to copper than cupric chloride and might prevent deactivation by lowering the rate at which chloride is abstracted from the catalyst. Potassium chlorocuprate[21] is known to exist as the dimer, $K_2Cu_2Cl_6$, which has a structure similar to that for cupric chloride. The bulky $Cu_2Cl_6^{2-}$ anion might lower the rate of deactivation by lowering the rate at which the copper chloride species sinters. It is noteworthy that potassium chlorocuprate is a chloride-bridged binuclear complex whose activity pattern for methanol oxidative carbonylation is similar to that for cupric chloride. These observations provide additional support for the mechanism proposed in the previous section.

SCHEME 1. VAPOR—PHASE PROCESS CHEMISTRY ON METHANOL OXIDATIVE CARBONYLATION

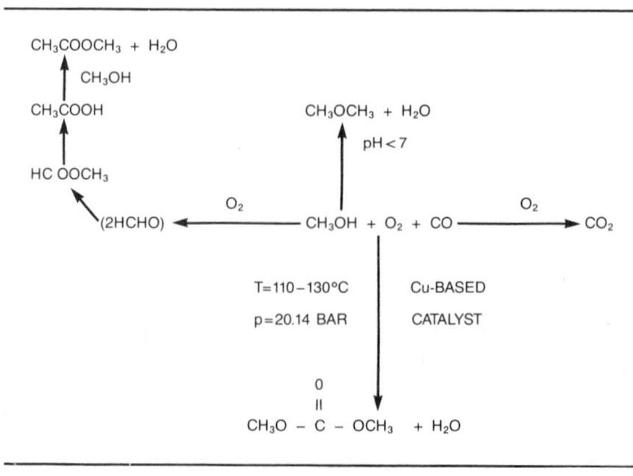

The results presented in Table I show that activated carbon is the best support for complex 1. The ability of activated carbon to catalyze the oxidation of metal cations[22] and supported metals[23] is well known. The same mechanism might be operating with Cu(I). Since carbon is an electrical conductor, it might enhance the rate of oxidation of Cu(I) by easily transferring the electron between the copper atom and atomic oxygen.

Deactivation Pathway and Regeneration

The results of catalyst characterization showed that cupric chloride is converted to copper hydroxychlorides during use. The formation of copper hydroxychlorides during methanol oxidative carbonylation has been previously reported by Romano et al.[2] for the case of the slurry reaction catalyzed by cuprous chloride. The mechanism involved in their formation might be the same in both cases. The byproduct water may hydrolyze chloride to hydrochloric acid which is converted to methyl chloride by reaction with methanol. In the case of heterogeneous supported copper catalysts, activity can be restored by converting the copper

CONCLUSIONS

This work has shown that $C_5H_5NCu(OCH_3)Cl$[14] or $CuCl_2$ supported on activated carbon are effective catalysts for oxidative carbonylation of methanol. DMC is formed with high selectivity. While the addition of potassium chloride was found to retard catalyst deactivation, magnesium chloride was also effective to some extent. Electron transfer between a Cu(I) atom and atomic oxygen via the activated-carbon support is assumed to be essential for high activity. Under conditions of oxidative carbonylation, copper catalysts gradually deactivate with time due to sintering of the copper chloride species and to loss of chloride from the copper forming Cu(OH)Cl and $Cu_2(OH)_3Cl$. Activity can be restored by treating the catalyst with hydrochloric acid. Multiple regenerations with HCl have been demonstrated.

ACKNOWLEDGMENTS

The authors gratefully acknowledge the contributions of the following people: R. E. Guerra - TEM, C. Crowder - XRD, W. Rigot - Neutron Activation, D. T. Doughty - assistance with DMC process work.

REFERENCES

1. M. M. Mauri, U. Romano, and F. Rivetti, Quad. Ing. Chim. Ital. 21:6 (1985).

2. U. Romano, R. Tesel, M. M. Mauri, and P. Rebora, Ind. Eng. Chem. Prod. Res. Dev. 19:396 (1980).
3. U.S. Patent 4 218 391 (1980).
4. U.S. Patent 3 952 045 (1976).
5. U.S. Patent 4 113 762 (1978).
6. U.S. Patent 4 360 477 (1982).
7. U.S. Patent 4 361 519 (1982).
8. U.S. Patent 4 370 275 (1983).
9. E.P. Application 71 286 (1983).
10. E.P. Application 90 977 (1983).
11. U.S. Patent 4 604 242 (1986).
12. U.S. Patent 3 980 690 (1976).
13. H. Finkbeiner, A. S. Hay, H. S. Blanchard, and G. F. Endres, J. Am. Chem. Soc. 31:549 (1966).
14. U.S. Patent 4 625 044 (1986).
15. P. Koch, G. Cipriani, and E. Perrotti, Gazz. Chim. Ital. 104:599 (1974).
16. T. Saegusa, T. Tsuda, and K. Isayama, J. Org. Chem. 35:2976 (1970).
17. R. A. Head, and M. I. Tabb, J. Mol. Catal. 26:149 (1984).
18. W. Scoeller, W. Schrauth, and W. Esser, Chem. Ber. 46:2864 (1913).
19. R. D. Willett and G. L. Breneman, Inorg. Chem. 22:326 (1983).
20. A. F. Wells, J. Chem. Soc. 1670 (1947).
21. D. W. Smith, Coord. Chem. Rev. 21:93 (1976).
22. V. S. Tripathi and P. K. Ramachandran, Def. Sci. J. 35:115 (1985).
23. K. Fujimoto, T. Takahashi, T. Kunugi, Ind. Eng. Chem. Prod. Res. Dev. 11:303 (1972); J.Catal. 43:234 (1976).

DEPENDENCE OF REACTION PATHWAYS AND PRODUCT DISTRIBUTION ON THE OXIDATION STATE OF PALLADIUM CATALYSTS FOR THE REACTIONS OF OLEFINIC AND AROMATIC SUBSTRATES WITH MOLECULAR OXYGEN

James E. Lyons

Applied Research and Development Department
Sun Refining & Marketing Company
P.O. Box 1135
Marcus Hook, PA

ABSTRACT - Palladium chemistry dominates the catalytic liquid phase oxidation of unsaturated hydrocarbons both in the breadth of commercial processes and in the large number of synthetic applications. Recent work has shown the potential for new routes via oxygen activation chemistry. Control of the oxidation state of Pd catalysts can provide new routes to allylic oxidation products. Routes to industrially important α,β-unsaturated alcohols, esters and acids occur via both Pd(IV) and Pd(0) intermediates. High oxidation state palladium complexes are implicated in selective catalytic aromatic ring oxidations, while Pd(II) intermediates give rise to oxidative coupling, and low oxidation state species are responsible for benzylic oxidations.

INTRODUCTION

Palladium(0) complexes react readily with molecular oxygen and activate it toward reaction with a number of unsaturated substrates(1,2). In the first section of this paper the activation and transfer of molecular oxygen by palladium complexes will be reviewed and the catalytic potential of this chemistry considered.

Although the industrially important production of vinylic oxidation products via Wacker chemistry has long dominated the palladium-catalyzed oxidation of unsaturated substrates, the equally important family of allylic oxidation products can now be generated via palladium catalyzed reactions. The oxidation state of the metal during these catalytic reactions can be a major determinant of whether allylic or vinylic products will be produced. The second section of this paper will deal with those factors which control and promote allylic oxidations in the presence of palladium catalysts.

Finally, in the third section the role of palladium complexes in aromatic oxidations will be discussed. Again, the oxidation state of the catalyst is critical. Regioselectivity in the oxidation of alkyl aromatics is a function of the oxidation state of the catalyst as is the pathway taken by palladium aryl intermediates in ring oxidation.

ACTIVATION AND TRANSFER OF OXYGEN <u>VIA</u> METALLACYCLIC INTERMEDIATES

It has long been known that the addition of molecular oxygen to a coordinatively unsaturated palladium(0) complex can give a peroxo complex, eq. 1(3,5) which can react with an electron-deficient olefin to produce a metallacycle, eq. 2(6). Olefins without strongly electron-withdrawing groups on at least one of the unsaturated carbon atoms failed to react in this manner. Thermolysis of the metallacycle in an attempt to produce an epoxide resulted in products of oxidative cleavage of the C-C bond instead, eq. 3(6).

Thus, the direct epoxidation of an olefin by molecular oxygen activated solely by a palladium center was not demonstrated. Protonation of the metallacycle, however, did lead to an epoxide, eq. 4a(6). One may speculate that such a protonation could give an intermediate, I, of the type that is known to produce olefin oxidation products. In addition to the epoxide, a smaller amount of olefin was also released during protonation of the macrocycle, 4b.

Protonation of a palladium peroxo complex can lead to an intermediate capable of converting an α-olefin to a methyl ketone, eq. 5(7). Presumably, the reason for production of a methyl ketone rather than an epoxide as in eq. 4a, is the migration of the β-H which is present in the intermediate metallocycle in this case.

$$L_2Pd^0 + O_2 \longrightarrow L_2Pd^{II}\langle{}^{O}_{O}| \qquad (1)$$

$$L_2Pd^{II}\langle{}^{O}_{O}| + (CH_3)_2C{=}C(CN)_2 \longrightarrow L_2Pd^{II}\langle{}^{O-O}_{C(CN)_2-C(CH_3)_2} \qquad (2)$$

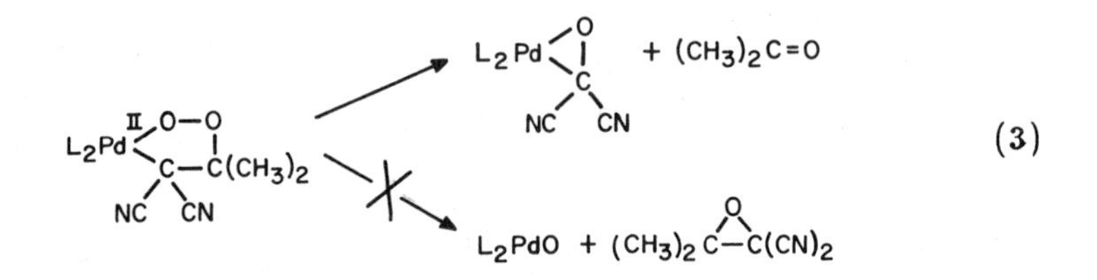

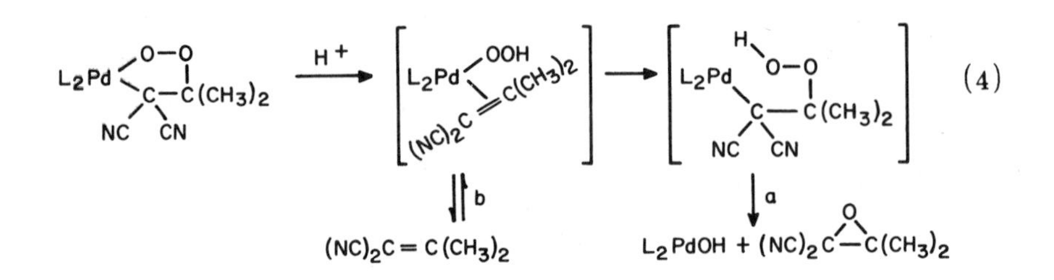

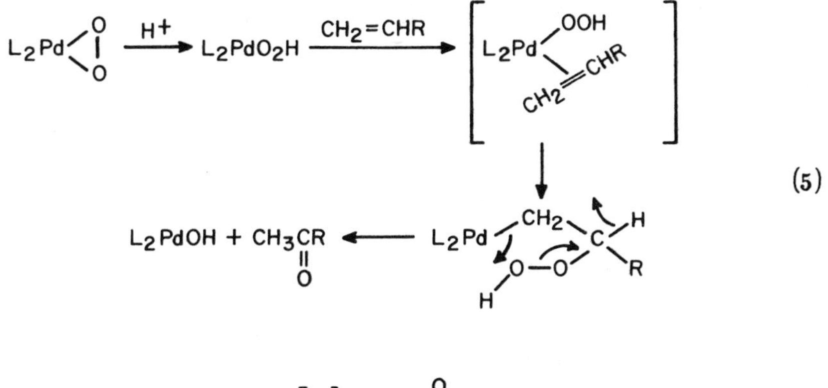

(5)

$$CH_2{=}CHR + R'O_2H \xrightarrow{[Pd]} CH_3\overset{O}{\overset{\|}{C}}R + R'OH \quad (R' = \text{Alkyl or H}) \qquad (6)$$

$$\begin{matrix} CH_2{=}CHR \\ Pd{-}OOR' \end{matrix} \longrightarrow \begin{matrix} CH_2{-}CHR \\ | \quad\quad \diagdown \\ Pd \quad\quad O \\ \quad R'O \diagup \end{matrix} \longrightarrow Pd^{II}OR' + CH_3\overset{O}{\overset{\|}{C}}R \qquad (7)$$

The catalytic conversion of an α-olefin can be accomplished using hydrogen peroxide or alkyl hydroperoxides as the oxidant eq. 6(7,8) and many of the same metallacycles have been postulated as reaction intermediates, eq. 7. Metallacyclic intermediates have been isolated in the catalytic oxidation of olefins catalyzed by palladium nitro complexes(9), eq. 8, 9. Principles similar to those outlined above appear to govern these reactions as well. If accessible β-hydrogens are available, migration will occur in the intermediate metallacycle and ketones will be the predominant product.

$$RCH{=}CHR' + O_2 \xrightarrow{[Pd(CH_3CN)_2ClNO_2]} R\underset{O}{\underset{\|}{C}}CHR' \qquad (8)$$

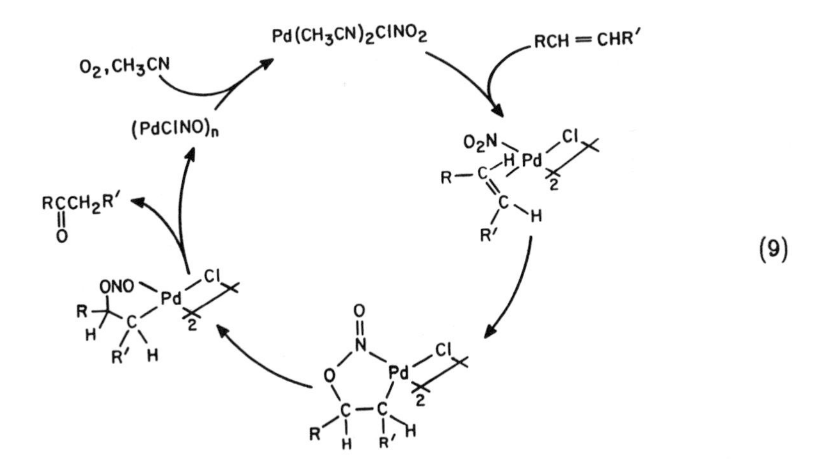

(9)

If, however, accessible β-hydrogens are not available, epoxides predominate as in the case of norbornene oxidation, eq. 10(9). In this case, the intermediate metallacycle was isolated at low temperature, characterized, and shown to decompose at room temperature to give the epoxide and a palladium nitrosyl complex, eq. 11(9). An important difference between the catalytic oxidations of olefins using palladium nitro complexes and the oxidations referred to previously is that the oxygen activation step occurs at the nitrogen atom and <u>not</u> at the palladium center, eq. 12(9).

In summary, therefore, it has been demonstrated that pathways involving oxygen transfer <u>via</u> metallacycles are important in both catalytic and stoichiometric oxidations of olefins using palladium complexes and that the structure of the olefin is critical in determining whether epoxides or carbonyl compounds are produced, eq. 13a, b.

Another strategy for catalytic oxidation of olefins which is related to that described in eq. 8-12, uses the palladium complex to bind only the olefin, and a cobalt center is used to carry the nucleophilic nitro group to the activated olefin,(10) eq. 14a-d. Interestingly, when the metal center that activates the olefin is changed from palladium(II) to thallium(III)--a metal center which does not readily participate in β-H migration--epoxides are formed even from olefins having accessible β-hydrogens.(10) This approach adds a new measure of control of oxidation reactions. Because of facile β-H migration, palladium catalyzed reactions are subject only to <u>substrate control</u>. By varying the metal center in the approach shown in equation 14a-d a measure of <u>metal control</u> over reaction pathway is achieved.

$$ + O_2 \xrightarrow{|Pd(CH_3CN)_2\ ClNO_2|} \qquad (10)$$

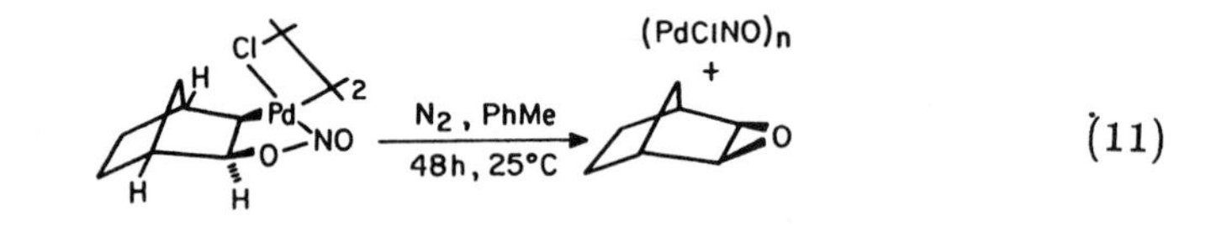

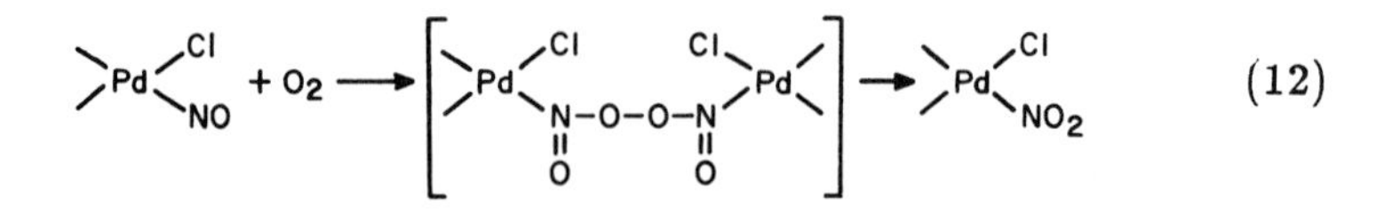

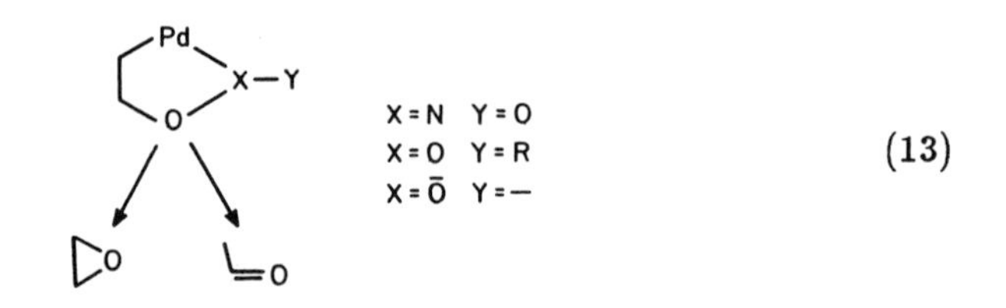

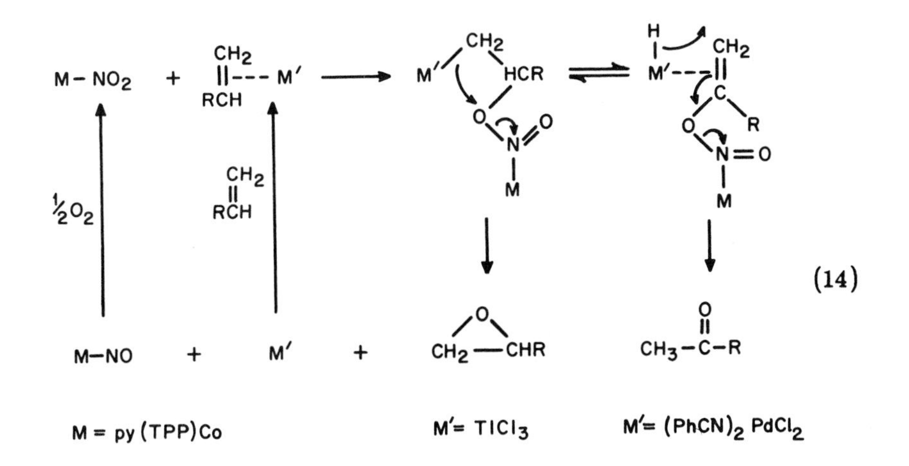

VINYLIC OXIDATIONS OF OLEFINS

The familiar Wacker reaction forms methyl ketones from α-olefins(1,2). It produces vinylic oxidation products or their stable tautomers via direct nucleophilic attack on olefin coordinated to a palladium center, eq. 15. The commercial utility of this reaction extends well beyond the production of acetaldehyde, eq. 15, to acetone, vinyl acetate and possibly a butane diol precurser as well, Table 1(11). The synthetic utility of Wacker Chemistry diminishes as the olefins become larger. While fairly good yields of methyl ketones are formed from 1-butene and 1-pentene, yields fall off markedly as the chain length grows so that 1-octene, 1-nonene and 1-decene produce methyl ketones in quite low yields, Table 2.(12) Only fair yields of ketones are obtained from cyclic olefins as well. Perhaps it is in the area of the production of specialty ketones from more complex olefinic structures that the new and selective methods of oxygen activation described in the first section could gain prominence.

TABLE 1. SOME INDUSTRIAL APPLICATIONS OF PALLADIUM CATALYZED OXIDATION OF OLEFINS

	CONDITIONS	CATALYSTS	YIELD AND/OR SELECTIVITY
(1) ACETALDEHYDE FROM ETHYLENE $C_2H_4 + O_2 \longrightarrow CH_3CHO$	100°C, 100 psia H_2O SOLVENT	$PdCl_2 + CuCl_2$	> 95 % ≃ 96 %
(2) VINYL ACETATE FROM ETHYLENE $C_2H_4 + O_2 + HOAc \longrightarrow CH_2{=}CHOAc + H_2O$	100 - 130°C 450 - 600 psia ACETIC ACID SOLVENT	$PdCl_2 + CuCl_2$	≃ 90 %
(3) ACETONE FROM PROPYLENE $C_3H_6 + O_2 \longrightarrow (CH_3)_2C{=}O$	50 - 120°C 50 - 100 ATM H_2O SOLVENT	$PdCl_2 + CuCl_2$	≃ 99 %
(4) DIACETOXYBUTENE FROM BUTADIENE $C_4H_6 + \frac{1}{2}O_2 \xrightarrow[-H_2O]{2HOAc} AcOCH_2CH{=}CHCH_2OAc$	80 - 100°C 20 psig	Pd-Te/C ; $Pd(OAc)_2$- $Cu(OAc)_2$- LiOAc	90 %

a) Data from reference 11.

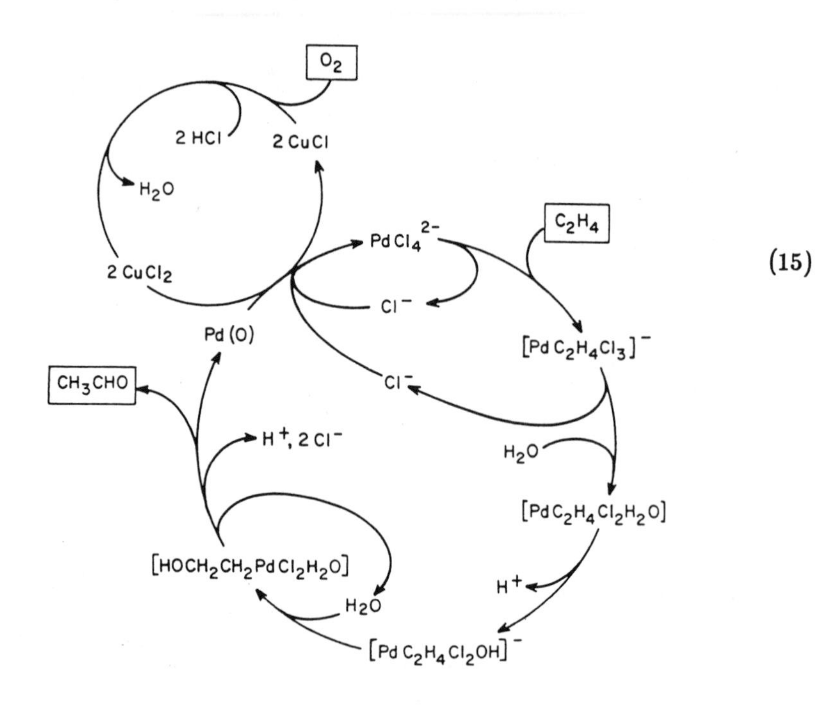

(15)

TABLE 2. OXIDATION OF HIGHER MONO-OLEFINS IN AQUEOUS SOLUTIONS OF THE $PdCl_2$-$CuCl_2$ CATALYST SYSTEM

Olefin	Temp.	Time(min)	Product	Yield(%)
1-Butene	20	10	2-Butanone	80
1-Pentene	20	20	2-Pentanone	81
1-Hexene	30	30	2-Hexanone	75
1-Heptene	50	30	2-Heptanone	65
1-Octene	50	30	2-Octanone	42
1-Nonene	70	45	2-Nonanone	35
1-Decene	70	60	2-Decanone	34
Cyclopentene	30	30	Cyclopentanone	61
Cyclohexene	30	30	Cyclohexanone	65
Styrene	50	180	Acetophenone	57

a) Data from reference 12.

Other synthetically important pathways are available to palladium(II) α-olefin complexes which produce not <u>vinylic</u> but <u>allylic</u> oxidation products.(13) The pathway which predominates can be controlled by the nature and environment of the Pd(II) complex which is used as the catalyst. For example, palladium acetate or palladium chloride can catalyze predominant vinylic oxidation, eq. 16a(1,2) or predominantly allylic oxidation if a poorly coordinating weak base is added to assist proton removal, eq. 16b(14-17). On the other hand, palladium trifluoroacetate gives predominant allylic oxidation eq. 16c. It has been suggested that Pd(II) complexes with strongly electron withdrawing ligands are highly electrophilic and oxidatively add an allylic C-H bond, resulting in Pd(IV) π-allyl intermediates(18). Thus, although isopropenyl acetate is the predominant product of the palladium acetate-catalyzed oxidation of propylene in acetic acid, addition of excess sodium acetate increases the allyl acetate yield markedly(16). When palladium trifluoroacetate is used as the catalyst, high yields of allyl acetate are formed(19).

Finally, another route from coordinated olefin to catalytically active π-allyl intermediates proceeds <u>via</u> oxidative addition to a coordinatively unsaturated palladium(0) center(20-23). Heterogeneous palladium catalysts are known to activate the C-H bond of α-olefins, and hence one can envision an efficient catalytic oxidation of α-olefins to produce allylic products, eq. 17, 18(13).

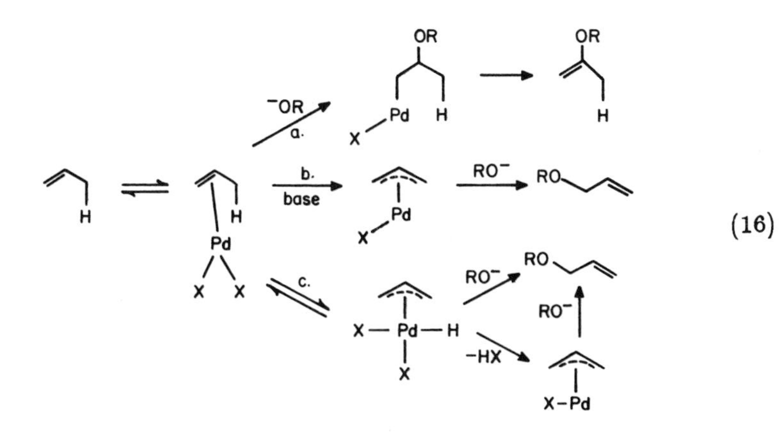

(16)

$$\tfrac{1}{2}O_2 + CH_2{=}CHCH_3 + HY \xrightarrow{Pd/S} CH_2{=}CHCH_2Y + H_2O \qquad (17)$$

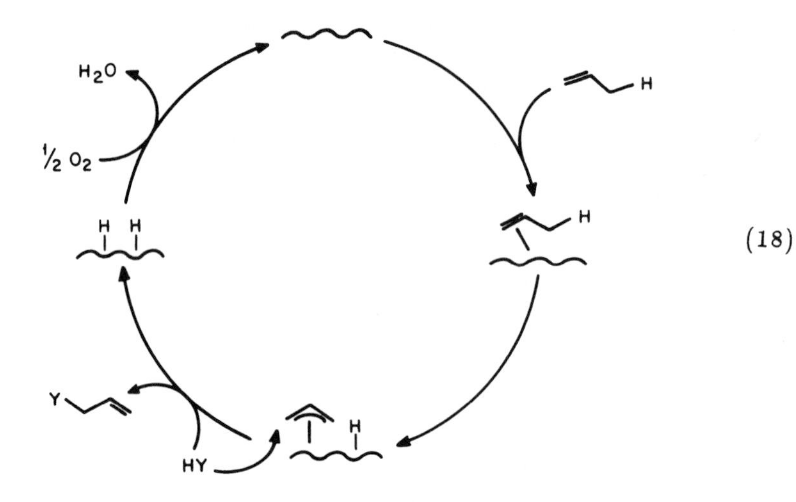

(18)

Reaction 17 (HY = HOAc, HOH) occurs over heterogeneous palladium catalysts under mild conditions in the liquid phase to give high yields of allylic oxidation products(13). An efficient catalyst for these reactions is 10% palladium on carbon, Table 3. When HY is acetic acid, allyl acetate is the major product, Table 4, and when HY is water, acrylic acid is produced in high yields, Table 5. Presumably in the latter case allyl alcohol is the initial product which is rapidly converted to acrylic acid on the surface of the catalyst, eq. 19.

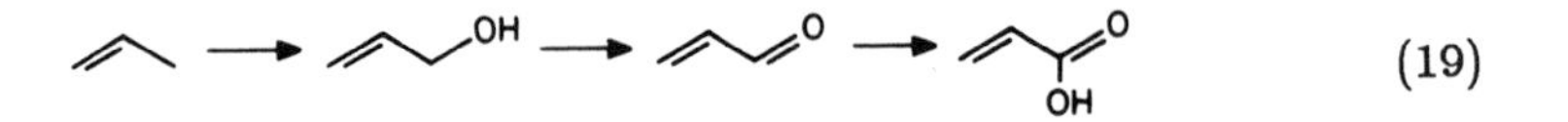

(19)

TABLE 3. PROPYLENE OXIDATIONS OVER SUPPORTED PALLADIUM CATALYSTS IN WATER [a)]

Catalyst	Gas Uptake Mol/g. atom Pd/hr	Sel. to Acrylic Acid Wt. % in Liquid
1% Pd-C	12.4	31
5% Pd-C	10.8	72
10% Pd-C	10.1	88
20% Pd-C	7.5	91
5% Pd-Al_2O_3	4.2	78
5% Pd-SiO_2	0.8	--

a) One gram of the catalyst and 30 ml of water were added to a Fisher-Porter aerosol tube, flushed 3 times with propylene and heated for 30 minutes at 65°C under 50 psig of propylene. The propylene was then replaced with 100 psi of a 60/40 $O_2/C_3^=$ gas mixture and stirred for ~5 hours. As gas was consumed, it was continually replaced with the 60/40 mix. Liquid products were analyzed by standardized GLPC.
b) Data from reference 13.

TABLE 4. PALLADIUM CATALYZED OXIDATION OF PROPYLENE TO ALLYL ACETATE[a)]

NaOAc moles/l.	Activation Time, Min.	Acetate Wt. % in Product Allyl	i-Propenyl	Allyl Acetate, %
0.0	0	1.69	1.70	29.9
0.167	0	1.89	1.75	39.1
0.333	0	5.79	1.65	69.7
0.0	15	4.94	0.83	85.6
0.167	30	16.04	1.71	90.6
0.333	30	17.57	0.98	94.7
0.500	30	17.40	0.38	97.9
0.667	30	18.57	TR	>99.0

a) Glacial acetic acid, 30 ml, and 10% Pd-C, 0.1 gram, were charged to a 100 ml Fisher-Porter aerosol tube, flushed 3 times with propylene and activated under propylene for the designated time. A 65/35 $C_3^=/O_2$ mixture was then added and the reaction allowed to proceed for 5 hours at 65°C.
b) Data from reference 13.

TABLE 5. EFFECTS OF ADDITIVES AND ACTIVATION TEMPERATURE ON RATE AND SELECTIVITY OF PALLADIUM-CATALYZED OXIDATIONS OF PROPYLENE IN WATER[a)]

Activation T°,C	Reaction T°,C	Additive mmoles	Reaction Products[b)], mmol						TON[h]
			CO_2	Acetaldehyde	Acetic Acid	Acetone	Acrolein	Acrylic Acid	
80	65	--	2.3	0.5	0.8	0.8	0.2	12.9	3.2
80	50	--	0.7	0.2	0.5	0.8	0.3	7.7	1.9
80	40	--	0.4	0.1	0.4	0.4	0.1	8.3	1.2
40	40	--	--	--	--	--	--	--	--
50	50	--	--	--	0.3	0.3	--	--	--
65	65	--	2.3	0.4	0.	1.0	0.3	14.4	3.5
65	65	BHT[c)]	2.4	0.4	1.0	0.6	0.2	17.3	4.3
--[d)]	65	--	2.5	0.4	1.1	1.1	0.3	9.7	2.4
65	65	LiCl[e)]	na	--	-	3.4	--	0.1	--
65	--[f)]	--	0.25	--	0.3	0.7	0.1	0.9	--

a) One gram 10% Pd-C and 30 ml H_2O were added to an aerosol tube, flushed 3 times with propylene, heated under 50 psi of propylene at "activation T" for 30 minutes, then brought to reaction T. The propylene was replaced with 100 psi of a 60/40 $O_2/C_3^=$ gas mixture and stirred for 4 hours. As gas was taken up, the 60/40 mix was periodically added. b) After cooling the reaction mixture, the gas was captured and analyzed by GC. The liquid was also analyzed by GC using propionic acid as an internal standard. All volatile reaction products were analyzed and the amounts (mmoles) reported in the table above. c) The radical inhibitor, 2,6-di-tert-butyl-p-cresol, 0.1 g, was added to the reaction mixture. d) No activationperiod--60/40:$O_2/C_3^=$ was admitted initially. e) LiCl, 1M, was present in aqueous solution. f) Product analysis directly after activation period prior to adding oxygen. g) Data from reference 13. h) TON = Turnover No. mmoles Acrylic/g.atom Pd/hr.

In order to achieve high reaction rates and selectivities the supported palladium catalyst is first treated with propylene in the absence of oxygen. A stoichiometric reaction ensues, likerating acetone ($HY=H_2O$) and presumably, creating an active Pd(0) center eq. 20 which can oxidatively add propylene to form a π-allyl complex, eq. 21(13).

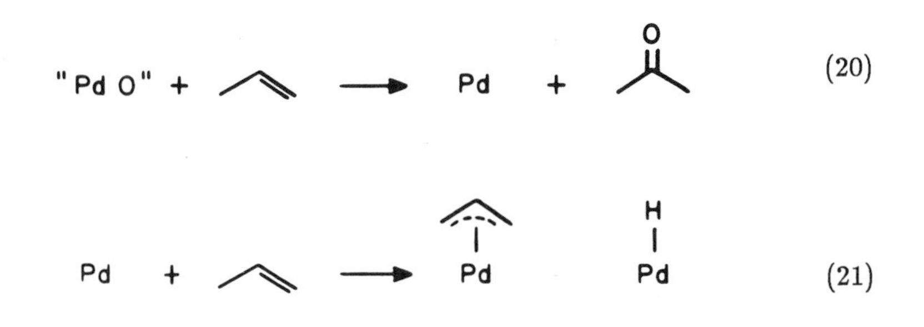

If the surface of the palladium catalyst is oxidized by contact either with a strong oxidant such as MnO_4^-, $S_2O_8^{=}$, NO_3^-, or even molecular oxygen at high temperature, prior to reaction, it is found that vinylic oxidation either competes or predominates over allylic oxidation. The presence of halide ion also promotes Wacker Chemistry to the virtual exclusion of allylic oxidation, Table 5. These observations suggest that the active catalyst for allylic oxidations may be low oxidation state palladium, and that conditions or reagents that cause higher oxidation state palladium to be formed on the catalyst surface lead to vinylic oxidations.

Thus, by properly activating the catalyst in water, by propene exposure prior to oxidation and by conducting the reaction in the presence of the radical inhibitor BHT which suppresses radical oxidation or polymerization of the reaction product, it is possible to smoothly oxidize propylene to acrylic acid in greater than 90% yield, eq. 22 (13).

Turnover numbers exceed 4 moles acrylic acid produced per gram atom of palladium per hour under these conditions. No leaching of catalyst into the aqueous phase was detected and a clear water-white solution was obtained on filtration which exhibited no activity toward oxidation under reaction conditions in the absence of added catalyst. Thus, palladium on carbon is an efficient, long-lived catalyst for the selective generation of acrylic acid from propylene in aqueous solution under very mild conditions.

$$CH_2{=}CH{-}CH_3 + \tfrac{3}{2}\,O_2 \xrightarrow[65^\circ C,\ 5\ BARS]{10\%\ Pd/C,\ BHT} \underset{>90\%}{CH_2{=}CH{-}CO_2H} \qquad (22)$$

In a similar manner, propylene was converted to allyl acetate, Table 4, and either cis- or trans-2-butene gave a mixture of branched and linear allylic acetates(13).

Butene-1, cis- and trans-butene-2 and isobutylene were oxidized over supported palladium catalysts, both in water as well as in acetic acid. Butenes gave crotonic acid in water but the *α,β*-unsaturated ketone and aldehyde were also major products, eq. 23. Isobutylene gave considerably more methacrolein than methacrylic acid, eq. 24. Perhaps the 2-methyl group causes desorption of the intermediate from the surface before it oxidizes further to the acid.

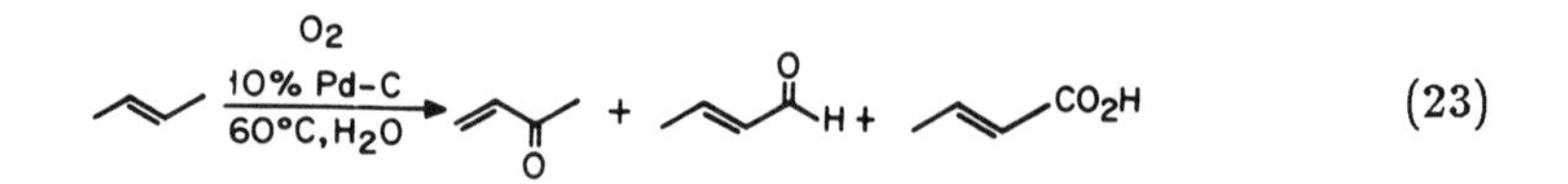

$$\text{isobutylene} \xrightarrow[60^\circ C,\ H_2O]{O_2,\ 10\%\ Pd\text{-}C} \text{CH}_2\text{=C(CH}_3\text{)-CHO} + \text{CH}_2\text{=C(CH}_3\text{)-CO}_2\text{H} \qquad (24)$$

In all cases it seems that although soluble palladium(II) complexes generally produce vinylic oxidation products, allylic oxidation products can be made to predominate under the following conditions: a) the presence of a weak base, b) highly electron withdrawing ligands on the palladium or c) the use of a supported metal catalyst capable of activating allylic C-H bonds, which is maintained in a low oxidation state.

OXIDATION OF AROMATIC SUBSTRATES

Just as the oxidation state of the palladium catalyst is important for alkene oxidations it is also a critical determinant of the reaction pathways of palladium-catalyzed oxidations of aromatic hydrocarbons. Palladium acetate catalyzes the oxidation of aromatics by strong oxidants in acetic acid to give aryl acetates and coupling products under mild conditions, eq. 25(24,25). It has been suggested that this reaction proceeds via electrophilic attack by $Pd(OAc)_2$ on the aromatic ring to give a palladium(II) aryl intermediate, eq. 26, which can either oxidize to give phenyl acetate or couple to give biphenyl (26,27). This mechanism is supported by the observations that: a) partial rate factors for oxidation of substituted aromatics are consistent with electrophilic palladation of the ring, b) isotope effects for forming both phenyl acetate and biphenyl are the same suggesting a common intermediate, and c) the phenyl acetate/biphenyl ratio is proportional to the concentration of strong oxidant used.

$$C_6H_6 \xrightarrow[\text{HOAc}]{[Ox],\ Pd(OAc)_2} C_6H_5OAc + C_6H_5{-}C_6H_5 \qquad (25)$$

$[Ox] = Cr_2O_7^{=}$, MnO_4^{-}, $S_2O_8^{=}$, etc.

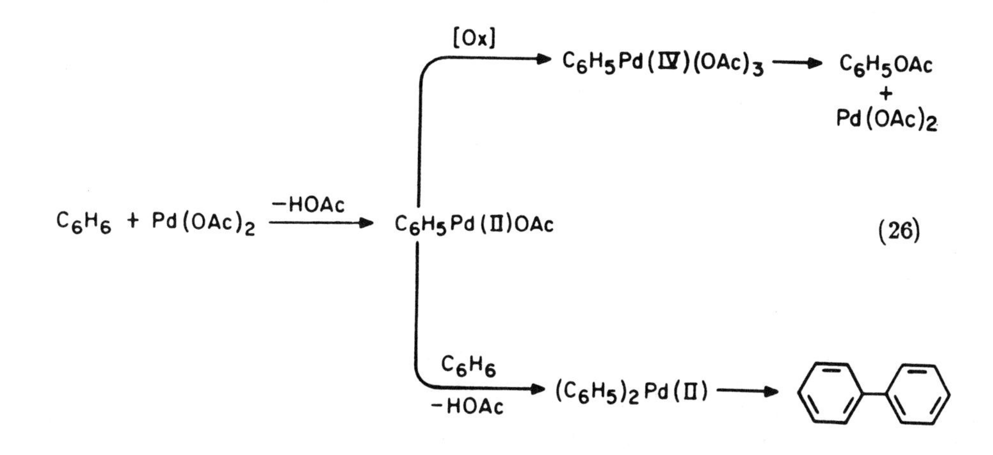

It would be of practical synthetic interest to carry out these reactions using air or oxygen as the oxidant rather than requiring stoichiometric consumption of expensive strong oxidants. Oxygen was successfully used as the oxidant for reaction 22 at high temperature but selectivity was poor(28). In this case both $Pd(OAc)_2$ and supported palladium catalysts were used.

Heteropolyacids, especially those containing vanadium are known to be rather strong oxidizing agents(29). The oxidation potential rises as the vanadium content increases and thus heteropolyacids such as $H_9PMo_6V_6O_{40}$ and $H_{11}PMo_4V_8O_{40}$ are good candidates for promoting palladium catalyzed oxidation of aromatics(29). Not only might they be able to oxidize Pd(II) aryls, but the reduced form can be re-oxidized with oxygen in aqueous solution eq. 27. Therefore they might provide a catalytic cycle for air oxidation of aromatics under mild conditions.

$$H_{3+n}PMo_{12-n}V_nO_{40} + Red + iH^+ \longrightarrow H_{i+3+n}PMo_{12-n}V_nO_{40} + Ox \quad (27)$$

(regenerated by $1/2\ O_2$, $-H_2O$)

Table 6 shows the attempt to use the heteropolyacid, $H_{11}PMo_4V_8O_{40}$ as a strong oxidant which is regenerable with molecular oxygen for the direct palladium catalyzed oxidation of benzene under mild conditions. It can be seen from this data that, although some ring oxidation occurs, selectivity is poor unless excess sodium acetate is added. Added acetate does not increase the yield of ring oxidation product, but rather inhibits the production of biphenyl. This may be due to increasing the coordinative saturation about the palladium which could have a far greater retarding effect on dimerization than oxidation eq. 28-30.

$$ArPdOAc + 2KOAc \rightleftharpoons K_2Pd(Ar)(OAc)_3 \quad (28)$$

$$K_2PdAr(OAc)_3 \xrightarrow{[Ox]} K_2Pd(OAc)_4 + ArOAc \quad (29)$$

$$K_2PdAr(OAc)_3 \xrightarrow{ArH} K_2Pd(Ar)_2(OAc)_2 + HOAc \quad (30)$$

In any case, in the absence of the heteropolyacid, only a small (sub-stoichiometric) amount of biphenyl was formed and no phenol or phenyl acetate was detected. Ring oxidation required the strong oxidant. Thus, it appears that in order for ring oxidation to occur, the palladium(II) aryl intermediate must be oxidized. Strong oxidants or forcing conditions are therefore necessary.

Little work has been done on the acetoxylation of substituted aromatics using molecular oxygen as the oxidant. We have explored the palladium-catalyzed oxidation of phenyl acetate as a route to derivatives of the industrially important dihydroxyaromatics: catechol, resorcinol and hydroquinone. We were interested in finding conditions under which O_2 could be used as the oxidant and in determining those factors which controlled acetoxylation and those which controlled dimerization.

Palladium acetate was found to catalyze the oxidation of phenyl acetate at 145°C using 800 psig of an oxygen-containing gas, eq. 31. Phenylene diacetates and diacetoxy biphenyls were formed as major reaction products. Although the three possible phenylene diacetates were formed, only two of the six possible diacetoxybiphenyls were produced in significant quantities: o,m'-diacetoxybiphenyl and o,p'-diacetoxybiphenyl.

TABLE 6. OXIDATION OF BENZENE USING PALLADIUM CATALYSTS PROMOTED WITH A HETEROPOLYACID[a]

Catalyst(mmoles)	NaOAc mmoles	Ring Oxidation [PhOH+PhOAc] mmoles	Biphenyl mmoles	Ring Oxid'n Selec.,%
$Pd(O_2CCH_3)_2$(0.44)	0	2.05	4.10	33
	12	1.98	1.37	59
	36	1.51	0.17	90
$Pd(O_2CCF_3)_2$(0.30)	36	1.58	0.23	87
$PdCl_2$ (0.56)	36	1.27	1.38	52
Pd(HPA-8) (0.48)[b]	36	2.63	0.30	90

a) Benzene, 50 mmoles, was added to a solution of the Pd complex and 0.5 mmoles of $H_{11}PMo_4V_8O_{40}$ in 3 ml water and 25 ml acetic acid. The solution was stirred at 120°C under oxygen, 100 psi, for 6 hours.

b) $PdH_9PMo_4V_8O_{40}$ was the catalyst; $H_{11}PMo_4V_8O_{40}$ was not added to this run.

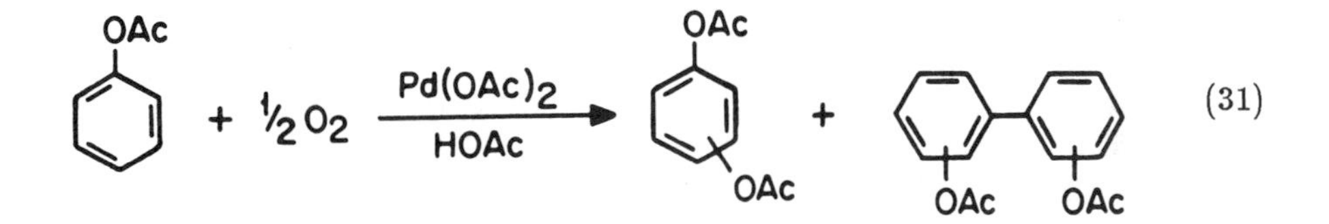

(31)

As expected, ring oxidation increased both with temperature and oxygen partial pressure. It is interesting to compare this pressure effect with the increase in ring acetoxylation on increasing oxidant concentration in the $Pd(OAc)_2$ catalyzed oxidation of benzene in acetic acid, Table 7. The fact that higher temperatures and greater oxygen partial pressures favor acetoxylation relative to oxidative dimerization is consistent with the requirement that palladium be in a higher oxidation state for acetoxylation than for oxidative dimerization in cases where molecular oxygen is the oxidant.

Just as was found in the case of benzene oxidations, the presence of alkali acetate had little effect on the rate of acetoxylation at low partial pressure of oxygen, but had a pronounced inhibiting effect on oxidative dimerization.

TABLE 7. PRODUCT DISTRIBUTIONS FOR THE $[Pd(OAc)_2]$-CATALYZED OXIDATION OF BENZENE BY DICHROMATE[a] AND PHENYL ACETATE BY OXYGEN[b]

Benzene Oxidation			Phenyl Acetate Oxidation		
$Na_2Cr_2O_7$, mmole	% Yield[c]		O_2 Partial Pressure, psig	% Yield[c]	
	C_6H_5OAc	$(C_6H_5)_2$		$C_6H_4(OAc)_2$	$(C_6H_4OAc)_2$
0	<1	22			
1	99	23	32	223	1953
3	190	33	168	726	1750
7.5	353	19	320	927	1747

a) Reactions conducted at 90°C for 16 hours using benzene (56 mmoles) and $[Pd(OAc)_2]$ (1 mmole) in in acetic acid (25 ml). Ref. 18.

b) Reactions conducted at 145°C for 3 hours under conditions given in Table 6.

c) Calculated on the basis of $[Pd(OAc)_2]$.

Thus, those conditions which favor acetoxylation over oxidative dimerization are high temperature, high oxygen partial pressure, high acetate ion concentration, and low phenyl acetate concentration. Under these conditions, (175°C, 320 psi O_2, 0.78M PhOAc) selective acetoxylation occurs. The molar ratio of the phenylene diacetates to biphenyl diacetates formed under these conditions was 94.5/1. Unfortunately, the low phenyl acetate concentration, coupled with high CsOAc concentration, depressed the overall rate to a rather low level.

Supported Palladium Catalyst for Aromatic Oxidation

Since benzene can be oxidized in acetic acid using both homogeneous and heterogeneous palladium catalysts(28), we compared supported palladium catalysts to $Pd(OAc)_2$ for oxidation of phenyl acetate. Supported catalysts required temperatures of >175°C for significant activity using O_2 as the oxidant. At these temperatures carbon was not a suitable support. Palladium (5%) on alumina catalyzed oxidation at high temperature (175°C) and high oxygen partial pressure (320 psig) to give predominant ring acetoxylation. Even at high temperatures over three-hour reaction periods, phenyl acetate conversion was low (5%); however, regioselectivity was considerably different from the homogeneous system. The major reaction product over 5% Pd/Al_2O_3 was p-phenylene diacetate, eq. 32, with very little o- and only traces of m- isomer.

$$\text{AcO}-\text{C}_6\text{H}_5 + O_2 \xrightarrow[\substack{175°\text{C}, 3\text{ hrs} \\ 500\text{ psi} \\ 40\%\ O_2 \text{ in } N_2}]{\substack{Pd/Al_2O_3 \\ \text{HOAc}}} \text{AcO}-\text{C}_6\text{H}_4-\text{OAc} \qquad (32)$$

It has long been known that $Pd(OAc)_2$ catalyzes the oxidation of toluene in acetic acid to give benzylic oxidation products(30). The catalyst appears to be more active when reactions are run in the presence of activated carbon(30). Low oxidation state palladium has been implicated in benzylic oxidations(31). We have found that palladium on carbon is effective for benzylic oxidations under mild conditions, eq. 32, under which little or no ring oxidation occurs. This encourages one to speculate, in close analogy with olefinic systems, that over supported metal catalysts, higher oxidation state palladium species lead to ring oxidation, whereas lower oxidation state species are responsible for benzylic oxidation.

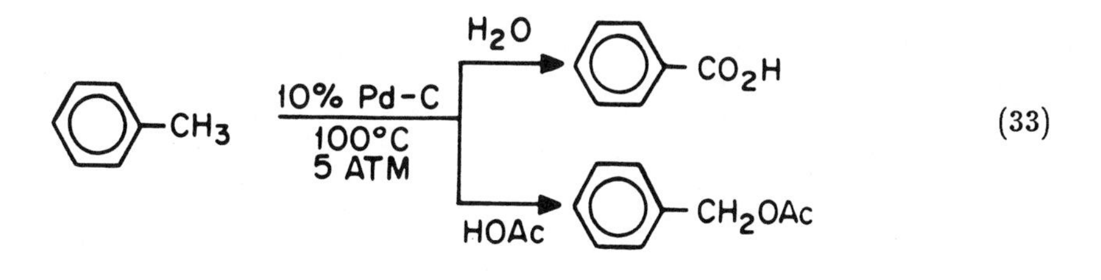

CONCLUSIONS

Although Pd(0) complexes can react readily with molecular oxygen, the peroxo complexes which form are not reactive toward un-activated olefins. Although olefins which are activated by electron withdrawing substituents react with Pd(II) peroxo complexes, generation of products in a catalytic manner does not occur. Palladium(II) nitro complexes catalyze the oxidation of olefins but it is the nitrogen atom and not the palladium that actually activates oxygen. Synthetic utility of these reactions may lie in the oxidation of higher molecular weight olefins to ketones or epoxides.

Nucleophilic attack on a Pd(II) π-complex gives an intermediate which can lead to a vinylic oxidation product or its stable isomer. If carried out in the presence of a suitable oxidant which is regenerable in air, the reaction is catalytic. Allylic oxidation products also arise via palladium catalyzed oxidation of olefins, but these catalytic reactions have not been extensively investigated. Several factors favor the allylic oxidation of olefins over vinylic oxidation. In homogeneous solutions of Pd(II) π-complexes, weak inorganic bases aid formation of π-allyl complexes which give rise to allylic oxidation products. Strongly electron withdrawing anionic ligands promote oxidative addition of C-H bonds of coordinated olefin to give π-allyl intermediates which can undergo nucleophilic attack to produce allylic, rather than vinylic oxidation products. Supported palladium catalysts which can be maintained in a low oxidation state appear to oxidatively add α-olefins to produce π-allyl surface species and give allylic oxidation products, whereas the addition of strong oxidants, chloride ion or use of forcing conditions promotes formation of vinylic oxidation products over these catalysts.

Palladium catalyzed aromatic oxidations proceed in two distinct directions--ring oxidation or oxidative coupling. In solution, oxidative coupling seems to be favored by low oxidation state [Pd(II) $\rightleftharpoons$ Pd(0)] catalysis, whereas ring oxidation is favored by high oxidation state [Pd(IV) $\rightleftharpoons$ Pd(II)] systems. Alkyl aromatic compounds may undergo competitive oxidation at the alkyl side chain. Benzylic oxidation is favorred by low oxidation state palladium.

By applying oxidation state considerations it is possible to direct both the oxidation of olefins and aromatic hydrocarbons in synthetically useful directions. Although much more refinement is needed, it is becoming possible to direct olefins along new and interesting allylic oxidation pathways and to promote selective aromatic oxidations using palladium catalysts.

REFERENCES

1. P. M. Henry, "Palladium Catalyzed Oxidation of Hydrocarbons", D. Reidel Publishing Company, Boston, (1980).
2. P. M. Maitlis, "The Organic Chemistry of Palladium, Vol. II Catalytic Reactions", Academic Press, New York, (1971).
3. S. Otsuka, A. Nakamura and Y. Tatsuno, Chem. Commun. 836 (1967).
4. S. Otsuka, A. Nakamura and Y. Tatsuno, J. Amer. Chem. Soc., 91: 6994 (1969).
5. J. J. Levison and S. D. Robinson, J. Chem. Soc. A., 762 (1971)
6. R. A. Sheldon and J. A. Van Doorn, J. Organometal. Chem., 94:115 (1975).
7. F. Igersheim and H. Mimoun, Nov. J. Chim., 4:711 (1980).

8. M. Roussel and H. Mimoun, J. Org. Chem. 45:5381 (1980).
9. M. Andrews and K. Kelley, J. Amer. Chem. Soc., 103:2894 (1981).
10. J. D. Solar, F. Mares, and S. E. Diamond, Catal. Rev. Sci. Eng., 27:1 (1985).
11. J. E. Lyons, "Applied Industrial Catalyst - Vol. 3" B. E. Leach, Ed., Academic Press, New York, 1984, p. 134.
12. E. Stern, Catal. Rev., 1:73 (1968).
13. J. E. Lyons, "Homogeneous and Heterogeneous Catalysis" Yu. Yermakov and V. Likholobov, Eds., VNU Science Press, Utrecht, The Netherlands, (1986) pp 117-138.
14. A. D. Ketley and J. Braatz, Chem. Commun., 169 (1968).
15. S. Winstein, J. McCaskie, H.-B. Lee and P. M. Henry, J. Am.Chem. Soc., 98:6913 (1978).
16. I. P. Stolyarov, M. N. Vargaftik, O. M. Nefedov and I. I. Moiseev, Kinetika i Kataliz, 23:376 (1982).
17. W. Swodenk and G. Scharfe, U.S. Patent 3,925,452 (1975).
18. B. M. Trost and P. J. Metzner, J. Am. Chem. Soc., 102:3572 (1980).
19. L. Saussine, J.-P. Laloz and H. Mimoun, French Patent 2,450,802 (1980).
20. T. Seiyama, N. Yamazoe, J. Hojo and M. Hayakawa, J. Catal., 739 (1974).
21. K. Fujimoto and T. Kunugi, J. Jpn. Petrol. Inst., 17:739 (1974).
22. R. David and J. Estienne, U.S. Patent 3,624,147 (1971).
23. J. A. Hinnenkamp, U.S. Patent 4,435,598 (1984).
24. P. M. Henry, J. Org. Chem., 36:1886 (1971).
25. a) L. Eberson and L. Gomez-Gonzalez, Act. Chem. Scand., 27:1162 (1973); 27:1249 (1973); 27:1253 (1973). b) L. Eberson and L. Jonsson, Act. Chem. Scand., 28:597 (1974); 28:771 (1974); 30:361 (1976). c) L. Eberson and L. Jonsson, Act. Chem. Scand., 28:771 (1974); Liebigs Ann. Chem., 233 (1977).
26. L. M. Stock, K. Tse, L. J. Vorvick and S. A. Walstrum, J. Org. Chem., 46:1759 (1981).
27. a) R. Van Helden and G. Verberg, Rec. Trav. Chim., 84:1263 (1965). b) M. O. Unger and R. A. Fouty, J. Org. Chem., 34:18 (1969). c) F. R. S. Clark, R. O. C. Norman, C. B. Thomas and J. S. Willson, J.C.S. Perkin I, 1289 (1974). d) H. Iataaki and H. Yoshimoto, J. Org. Chem., 38:76 (1973). e) H. Itatani and H. Yoshimoto, Chem. Ind., 674 (1971). f) A. I. Rudenkov, G. U. Mennenga, L. N. Rachkovskaya, K. I. Matveev and I. V. Kozhevnikov, Kinet. i Katal., 18:915 (1977). g) Y. Itahara, Chem. Ind., 330 (1982).
28. V.H.-J. Arpe and L. Hornig, Erdol und Kohle, 23:79 (1970).
29. I. V. Kozhenikov and K. I. Matveev, Applied Catalysis, 5:135 (1985).
30. D. R. Bryant, J. E. McKeon and B. C. Ream, Tetrahedron Lett., 3371 (1968); J. Org. Chem., 33:4123 (1968).
31. M. K. Starchevskii, M. N. Vargaftik and I. I. Moiseev, Kinetika i Kataliz, 20:1163 (1979).

OXYGEN ACTIVATION AND OXIDATION REACTIONS

ON METAL SURFACES

Robert J. Madix

Department of Chemical Engineering
Stanford University
Stanford, CA 94305

Over the years we have developed ways of dissecting the kinetics and mechanisms of increasingly complex reactions using metal single crystals in ultra high vacuum as model reaction systems, and I would like to discuss that work today, particularly with regard to oxidation processes. The strengths of this approach primarily lie in the ability to dissect mechanistic processes and determine the activation energies, frequency factors, and rate-limiting steps for reactions of metastable intermediates. The major difficulties arise in attaching relationships between the structure of these intermediates to their reactivity; structure of surface species is currently very difficult to determine. We do not have the equivalent of NMR, so it is painstaking, as I think you will see, to begin to understand the relationship between the disposition of bonds with respect to the metal atoms on the surface and the reactions taking place. We are, however, beginning to make some progress there, as well, and I find that most exciting.

Oxidation reactions on surfaces can, in principle, proceed via reactions with either adsorbed dioxygen or atomic oxygen. There is relatively little work that has been done with molecular oxygen on metal surfaces. On several metal surfaces, states of dioxygen have been identified at temperatures near 100 K [1-8]. On silver at lower temperatures a very weakly perturbed form of dioxygen forms with very high collisional efficiencies [4]; it is effectively a physically absorbed species with a very weak binding energy, and the moment of inertia of the molecule as measured by rotational spectroscopy is essentially the same as the gas phase. This species is very weakly bound and probably has no chemical significance except that it may act as precursor for dissociation, in which a molecule colliding with the surface becomes trapped in this state and hops around the surface before ultimately dissociating.

At approximately 100 K, it is possible to form a state of dioxygen on palladium, platinum and silver in which the O-O bond remains intact, as shown by oxygen isotope exchange experiments. This surface dioxygen complex has a very low O-O stretching frequency compared to the gas phase [2,3, 5,8]. Clearly electron charge transfer from the metal into the species occurs which strongly reduces the O-O bond strength. In fact, the frequencies correlate with a bond order of roughly one on Ag(110) and 1.5 on Pt(111); the bond O-O order is thus substantially lowered in these surface complexes.

On the more reactive metals, e.g., palladium, in order to form this dioxygen state, one must first dissociate the dioxygen; in the presence of adsorbed atomic oxygen the dioxygen species is stable [8]. All of these species show a relatively low activation energy for conversion into the dissociated form or for desorption. This value is about 8 kilo calories, and the complexes are, therefore, unstable above 140 K in ultra high vacuum. It is notable that to date, although there have been very few reactions tried, no reactions of these species have been observed for a variety of molecules, including formic acid and ethylene [9]. They seem to be relatively unreactive species. There is an experimental problem with such studies, however, because the dioxygen species is only stable up to 140 K and these reactions must therefore be conducted at very low temperature. Apparently there is an activation barrier for the reaction of dioxygen so that reactivity cannot be accessed under normal laboratory conditions with room temperature gases and a cold surface in ultra high vacuum. One simple displacement reaction has been studied; namely, the displacement of O_2 by carbon monoxide on Pd(100), which is more strongly held than the dioxygen [7]. This displacement occurs at 80 K, again reflective of the very weak surface bond energy of the dioxygen species.

It is of interest to understand as much as possible about the structure of the species. Dioxygen has been proposed as the unique intermediate for the selective oxidation of ethylene to ethylene oxide on silver surfaces. The mechanism for this reaction is still not understood. This question has been discussed in the literature for years, and the issue is whether dioxygen or atomic oxygen, or both, is responsible for the selective oxidation. The answer is not yet known. Since adsorbed dioxygen may be an important reaction intermediate, I want to present the results of recent studies of the structure of dioxygen species on Ag(110) and Pt(111) utilizing synchrotron radiation [10]. Synchrotron radiation is, of course, tuneable in energy; a soft X-ray photon beam can be tuned over a rather wide range. Further, the electric field vector of the light is fixed in space, since it is a linearly polarized source, and the relative orientation of the electric field vector and the surface can be varied at will. When the electric field vector points along the internuclear axis, electronic transitions from the oxygen 1s core level to σ^* orbitals are produced at an energy which is characteristic of the bond length. Furthermore transitions from the carbon 1s core level to the π^* system occur when the electric field vector is perpendicular to the internuclear axis. These effects can be used to probe the orientation of the internuclear axis to obtain the bond length and to see whether or not the π^* orbitals are fully occupied [11.12]. If the adsorbed dioxygen posesses an O-O single bond due to charge transfer from the metal to fill the π^* orbitals, a transition to the π^* orbitals would not be observed; only a σ^* resonance is expected. The feature in the spectrum would be due solely to a transition from the oxygen 1s to the σ^* along the internuclear axis. This method is referred to as near edge X-ray absorption fine structure (NEXAFS).

The silver(110) surface is a surface with close-packed rows of silver atoms in one direction separated by a space. The surface looks like a plowed field near Texas A&M with close-packed rows of atoms running in one direction with furrows between. This arrangement is shown schematically in figure 1, which gives the NEXAFS results for dioxygen on Ag(110) for several orientations of the electric field vector of the light with respect to the surface [10]. When the electric field vector, E, is in the plane of the surface, θ is 90°; when it is nearly perpendicular to the surface it is 10°. The transition observed is from O(1s) to σ^*. In the 10° spectra with E along the close-packed direction there is essentially no resonance, and, therefore, no transition occurs to the σ^* orbital along the internuclear axis. In some cases shown there is a small transition observed when the electric field vector is perpendicular to the surface (θ = 90°). However, the largest transition intensity occurs with the electric field vector in the plane of the surface pointing along the close-packed direction. The small peaks shown at different orientations of the electric field vector are due to non-perfect polarization of the photon beam; that is, there is a small component of the field perpendicular to the primary component; there is not 100% polarization. The conclusion is that the dioxygen species lies in a plane parallel to the plane of the Ag(110) surface with its

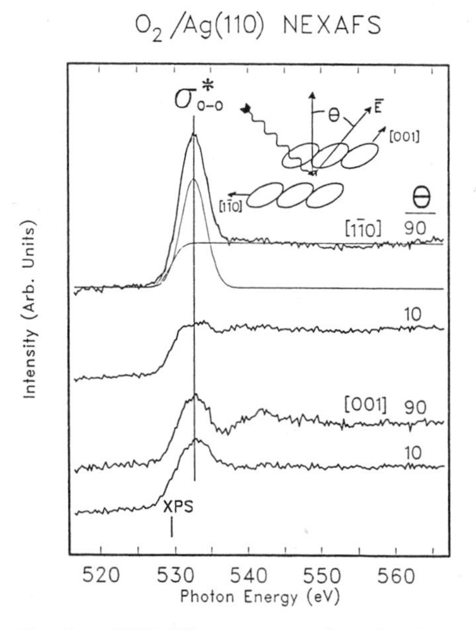

Figure 1. Oxygen K-edge NEXAFS spectra for O_2 on Ag(110) at 90 K as a function of polar and azimuthal E orientations. The O-O σ^* peak at 532.6 eV is strongest when E lies along the O-O bond direction which occurs when E is along the [1$\bar{1}$0] asimuth and parallel to the surface (θ = 90°). The line at 529.3 eV marks the O(1s) binding energy relative to the Fermi level for O_2 on Ag(110).

internuclear axis along the close-packed direction, most likely sitting in the furrows. The O-O bond distance can be obtained by correlating the transition energy with that observed for dioxygen and peroxy compounds in the gas phase [10]. The transition energy clearly shows that this is an O-O single bond, very much like peroxides. The absence of a transition to the π^* system is consistent with the π^* orbitals being full. For the dioxygen species on platinum, however, there is a sharp π^* transition, and the position of the σ resonance indicates a bond order of about 1.5 [10].

I would now like to turn my attention to the reactions of atomic oxygen on surfaces and to the general state of understanding of the mechanism of heterogeneous oxidation reactions. I will rely considerably on a temperature programmed method that we have developed which determines the reaction energetics. The concept is really very straightforward [13]. Initially the surface is held at some initial temperature, T_o, and a reactant is adsorbed. T_o is intentionally lower than the temperature of the reaction to be studied. As the surface is then heated linearly in time, temperatures are reached at which certain reaction channels "ignite." Basically, the evolution of products from the surface due to the reaction rates for different reactions is observable at different temperatures as the temperature is ramped, since they possess different rate constants, and the products resulting from different elementary reactions are separated in time by the programmed heating. With the crystal in the evacuated region, the products are observed directly in line of sight with a mass spectrometer. A quadrupole mass spectrometer is driven and processed by a microcomputer to multiplex up to 200 masses simultaneously with a one second time resolution to record all the products during the temperature ramp. This multiplexing is essential for reactions of any complexity, because a cracking fraction analysis must be done to identify the separate products. The temperature at which products appear must then be compared to the _desorption_ temperature expected for that molecular species itself in order to discern whether the rate of reaction _on_ the surface or the simple detatchment of the product molecule itself _from_ the surface is being measured. Let us suppose, for example, that a reaction product being observed is carbon monoxide. The temperature at which it appears from the reaction must be compared to the temperature at which it desorbs if it itself is adsorbed alone. In the latter experiment, the binding energy of carbon monoxide to the surface is measured, whereas in the former experiment the activation for its appearance due to reaction is determined. If the carbon monoxide evolves at a temperature higher than expected from its characteristic binding energy with the surface, the rate of a surface reaction, and not simply the rate of its desorption, is being measured [14].

Obviously, isotopically labelled molecules can be employed to reveal mechanistic details. For example, with an oxygen-deuterated phenol (OD), deuterium and hydrogen would be evolved at different temperatures, each of these being characteristic of a different reaction channel. For example, the rate of D_2 evolution could be characteristic of the barrier to O-D bond breaking or of the D atom recombination on the surface to form D_2 subsequent to that bond breaking. The temperature of H_2 evolution could correspond to C-H bond breaking. By quantitatively determining the products in each channel the surface species can be identified and then studied with a variety of electron spectroscopies to refine the understanding of their structure and bonding.

An example that is now well understood is illustrated in figure 2. Deuterated formic acid was deposited on a copper(110) surface at approximately 150 K and heated, and the temperature programmed reaction spectrum is shown here [15]. The first species to appear is the formic acid molecule which simply desorbs intact. This peak temperature can be used to calculate the binding energy of that species to the surface. The next product to appear is hydrogen (H_2) which is the result of a desorption-limited step; in other words, the rate at which it evolves is determined by recombination of hydrogen atoms, not O-H bond cleavage. The rate of H_2 evolution from DCOOH is identical to that obtained if H atoms alone are adsorbed on the surface. Knowing this step we can clearly deduce that initially formic acid was activated by O-H bond breaking to leave an intermediate on the surface. At higher temperatures (475 K) a reaction channel is apparent in which solely deuterium and CO_2 are evolved. These two products have exactly the same peak shape and peak position, signifying they have precisely the same rate vs. temperature behavior. They are precisely simultaneously evolved, and they must come from a common rate-limiting step. From that fact and by quantitatively determining the amount of these two products formed, it can be very straightforwardly deduced that the intermediate is a surface formate. For the spectroscopists who doubt the power of such techniques, other spectroscopies can be used to interrogate the nature of the intermediate. Also shown in the figure is the vibrational spectrum which was taken after annealing the surface to about 400 K [16]. The spectra clearly show the C-H stretch, the symmetric O-C-O stretch, the bending mode for the O-C-O linkage and the vibration of the surface bond between copper and oxygen. The asymmetric stretch is very small because in electron energy loss vibrational spectroscopy dipole moments parallel to the surface are screened, and consequently the asymmetric stretch parallels the surface. The general orientation of the formate intermediate on the surface can be deduced from such information; it is a bidentate rather than a monodentate form. A significant amount can be learned about the nature of surface reaction intermediates by combining these methods. It must be emphasized that the temperature of these peaks allows one to directly calculate the activation energy and frequency factors for all the reaction-limiting steps.

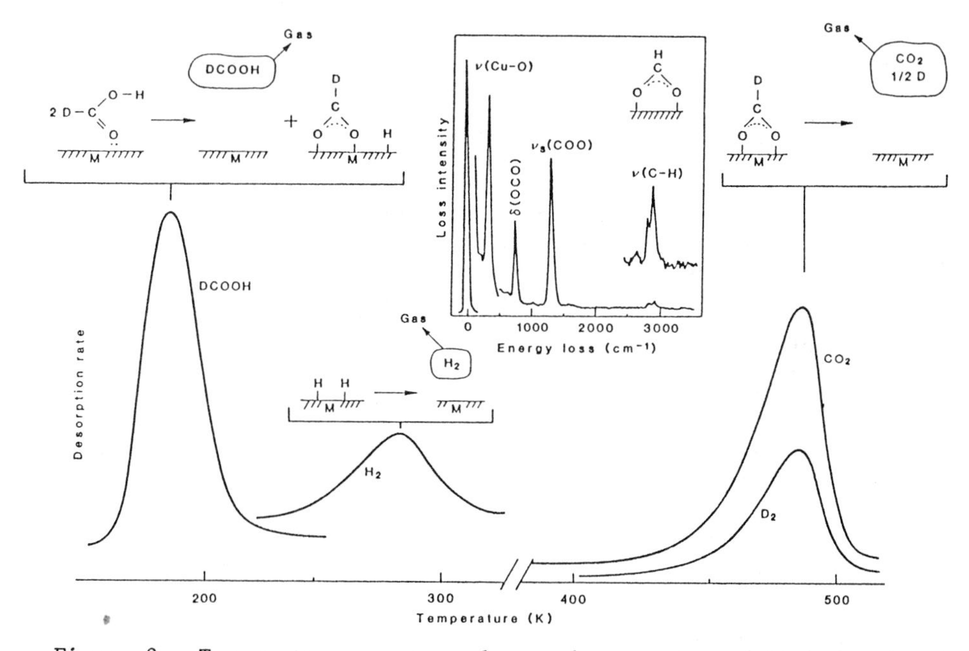

Figure 2. Temperature programmed reaction spectrum (TPRS) for formic acid reaction with Cu(110). Each step in the reaction DCOOH = $H_{(a)}$ + $D_{(a)}$ + CO_2 is revealed by product evolution into the evacuated gas phase as the surface complex is heated. Simultaneous formation of CO_2 and D_2 at 480 K is clear indication of the $DCOO_{(a)}$ intermediate. The vibrational spectrum for the formate existing between 300 to 480 K is shown in the inset. The absence of the asymmetric O-C-O stretch indicates bidentate or chelating bonding to the copper atoms on the surface. The combination of these spectroscopies makes it possible to determine both chemical and structural identity.

Considering oxidation reactions on metals at relatively modest temperatures, it is nearly obvious that metals which form strong bonds with oxygen are unlikely to be effective oxygen transfer agents, since in these cases the oxygen would prefer to bind to the metal to form an oxide. Rather, surfaces which reversibly bind oxygen, in other words, that dissociate the molecule but give it up via recombination relatively easily, are preferable. For example, cobalt, nickel and copper all form very stable oxides. The oxides of rhodium, palladium, and silver become less stable in the direction from left to right in the periodic table, and similarly for iridium, platinum and gold from left to right the oxides are less stable. The oxygen atoms bound to single crystals of these materials in ultra high vacuum yield dioxygen by recombination and the respective temperatures for this reaction agree roughly with the temperatures at which the corresponding bulk oxides decompose. Thus, the bulk oxide stability can be used as a rough measure of the reversibility of oxygen adsorption on metals. The metals that exhibit significant reversibility are useful as oxidation catalysts; platinum and palladium are extremely useful for complete oxidation. Platinum is used in the auto exhaust emissions control muffler systems, because complete combustion is desired, whereas silver is commonly used for with partial and selective oxidation. Silver has the weakest oxygen-metal bond energy except for that of gold.

It is possible to identify general chemical properties of adsorbed atomic oxygen on metals and to relate these properties to the mechanism of oxidation [13,17,18]. Basically, oxygen has three very important characteristics on silver. First of all it acts very much like a strong base. In general, consider a protypical hydrogen-containing molecule, BH, as a gas phase acid. Atomic oxygen on silver very readily accepts the proton from the acid to form hydroxyl groups and the adsorbed conjugate base. There is also definite charge transfer from the metal to the adsorbed conjugate base; they are not neutral. This generic class of reactions is an absolute predictor of activation of a variety of substrates.

The second important property is that adsorbed atomic oxygen is a strong nucleophile; it attacks electron deficient centers in molecules to form the corresponding intermediate [19,20]. For example, aldehydes are attacked to form $RCHO_2$. These intermediates are very metastable, and the C-H bond is very facilely broken. Through this mechanism surface carboxylates are formed from aldehydes; in fact, this reaction is partially responsible for degradation of selective oxidation processes on silver when the desired product is an aldehyde.

The third property, which is somewhat more subtle but has a very important effect in directing selective oxidations, is that oxygen imbues the neighboring metal atom with a Lewis acidity [21,22]. The adsorbed oxygen is, of course, electronegative, withdrawing electrons from its immediate vicinity. At the silver atoms neighboring the oxygen, the electron charge density is somewhat lower than it would be were the oxygen not there. This has the effect of increasing the bonding energy of electron donors to the metal in the vicinity of oxygen. Quite often the gas phase acids referred to earlier possesses an electron-rich center which helps direct the gas phase acid onto a binding site that is favorable for direct proton transfer to the surface oxygen. These two acidity properties work in a cooperative way to produce very selective proton transfer. In the absence of the oxygen, hydrogen is not be transferred to the surface; the oxygen is the activator for these reactions.

The oxidation of acetonitrile on Ag(110) is an interesting example of the acid/base properties of oxygen [14]. This study was motivated by our desire to see whether or not we could abstract protons from methyl groups via adsorbed oxygen atoms. This reaction does not occur with ethane since ethane is a very bad gas phase acid, whereas acetonitrile is a rather strong gas phase acid because of the presence of the nitrile group. The oxygen activity for activation of the methyl grous in ethane and acetonitrile thus should offer an interesting contrast if the gas phase acidity is the important correlating property for reactivity.

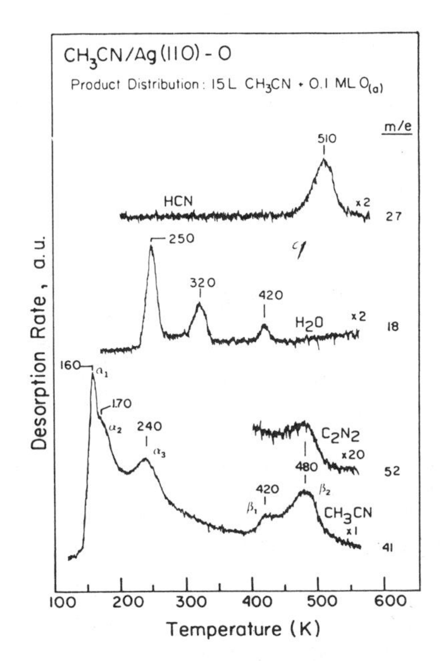

Figure 3. TPR spectrum for a 15 L dose of CH_2CN at 140 K to the Ag(110) surface covered with 0.1 ML oxygen atoms. Products were CH_3CN (m/e = 41), H_2O (m/e = 18), and HCN (m/e = 27). A heating rate of 5 K s^{-1} was used.

An Ag(110) surface was covered with one tenth of a monolayer of oxygen and then enough acetonitrile was deposited at 100 K to actually build up two or three layers condensed on the surface. The evolution of different products was then followed while heating (figure 3). There are clearly several channels for the evolution of acetonitrile from the surface. At low temperature molecular acetonitrile is evolved (the α and α_2 states) and no bonds have been broken, it simply desorbs when heated. This α_3 state is a molecular state of acetonitrile which is stabilized by the presence of surface oxygen and is indicative of the Lewis acid inducing capacity of atomic oxygen. It produces a stronger binding state for the acetonitrile, presumably due to electron donation from the π-system. At higher temperatures a number of products form. The water peaks are indicative of the extent of the initial reaction between acetonitrile and the surface oxygen. The feature at

250 K is the result of desorption of molecular water and signifies that initially acetonitrile reacts with surface oxygen to form molecular water on the surface. There is another water peak at 320 K which is due to hydroxyl disproportionation [23] indicative of direct proton transfer. Water is also evolved at 420 K with some acetonitrile due to reaction with oxygen left behind in hydroxyl disproportionation. HCN is formed at 510 K as follows. First partial dehydrogenation of the methyl group occurs to form a $-CH_2CN$ which disproportionates to make HCN and CH_3CN near 500 K. It is clearly easy to activate that methyl C-H bond by preabsorbed oxygen whereas that is not possible with ethane. In detail, the $-CH_2CN$ dehydrogenates at 480 K to form =CHCN and adsorbed hydrogen atoms. The H atoms react with $-CH_2CN$ to form CH_3CN. The CHCN reacts to form HCN and adsorbed carbon atoms. This mechanism is verifiable with vibrational spectroscopy and acid displacement reactions. In the vibrational spectrum the methylene CH_2 wag appears as a very strong vibrational loss in the spectrum. In fact it is so large that it suggests that the CN group is lying parallel to the surface with the two C-H bonds protruding at an oblique angle from the surface.

The surface $-CH_2CN$ group can be isolated by exposing the oxygen predosed surface to acetonitrile and then heating to 350 K to drive off the molecular acetonitrile and all of the water. It can then be exposed to deuterated formic acid. The carboxylic acid is a stronger acid than is acetonitrile and deuteron transfer from the carboxylic acid to the surface intermediate occurs. Heating the surface then yields the singly deuterated molecular acetonitrile and the formate; the formate actually falls apart to give CO_2 at 410 K, a well characterized reaction. It is clear that the intermediate is displaced by the stronger acid and that it is $-CH_2CN$.

Table 1. Relative stabilities of surface intermediates as determined by the displacement reaction $B_{(a)} + B'H_{(g)} + B'_{(a)}$ following formation of $B_{(a)}$ by quantitative titration of $O_{(a)}$ by $BH_{(g)}$; $pK_{(a)}$ negative logarithm of the acidity constant.

Order of Stability of $B_{(a)}$ on Ag(110)	$B_{(a)}$	$BH_{(g)}$	H^o_{acid} (gas phase) (kcal/mol)	pK_a	D^o(B-H) (kcal/mol)	Identifying products in TPRS (characteristic temperature, K)
1, 2	HCOO	HCOOH	345.2	3.7	112	CO_2 (420)
	CH_3COO	CH_3COOH	348.5	4.8	112	CO_2 (650)
3	C_2H_5O	C_2H_5OH	376.1	17	104	CH_3CHO (275)
4	C_2H	C_2H_2	375.4	26	120	C_2H_2 (270)
5	CH_3O	CH_3OH	379.2	15.5	104	H_2CO (300)
6, 7	C_3H_5	C_3H_6	390.8	35	89	
	OH	H_2O	390.8	15.7	119	H_2O (320)

Titrations of this sort are stoichiometric displacements of a weaker acid by a stronger acid. Generally, one can utilize a wide variety of these gas phase acids, including amines, to make corresponding surface intermediates by selective oxygen activation of that species. This intermediate can then be exposed to another acid, etc., to see whether forward and/or reverse reactions occur in order to assign a relative stability scale of the intermediates on the surface [13,24,25]. All of these species have their own characteristic sets of products when they are heated, and it is easy to discern whether or not one has displaced the other. When such a series of experiments is performed, the hierarchy of stability shown in table 1 is revealed. The carboxylic acids indeed are the strongest, propylene and water are two of the weakest. A number of species do not react at all; hydrogen, ethylene, and ethane are thus clearly weaker acids than even the weakest of the group that react. The relative stabilities of the intermediates do not correlate very well with the overall rankings of their aqueous acidities; in fact, they correlate well with the gas phase acidities. Neither do they correlate with bond energies. The message here is that the formation of the surface species involves charge acceptance from the surface; they are anionic to some degree and thus their relative stabilities correlate well with gas phase acidities. Of course, a large contributing factor to the gas phase acidity is the electron affinity of the conjugate base in the gas phase. The electron affinities can run from 2 to almost 80 kcal/mol, producing acidities compared to the relative bond energies, for example [26].

It is important to examine how these general chemical characteristics of adsorbed atomic oxygen carry over to other metals. The first question is how they carry over to the other group IB metals, copper and gold. The extension to these metals is really quite straightforward with some subtle differences. First of all, copper itself activates some bonds without oxygen. It activates O-H bonds in carboxylic acids and alcohols, so it already has an inherent reactivity which is higher than that of silver. On the other hand, the presence of oxygen on the surface facilitates direct proton transfer and enhances the reactivity. The reactivity of adsorbed oxygen has the same qualitative characteristics on copper, but competing reaction channels take place on the metal itself. However, in the absence of surface oxygen, N-H bonds are not activated on copper, whereas in its presence they are very facilely cleaved [27]. At the other end of this column in the periodic table is gold. Because it does not dissociate oxygen, gold is a poor oxidation catalyst. In fact, when exposed to dioxygen above one atmospheric pressure at 600 K, pristine gold does not dissociate oxygen [28]. Oxygen does dissociate if there is silicon at the surface, but it does not if the gold is clean [29]. For this reason obviously gold is a poor oxidation catalyst. If, however, oxygen atoms from the gas phase are adsorbed on the surface, they have all the same chemical properties as on silver. They act as a strong nucleophile, a Bronsted base, and so forth. Exactly the same kind of oxygen-activated chemistry is exhibited as far as we can tell from our observations to date.

Little is yet known about oxygen activated processes on the group VIII metals, palladium and platinum. On these surfaces oxygen is very easily dissociated, and it is more difficult to determine whether it has these special chemical properties, because these metals are very reactive, and they all activate C-H bonds, O-H bonds and N-H bonds. Since, however, water is evolved from palladium at two distinct temperatures for OH disproportionation and hydrogen-oxygen recombination [32], it is possible to detect direct proton transfer to adsorbed oxygen. Palladium does facilitate the proton transfer reactions. OH groups are formed by reaction of water or alcohols with surface oxygen [32]. There is therefore an indication that the oxygen on palladium has these same characteristics; it is a Bronsted base and strong nucleophile, but the surface itself shows competing reactivity.

With these concepts in mind it is possible to formulate two rather general mechanisms for catalytic oxidation on metals. The first is a rather simple cycle. The metal itself activates substrates by bond cleavage to produce molecular fragments on the surface. It also dissociates oxygen facilely, so that various intermediates coexist on the surface with oxygen. The oxygen then reacts with these intermediates to form oxidation products. In the limit of complete oxidation, the oxidation products are CO_2 and water. In fact it is very difficult to conduct selective oxidation over such a reactive metal surface. I call this particular catalytic cycle the <u>scavenger mechanism</u>, because oxygen is effectively scavenging intermediates from the surface. The other mechanism, which is not so obvious, I call oxygen-activated. Examples of such reactions were discussed above. For example, substrates can be directly activated by proton transfer to adsorbed oxygen to form water and a surface partially covered with an intermediate which then reacts by some other route to form the partially oxidized products. In excess oxygen, some scavenging may also occur to yield secondary reaction products of these species as well, but in lean oxygen conditions scavenging reactions are minimized.

With studies of the type described above in ultra high vacuum, it is possible to elucidate catalytic cycles. For example, in the oxidation of methanol on copper, methanol reacts with surface oxygen to form methoxide groups and hydroxyls. A second molecule of methanol reacts away the hydroxyl group to form water, leaving the methoxyl species on the surface, which then dehydrogenate to make formaldehyde and hydrogen [33]. This mechanism was not understood for many years until these reaction steps could be isolated. When isolated, the steps are very clear. There are also side reactions. One is a simple equilibrium between water and surface oxygen to make hydroxyl groups [34], which gives a side loop; water affects the surface oxygen concentration and it introduces another time constant in the overall reaction process. An important side reaction which leads to degradation of the product and lowering of the selectivity is the direct oxidation of formaldehyde [35]. This reaction proceeds by nucleophilic attack of the oxygen on the formaldehyde itself to form H_2CO_2, which dehydrogenates to make the formate; the formate then decomposes to yield CO_2 and hydrogen. Overall this cycle produces CO_2, the dilatorious side product. By such studies these selective oxidation processes can be rather completely understood. Furthermore, in each case we can identify the rate limiting step and measure its rate constant. This rate constant can be used as a predictor of the overall kinetic behavior for these catalytic cycles.

Recently we have begun to examine competitive oxidations that involve more than one functional group in a molecule. A simple example is the oxidation of ethylene glycol [36]. First, selective oxidation of the glycol occurs to form a surface dialkoxide. When this intermediate is heated, it dehydrogenates at one end to yield $-OCH_2CHO$, which subsequently dehydrogenates to give the dialdehyde. The reactivity pattern is predictable on the basis described above. There is direct evidence for the formation of $CHOCH_2O-$ in the reaction spectrum. The initial reaction of the dialkoxide is the dehydrogenation at one end of the intermediate; the second step is the dehydrogenation of the second functional group to give the dialdehyde. These steps can be also traced spectroscopically. Vibrational spectra taken sequentially with annealing from low to medium to high temperature reveal first the characteristic features of the diol (125 K), including the O-H stretch. Heating to 175 K drives off water and the O-H bond is lost in the spectrum. Heating further to partially dehydrogenate the dialkoxide yields the carbonyl stretch and an additional C-H stretch characteristic of the lower C-H vibrational stretch in aldehydes. It is clear that we can identify the $-OCH_2CHO$ intermediate and that our postulated intermediate to the dialdehyde is correct.

A most interesting C-C bond cleavage reaction occurs in the presence of excess oxygen. The experiments described above were conducted by first adsorbing approximately 0.1 monolayer of oxygen and then adding an excess of alcohol. In this fashion secondary oxidation reactions are suppressed. Different behavior could be expected were oxygen in excess; scavenging reactions might be expected. In fact, the excess oxygen does abstract hydrogen from the dialkoxide intermediate initially formed. It is very clear from the magnitude of the kinetic isotope effect that the reaction is rate-limited by C-H bond breaking. The intermediate formed, presumably $-OCH_2CHO$, is then very rapidly attacked at the carbon by oxygen to cleave the C-C bond, simultaneously liberating formaldehyde and forming a surface formate.

These reactions can be tracked either with temperature programmed reaction spectroscopy or vibrational spectroscopy. In the presence of excess oxygen with the alcohol, the primary water peak at low temperature (in this case O^{18}), instead of originating from adsorbed molecular water, is due to hydroxyl recombination. At higher temperatures formaldehyde is evolved, followed by even higher temperature products which are characteristic of surface formate. The surface formate contains the labelled oxygen, whereas we find no labelled oxygen in the formaldehyde. In summary, the selective oxidation to the dialdehyde dakes place particularly in an oxygen-lean situation. With excess oxygen the scavenging processes begin, leading to non-selective oxidation.

SUMMARY

Heterogeneous oxidation processes can be understood in terms of specific acid/base properties of oxygen atoms adsorbed on metal surfaces. Oxygen-activated processes are responsible for selective oxidation on silver and gold and also play a role on copper, palladium and platinum. Two general mechanistic routes can be identified for these oxidation processes. The scavenger mechanism involves reactions of coadsorbed molecular fragments and oxygen and generally is very non-selective. The oxygen-activated mechanism leads to selective partial oxidation.

ACKNOWLEDGEMENT

Support of the National Science Foundation and the Donors to the Petroleum Research Fund of the American Chemical Society is gratefully acknowledged.

REFERENCES

1. M.A. Barteau and R.J. Madix, Surf. Sci. 97 (1980) 101.

2. B.A. Sexton and R.J. Madix, Chem. Phys. Lett. 76 (1980) 294.

3. C. Backx, S.P.M. de Groot and P. Biloen, Surf. Sci. 104 (1981) 300.

4. K.C. Prince, G. Paolucci and A.M. Bradshaw, Surf. Sci. 175 (1986) 101.

5. J.L. Gland, B.A. Sexton and G.B. Fisher, Surf. Sci. 95 (1980) 587.

6. N.R. Avery, Chem. Phys. Lett. 96 (1983) 371.

7. E.M. Stuve, R.J. Madix and C.R. Brundle, Surf. Sci. 146 (1984) 155.

8. C. Nyberg and C.G. Tengstal, Surf. Sci. 126 (1983) 163.

9. R.J. Madix, unpublished results

10. D.A. Outka, J. Stohr, W. Jark, P. Stevens, J. Solomon and R.J. Madix, Phys. Rev. B 35(8) (1987) 4119.

11. J. Stohr and R. Jaeger, Phys. Rev. B 26 (1982) 4111.

12. F. Sette, J. Stohr and A.P. Hitchcock, J. Chem. Phys. 81 (1984) 4906.

13. R.J. Madix, Science 233 (1986) 1159.

14. For a recent example of the application of this method to surface reaction mechanisms, see Armand J. Capote, Alex V. Hamza, Nicholas D.S. Canning and Robert J. Madix, Surf. Sci. 175 (1986) 445.

15. David H.S. Ying and Robert J. Madix, J. Catal. 61 (1980) 48.

16. B.A. Sexton, Surf. Sci. 88 (1979) 319.

17. M.A. Barteau and R.J. Madix, Surf. Sci. 115 (1982) 355.

18. M.A. Barteau and R.J. Madix, J. Am. Chem. Soc. 105 (1983) 344.

19. M.A. Barteau and R.J. Madix, J. Chem. Phys. 74 (1981) 4144.

20. D.A. Outka and R.J. Madix, Surf. Sci. 137 (1984) 242.

21. C. Backx, C.P.M. de Groot, P. Biloen, Appl. Surf. Sci. 6 (1980) 256; M.A. Barteau and R.J. Madix, Surf. Sci. 103 (1981) L171; E.M. Stuve, R.J. Madix, B.A. Sexton, Chem. Phys. Lett. 89 (1982) 48.

22. C.H. Duboise and B.R. Zegarski, Chem. Phys. Lett. 120 (1985) 537.

23. E.M. Stuve, S.W. Jorgensen and R.J. Madix, Surf. Sci. 146 (1984) 179.

24. M.A. Barteau and R.J. Madix, Surf. Sci. 115 (1982) 355.

25. Scott W. Jorgensen and R.J. Madix, Surf. Sci. 130 (1983) L291.

26. J.E. Bartines and R.T. McIver, Jr., in Gas Phase Ion Chemistry, M.T. Bowers, Ed. (Academic Press, New York, 1979) vol. 2, pp. 87-121.

27. M.H. Matloob and M.W. Roberts, J. Chem. Soc. Far. Trans. 1 73 (1977) 1393.

28. A.G. Sault, R.J. Madix and C.T. Campbell, Surf. Sci. 169 (1986) 347.

29. N.D.S. Canning, D. Outka and R.J. Madix, Surf. Sci. 141 (1984) 240.

30. Duane A. Outka and R.J. Madix, Surf. Sci. 179 (1987) 361.

31. Duane A. Outka and R.J. Madix, J. Am. Chem. Soc. 109 (1987) 1708.

32. Scott W. Jorgensen and R.J. Madix, Surf. Sci., in press.

33. Israel E. Wachs and R.J. Madix, J. Catal. 53 (1978) 208.

34. K. Bange, D.E. Grider, T.E. Madey and J.K. Seiss, Surf. Sci. 136 (1984) 38.

35. Israel E. Wachs and Robert J. Madix, Surf. Sci. 84 (1979) 375.

36. A.J. Capote and R.J. Madix, to be published.

METHANE OXIDATION AT METAL OXIDE SURFACES

J.H. Lunsford

Department of Chemistry
Texas A&M University
College Station, Texas 77843

INTRODUCTION

The presence of large reserves of natural gas, located mainly in remote regions, has prompted extensive research on the partial oxidation of methane. Research on catalytic oxidation has been directed mainly toward the development of processes which will produce either oxygenates (methanol and formaldehyde) or coupling products (ethane and ethylene). The latter approach has been somewhat more successful and will be the subject of this paper.

The more promising catalysts for oxidative dimerization include certain members of the lanthanide oxide series,[1,2] as well as a number of metal oxides promoted with Group IA ions.[3-8] Some of the highest steady-state yields of C_2 products (ethane plus ethylene) have been achieved using lithium-promoted magnesium oxide (Li/MgO),[9] and some typical results are shown in Table 1. Although pure MgO is neither active nor selective for formation of C_2 products, the presence of Li results in C_2 yields of approximately 15-20%. The ethylene/ethane ratio depends upon the severity of the conditions; however, as indicated in the table, ratios up to 2 can be easily achieved. The condition where the pressures of CH_4 and O_2 are 303 torr and

Table 1. Conversion and Selectivity During Methane Oxidation

Catalyst (Mass, Temp.)	Partial Pressure, Torr: Reactants		Products				Conv.,%	C_2-Sel.,%
	CH_4	O_2	C_2H_4	C_2H_6	CO	CO_2		
MgO (4g, 720°C)	72	39	0.0	0.3	1.2	1.4	4	10
3% Li/MgO (4g, 720°C)	55	27	3.1	1.9	0.2	10.0	37	50
	303	157	20.2	9.7	3.9	75.2	38	43
7% Li/MgO (4g, 720°C)	59	29	3.5	2.2	0.0	11.3	38	50

Reactant flow rate = 0.83 mLs^{-1}.

157 torr, respectively, were chosen to simulate a mixture of CH_4 and air at a total pressure of 760 torr with a $CH_4:O_2$ ratio of 2. The yields under these conditions were comparable to those obtained at much lower partial pressures of CH_4.

Other combinations of Group IA/Group IIA oxides yield similarly active and selective catalysts, provided there is a reasonable size match between the alkali metal ion and the alkaline earth ion. As shown in Table 2, Na/CaO is as effective as Li/MgO in forming C_2 products, but Na/MgO is a considerably poorer catalyst under these conditions. At somewhat higher temperatures Aika and co-workers[6] have demonstrated that Na/MgO also promotes good C_2 yields. The origin for these phenomena will be discussed in a subsequent section.

As reported initially by Otsuka et al.[1] and more recently by our group,[2] certain members of the lanthanide oxide series are effective catalysts for the oxidative dimerization of methane. Otsuka et al. found that Sm_2O_3 was the best catalyst; however, we have observed recently that Nd_2O_3 is superior.[2] Our results showed that the pretreatment of the catalyst has a marked effect on its behavior, which may explain the different results obtained by the two groups. The effective lanthanide oxides which fall into the activity sequence $Nd_2O_3 > La_2O_3 > Sm_2O_3$ are all considerably more active than Li/MgO for methane oxidation, as well as for the oxidation of the products, ethane and ethylene. Thus, over the lanthanide oxides high selectivity can be achieved only at relatively low conversions (<10%) and under oxygen-limiting conditions. The C_2 selectivity can be improved considerably by adding a lithium salt to the catalyst.[4] The least effect of the lanthanide oxides is cerium oxide, which is characterized by multiple oxidation states. This observation suggests that metal oxides having multiple oxidation states are not necessary for selective methane oxidation, and, in fact, such materials may be poor catalysts.

PROPOSED MECHANISM

A matrix-isolation electron spin resonance (MIESR) system has been used to detect gas phase radicals which emanate from surfaces of catalytic interest.[10] This technique has been used to show that those catalysts which are effective in the oxidative dimerization reaction are also effective in the

Table 2. Comparison of Different Group IA/Group IIA Catalysts

	Li/MgO	Na/MgO	Li/CaO	Na/CaO
Wt %[a]	7	20	5	15
SA, m^2g^{-1}	4.1	6.3	1.6	1.5
Conv., %[b] (CH_4)	22.6	11.9	10.8	22
C_2 Sel., %	56.7	17.8	67.2	51.1
=/-	1.0	0.5	0.7	0.9

[a] Group IA/Group IIA atomic ratio = 0.33.

[b] 1.0 g catalyst, 700°C, CH_4/O_2 = 2, 55 $mLmin^{-1}$, 1 atm, all catalysts pretreated at 700°C.

generation of gas phase methyl radicals. Conversely, materials such as cerium oxide do not give rise to any gas phase radicals.[2] In fact, this oxide actually scavanges methyl radicals. Recent quantitative experiments have demonstrated that the flux of $CH_3\bullet$ radicals emanating from a Li/MgO catalyst is approximately equal (within a factor of 2) to the formation of C_2 products.[11]

We have concluded, therefore, that ethane is formed by a methyl radical coupling reaction which occurs primarily in the gas phase. Even so, up to half of the coupling reactions may occur within the void volume of the catalyst particle. Ethylene is formed by the subsequent homogeneous and heterogeneous dehydration of ethane.[9]

Scheme I

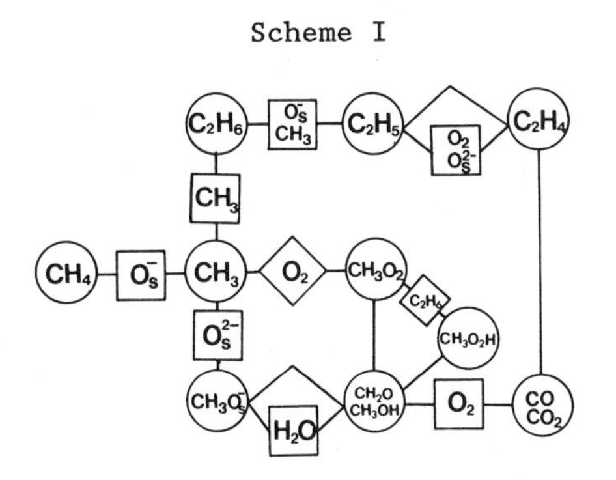

These reactions, as well as the nonselective oxidation reactions are graphically shown in Scheme I. At least two nonselective pathways are of importance: secondary reactions of methyl radicals with the surface and gas phase chain branching reactions which involve $CH_3O_2\bullet$ radicals. The selective coupling reaction and the nonselective oxidation reactions are

$$2CH_3\bullet \rightarrow C_2H_6 \qquad (1)$$

$$CH_3\bullet + O^{2-} \rightarrow CH_3O^- \rightarrow CO, CO_2 \qquad (2)$$

$$CH_3\bullet + O_2 \rightleftharpoons CH_3O_2\bullet \qquad (3)$$

The rates of the selective and nonselective reactions are given by

$$dC_2/dt = k_1[CH_3\bullet]^2 \qquad (4)$$

$$dC_1/dt = k_2[CH_3\bullet]\{[S.A.] + P(O_2)\} \qquad (5)$$

where the amount of surface O^{2-} is proportional to the surface area (S.A.). Upon taking the ratio of equations 4 to 5 the ratio of the selective to nonselective reactions is given as

$$(dC_2/dt)/(dC_1/dt) = \frac{k_1}{k_2}[CH_3\bullet]/\{[S.A.] + P(O_2)] \qquad (6)$$

from which it can be seen that C_2 selectivity may be improved by operating at low O_2 pressures with low surface area catalysts. Aika and co-workers[6] have convincingly demonstrated the latter point.

ACTIVE FORMS OF SURFACE OXYGEN

Of the potentially active forms of surface oxygen one must seriously consider the ions O^-, O_2^-, O_3^- and O_{LC}^{2-}, where the latter refers to oxide ions in a state of low coordination. Normal surface oxide ions are not very active, as demonstrated by the behavior of MgO reported in Table 1. In the pure oxides O_{LC}^{2-} ions may give rise to limited activity, but they would increase in concentration with increasing surface area. We have already seen, however, that high surface areas result in low selectivity.

The O^-, O_2^- and O_3^- ions are all paramagnetic and have been studied in detail using esr spectroscopy. The formation, thermal stabilities and activity of these ions have been reviewed,[12,13] and the latter two properties are briefly summarized in Table 3. The sequence of activity is $O^- \gg O_3^- \gg O_2^-$. One should note that O^- on MgO reacts with CH_4 at temperatures <-150°C, which is consistent with the high reactivity of O^- with alkanes in the gas phase as reported by Bohme and Fehsenfeld.[14] Methane has been observed to react with O^- on supported $Mo^{VI}O^-$ at -196°C, and in this case the resulting $CH_3\bullet$ radicals were observed on the surface.

The formation of the oxygen species described in Table 3 required irradiation of the MgO at room temperature. For these ions to be involved in the activation of CH_4 at elevated temperatures there must be another mechanism by which they are thermally generated. In fact, Abraham and co-workers[15] have shown that O^- ions in the form of $[M^+O^-]$ centers exist in single crystals of Group IIA oxides doped with Group IA ions, provided the samples are heated in O_2. These centers may be detected by esr after quenching the oxide which was at elevated temperatures (T > 1000°C).

Similar centers have been detected in the used Li/MgO catalysts, as well as in Li/CaO and Na/CaO (Table 2).[9,16,17] The spectrum observed after quenching the used Li/MgO catalyst from 650°C is depicted in Fig. 1b. An even larger signal is obtained upon irradiating the sample with ultraviolet light (Fig. 1c). In addition to the $[Li^+O^-]$ spectrum one can also detect the spectra of O_2^- and O_3^- ions. It is unlikely that these two forms of

Table 3. Thermal Stability and Activity of Oxygen Ions on MgO

Oxygen Ion	Maximum Temp. for Stability	Activity with Hydrocarbon
O^-	$T \simeq 25°C$	Reacts with CH_4 at $T < -150°C$
O_2^-	$T \simeq 175°C$	No reaction with C_1 or C_2 alkanes at $T < 175°C$; reacts with propylene
O_3^-	$T \simeq 25°C$	Reacts with alkanes at $T = 25°C$

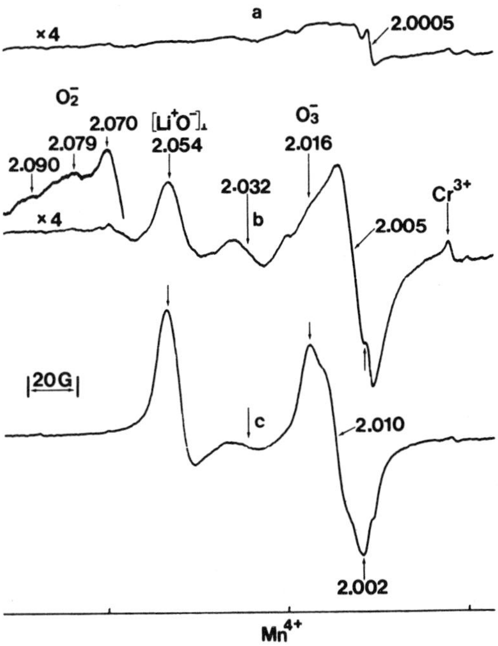

Figure 1. EPR spectra of 7 wt % Li/MgO after heating in 192 Torr of O_2 at 923 K for 1 h: (a) sample cooled slowly to 298 K, (b) sample quenched in liquid O_2 at 77 K, and (c) sample (a) irradiated in 15 Torr of O_2 for 30 min.

oxygen are important in the activation of CH_4 at 700°C since (a) they would be thermally unstable under these conditions (Table 3) and (b) we have shown that the catalyst retains its ability to generate $CH_3\cdot$ radicals for several minutes after the O_2 has been removed from the system.[10]

In addition to the results of Table 2 there is considerable evidence which suggests that centers of the type $[M^+O^-]$ are responsible for the activation of CH_4. The formation of $CH_3\cdot$ radicals and the presence of $[Li^+O^-]$ centers have a similar functional relationship with respect to the level of Li addition. Moreover, as shown in Fig.2 the overall CH_4 conversion and the concentration of $[Li^+O^-]$ have a similar response to the O_2 partial pressure.[9] It is significant to note that most of the $[Li^+O^-]$ centers observed by esr are in the MgO bulk, as determined by broadening experiments with molecular O_2, yet at the elevated temperatures these centers are believed to be in communication with the surface oxide ions via the reaction

$$[Li^+O^-] + O_s^{2-} \rightleftharpoons [Li^+O^{2-}] + O_s^- \qquad (7)$$

We have shown, for example, that CH_4 at -60°C affects the rate of decay of the $[Li^+O^-]$ centers, presumably through reacting with O_s^- centers at the surface.

The catalytic cycle for the activation of CH_4 and the regeneration of $[Li^+O^-]$ center is shown in Scheme II. As noted previously, the hydrogen atom abstraction must be quite rapid under the reaction conditions, thus the

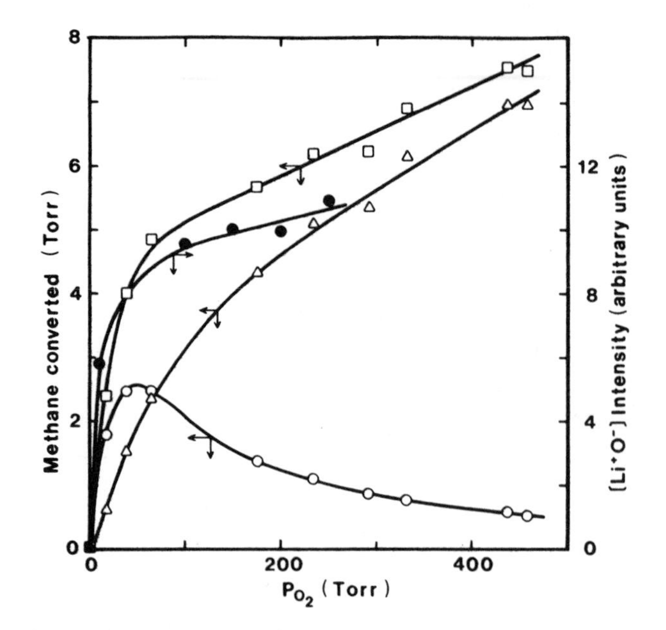

Figure 2. Changes in amount of CH_4 converted and concentration of $[Li^+O^-]$ centers with respect to an increase in O_2 pressure. A reactant mixture containing 300 torr of CH_4 was fed over 1 g of 7% Li/MgO at 620°C and at a flow rate of 0.83 mLs^{-1}: □, total; ○ , to C_2H_6 plus C_2H_4; Δ , to CO plus CO_2.

loss of hydroxyls or the reincorporation of oxygen into the lattice must be the slow step. The rapid increase in activity at low pressures of O_2, as described in Fig.2, may result from the slow incorporation of oxygen; whereas, at high O_2 pressures the loss of hydroxyls may be rate limiting.

Although $[M^+O^-]$ centers probably are responsible for the oxidative dimerization of CH_4 over Li/MgO, Li/CaO and Na/CaO there is no positive evidence to suggest that O^- centers exist on the lanthanide oxides. Rather, a study of quenched La_2O_3 revealed the presence of O_2^-, bound such that the two oxygen atoms were not equivalent.[18] For this catalyst it was found that the ability to generate $CH_3\cdot$ radicals decreased rapidly upon removal of gas phase O_2, which is in contrast to the Li/MgO system.[2,10] The origin of the unpaired electron for the formation of O_2^- is not certain; however, Loginov et al.[19] have suggested that the reaction

$$O_2 + O_2^{2-} \rightarrow 2O_2^- \qquad (8)$$

may occur on these oxides. The superoxide ion was not capable of attacking CH_4 at 175°C (Table 3), but it may do so at 700°C.

At sufficiently high temperatures ($T \geq 750$°C) it appears that the alkali metal oxides themselves may become effective for the activation of CH_4. As noted above Na/MgO becomes a reasonably good catalyst at 750 C even though no $[Na^+O^-]$ centers were detected.[6,17] Likewise K/CaO is effective for methane oxidation, although K^+ ions are too large to substitute for Ca^{2+} ions and therefore no $[K^+O^-]$ centers were observed by esr.[17] At elevated temperatures the most stable form of sodium oxide is Na_2O_2 and that of

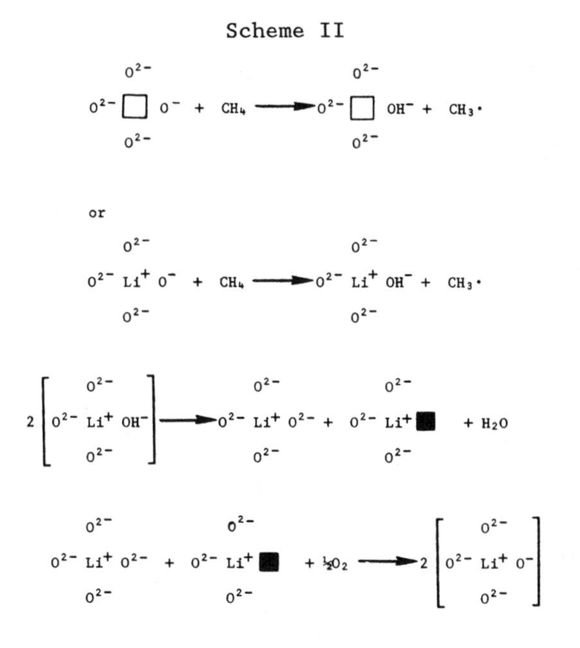
Scheme II

potassium oxide is KO_2. Thus, with K/CaO the superoxide ion would be expected in the supported phase.[20] In the presence of excess oxygen Na_2O_2 is oxidized to NaO_2 which yields the superoxide ion. Both the potassium and sodium oxides are undoubtedly in equilibrium with the respective carbonates which would decrease the amount of the oxide available for catalysis.

SUMMARY

1. If the cation size match is favorable in Group IA/Group IIA oxides, centers of the type $[M^+O^-]$ are involved in the activation of CH_4 for the catalytic oxidative dimerization reaction.

2. Other oxygen species such as O_2^- may be important for the activation of CH_4 on the lanthanide oxides or on supported alkali metal oxides.

ACKNOWLEDGMENT

The author wishes to acknowledge the contributions of graduate students and research associates including K.D. Campbell, D.J. Driscoll, T. Ito, C.-H. Lin, J.-X. Wang and H. Zhang. The research was supported, in part, by the National Science Foundation under Grant CHE-8405191.

REFERENCES

1. K. Otsuka, K. Jinno, and A. Morikawa, Chem. Lett. 499(1985); J. Catal. 100:353 (1986).
2. C.-H. Lin, K.D. Campbell, J.-X. Wang, and J.H. Lunsford, J. Phys. Chem. 90:534 (1986); K.D. Campbell, H. Zhang, and J.H. Lunsford, submitted to J. Phys. Chem.
3. W. Hinsen, W. Bytyn, M. Baerns, Proc. 8th Intl. Congr. Catal. Berlin, July 1984, Vol.III, 581 (1984); W.Bytyn and M. Baerns, Appl. Catal. 28:199 (1986).

4. K. Otsuka, Q. Liu, M. Hatano, and A. Morikawa, Chem. Lett. 903 (1986); K. Otsuka, K. Liu, M. Hatano, and A. Morikawa, Chem. Lett. 467 (1986); K. Otsuka, Q. Liu, and A. Morikawa, J. Chem. Soc. Chem. Commun. 586 (1986).
5. I.T. Ali Emesh and Y.J. Amenomiya, J. Phys. Chem. 90:4785 (1986).
6. T. Moriyama, N. Takasaki, E. Iwamatsu, and K. Aika, Chem. Lett. 1165 (1986); E. Iwamatsu, T. Moriyama, N. Takasaki, and K. Aika, J. Chem. Soc. Chem. Commun. 19 (1987).
7. I. Matsuura, Y. Utsumi, M. Nakai. and T. Doi, Chem. Lett. 1981 (1986).
8. C.A. Jones, J.J. Leonard, and J.A. Sofrenko, J. Catal. 103:311 (1987).
9. T. Ito, J.-X. Wang, C.-H. Lin, and J.H. Lunsford, J. Am. Chem. Soc. 107:5062 (1985).
10. D.J. Driscoll, W. Martir, J.-X. Wang, and J.H. Lunsford, J. Am. Chem. Soc. 107:58 (1985).
11. K.D. Campbell and J.H. Lunsford, to be published.
12. J.H. Lunsford in "Catalytic Materials: Relationship Between Structure and Reactivity," T.E. Whyte, R.A. Dalla Betta, E.G. Derouane, and R.T.K. Baker, eds, American Chemical Society, Washington, 1984; ACS Symposium Series 248:127.
13. M. Che and A.J. Tench, Adv. Catal. 31:77 (1982); 32:1 (1983).
14. D.K. Bohme and F.C. Fehsenfeld, Can. J. Chem. 47:2717 (1969).
15. Y. Chen, R.H. Kernohen, J.L. Boldu, and M.M. Abraham, Solid State Commun. 33:441 (1980); D.N. Olson, V.M. Orera, Y. Chen., and M.M. Abraham, Phys. Rev. B 21:1258 (1980); Y. Chen, H.T. Tohver, J. Narayan, and M.M. Abraham, Phys. Rev. B 16:5535 (1977).
16. J.-X. Wang and J.H. Lunsford, J. Phys. Chem. 90, 5883 (1986).
17. C.-H. Lin, J.-X. Wang, and J.H. Lunsford, J. Am. Chem. Soc., in press.
18. J.-X. Wang and J.H. Lunsford, J. Phys. Chem. 90:3890 (1986).
19. A.Y. Loginov, K.V. Topchieva, S.V. Kostikov, and N.S. Krush, Dokl. Akad. Nauk SSSR 232:135 (1977).
20. T.P. Whaley in "Comprehensive Inorganic Chemistry," J.C. Bailar, H.J. Emeleus, R. Nyholm, and A.F. Trotman Dickenson, eds, Pergamon Press, Oxford, 1973, pp. 369-529.

THE ACTIVATION OF OXYGEN BY METAL PHOSPHORUS OXIDES - THE VANADIUM PHOSPHORUS OXIDE CATALYST

Jerry R. Ebner* and John T. Gleaves

Monsanto Chemical Company

St. Louis, MO 63167

ABSTRACT

The oxidation of C_4 hydrocarbons has been studied over the active phase for butane oxidation to maleic anhydride, $(VO)_2P_2O_7$. The activation of oxygen by $(VO)_2P_2O_7$ and the reactivity of the various oxygen surface species was examined by thermogravimetric techniques and transient reaction studies. A new transient reactor system, TAP for Temporal Analysis of Products, was utilized to identify the sources of active oxygen for selective and non-selective reaction pathways. Double pulsed TAP experiments provided unique information on product formation as a function of the surface lifetime of oxygen species. Spectroscopic studies were conducted to determine the nature of surface hydrocarbon species. The results from this study illustrate the multiple oxygen activation methods employed by this catalyst in selective alkane oxidation.

INTRODUCTION

Transition metal phosphorus oxide compounds have been identified as useful systems for selective oxidation of hydrocarbons, particularly for oxydehydrogenation reactions (1). Previously, we reported on a manganese phosphorus oxide system which selectively ammoxidizes methanol to hydrogen cyanide (2). The source of active oxygen in this catalyst system was shown to be chemisorbed oxygen, in contrast to the lattice oxygen commonly used by many selective oxidation/ammoxidation metal oxide catalysts (3,4). Certainly one of the most scientifically fascinating selective oxidation systems in commercial use today is the vanadium phosphorus oxide system used for the oxidation of butane to maleic anhydride. This 14 electron oxidation reaction requires the abstraction of eight hydrogen atoms and the insertion of three oxygen atoms. It has been the subject of extensive study in the literature, in which questions of catalyst structure (5-13) and reaction mechanism (14-25) have been addressed. We have investigated this reaction as well, and have attempted to identify reaction pathways and the sources of selective and non-selective oxygen.

In order to establish a framework for discussion of the types of active oxygen, it is useful to consider the possible forms of active oxygen known to be available with metal oxide systems. This subject has been recently reviewed (26,27). One general mechanism considered to be of minor importance at high temperatures involves the simultaneous activation of dioxygen and hydrocarbon on the surface of the metal oxide. The more prevalent mechanism involves a stepwise process in which the electrophilic dioxygen molecule is first chemisorbed to form an activated species (hereafter symbolized as O*). This species may then react with the hydrocarbon, or it may replenish the surface lattice oxygen [O_{SL}] which in turn reacts with the hydrocarbon. The selective oxygen in the surface lattice is often a covalently bonded metal oxo species. The O* species can be molecular surface species such as O_2^{-2} or O_2^{-1}, or dissociatively adsorbed monoatomic anions such as O^- or O^{-2}. In the Mars van Krevelen mechanism (28) the surface lattice oxygen provides the oxidizing equivalents to the hydrocarbon and replenishment occurs through the bulk. Determining the mode of oxygen activation and reaction of dioxygen in any particular system requires an understanding of the pertinent reactions between [O*], [O_{SL}] and subsurface lattice oxygen [O_L].
Scheme 1, where R indicates hydrocarbon and V_{surf} the surface oxide

$$O_2 \rightleftharpoons [O^*]$$

$$[O_{SL}] \rightleftharpoons [O_L]$$

$$R + [O_{SL}] \rightleftharpoons RO + V_{surf}$$

$$R + [O^*] \rightleftharpoons RO$$

$$V_{surf} + [O^*] \text{ or } [O_L] \rightleftharpoons [O_{SL}]$$

Scheme 1

vacancy, gives a simplified summary of these relationships. We will be discussing the activation of oxygen by vanadium phosphorus oxide within this framework.

THE CATALYST

The vanadium phosphorus oxide catalysts used in this study were samples prepared in organic solvent according to patent literature procedures (29, 30). The preparation steps include reduction of V_2O_5 by the isobutyl alcohol solvent in the presence of phosphoric acid and a solvent additive (eg. HI, oleum, HCl, lactic acid, benzyl alcohol, etc.) to form the vanadium hydrogen phosphate hemihydrate precursor $(VOHPO_4)\cdot 0.5H_2O$, calcination of the precursor in air to remove residual organics and dehydrate the precursor phase and an in situ treatment in air/butane to form active catalyst. In general, the stable active catalyst is not fully realized until the butane oxidation reaction has been carried out for a number of hours. In this study catalysts were on stream for at least 200 hours, and are designated equilibrated catalysts. Extensive analysis of equilibrated catalyst samples shows the P/V ratio is approximately 1.0 and the average vanadium oxidation state is +4 across the entire length of a fixed bed reactor tube (31). Single crystals of the catalyst were obtained by melting a finely ground powder of equilibrated catalyst in a Pt crucible and then cooling at about 6°C per hour from 975°C to 600°C in a N_2 purged oven. A range of crystals of varying color and phase composition were obtained. Figure 1 compares the FTIR and laser Raman microprobe results for a green crystal and the active catalyst, and illustrates the similarity between the crystal and catalyst structure. The vibrational spectra match those reported for vanadyl pyrophosphate, $(VO)_2P_2O_7$ (32). Single crystal x-ray studies are also in essential agreement with the structure reported in the literature (33) for $(VO)_2P_2O_7$. However, accounting for a disordering in the vanadium positions results in normal vanadyl distances of 1.58-1.63 angstroms and lowers the R values to 4% or better. These results will be reported in detail in a separate publication. In the structure of $(VO)_2P_2O_7$ there are edge shared VO_6 octahedra with the vanadyls of each dimer pair in an *anti* arrangement, and the edge shared vanadium octahedra are linked along the axial direction by V-O-V chains and layer bridging pyrophosphate groups. The V to V separation within a dimer ranges from 3.15 - 3.20 angstroms. In general, $(VO)_2P_2O_7$ is the only crystalline phase found in equilibrated vanadium phosphorus oxide catalysts, and we consider it the active structure for selective oxidation of butane to maleic anhydride.

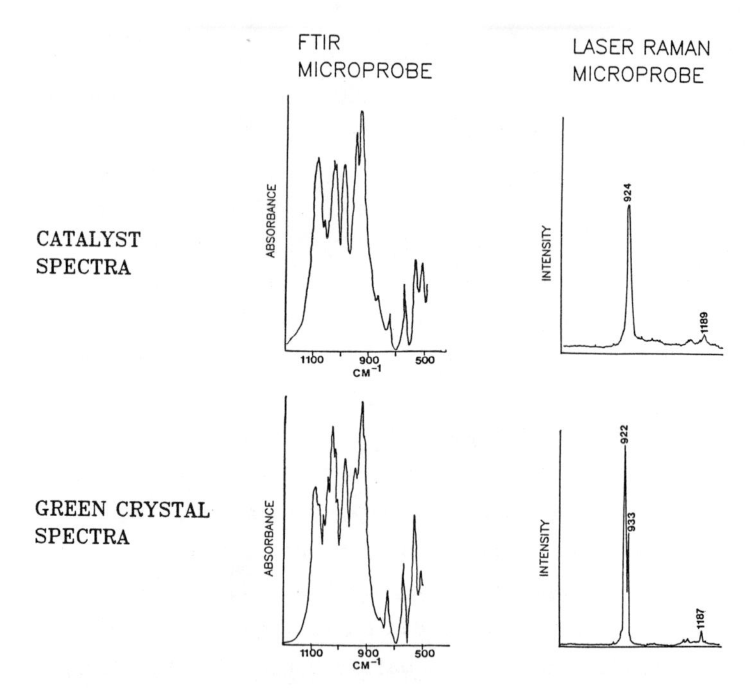

Fig. 1. FTIR and Raman microprobe spectra for active catalyst and green crystal formed from catalyst powder.

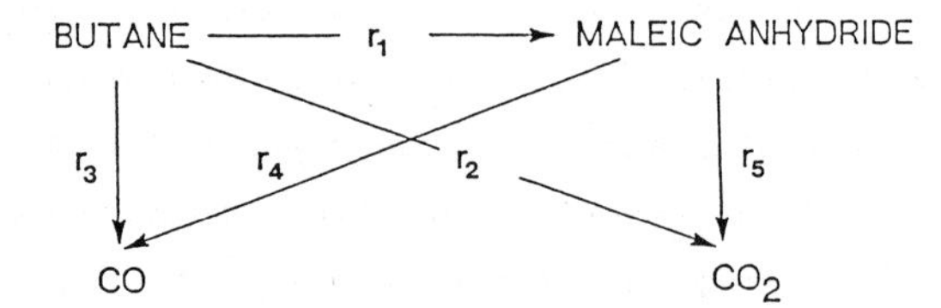

AUTHOR	CATALYST(P/V)	r_1 EXPRESSION	E_a (kcal/mole)
Escardino, 1973	oxides,(.8)	$k_1 P_B P_{O_2}$	14 (400–480°C)
Hoffman, 1980	aqueous,(1)	$k_1 K_B P_B P_{O_2}^{.28}/[1+K_B P_B]$	21 (446–504°C)
Cresswell, 1984	commercial	$k_1 P_B^{.54}/[1+K_M P_M]$	22 (300–380°C)
Trifiro, 1985	organic,(1.01)	$k_1 K_B P_B P_{O_2}^{.29}/[1+K_B P_B]$	(300–340°C)
Sundareson, 1986	organic,(1)	$k_1 P_B/[1+K_B P_B/P_{O_2}+K_M P_M/P_{O_2}]$	30 (390–440°C)
Larou, 1986	organic,(1)	$k_1 K_B P_B P_{O_2}/[1+K_B P_B+K_M P_M+K_W P_W]$	25 (330–450°C)

Fig. 2. Summary of literature kinetic studies on butane to maleic anhydride over vanadium phosphorous oxides.

REVIEW OF KINETIC STUDIES

The kinetics of the oxidation of butane to maleic anhydride have been studied by a number of research groups and there is substantial agreement in the literature that the kinetics best fit a triangular reaction network with reaction rates r_1 and r_2 to maleic anhydride and CO_x, respectively, and a combustion rate r_3 for burning of maleic anhydride to CO_x (20-25).

As shown in Figure 2, there is considerable variance in the rate expressions for maleic anhydride formation developed from the kinetic data. We propose this reflects the differing surface states of the particular catalyst due to its method of preparation and its reaction history. Only the Cresswell rate expression contains less than a first order dependence on butane, and the oxygen order dependencies range from 0 to 1. Sundareson's expression is developed using a redox model (reduction of the catalyst with hydrocarbon and reoxidation with molecular oxygen), and the data can be fit equally well using a half or first order oxygen dependence. A maleic inhibition term appears in several of the rate expressions, and it only becomes important at high conversions on equilibrated catalyst. Lerou (23) has reported a suppression of oxidation rate due to water concentration, as well as an increase in maleic anhydride selectivity. Sundareson also mentions this effect, but lumps it into the maleic term. A butane inhibition term appears in several of the rate expressions. Finally, Pepera et al. (14) have demonstrated a kinetic isotope effect using deuterated butane, and conclude the rate determining step involves breaking of the methylene C-H bonds. Maleic anhydride reaction data from the catalyst of this study can be fit well using a rate expression like that of Lerou.

The kinetics studies have led to a relatively simple reaction network in which maleic is formed directly from butane, and the rate of this reaction is clearly dependent on hydrocarbon and oxygen concentration. However, the exact order of the reaction in oxygen is uncertain, and the nature of the oxygen species important in the selective and non-selective pathways is unclear. Also, the question of whether or not intermediate unsaturated hydrocarbons and oxygenated hydrocarbons form and desorb in a stepwise path to maleic has also been the subject of much discussion in the literature. Pepera et al. (14) in pulse reactor studies report only maleic and CO_x formation, whereas intermediates have been detected by Trifiro (34) at high hydrocarbon concentrations. Pepera et al. (14) also point out the selectivity controlling steps occur after the rate determining step. In order to study the very fast steps that occur after the rate determining step it is helpful to do transient experiments. These studies have been conducted in our lab using a new type of transient reactor system called Temporal Analysis of Products (TAP). Results from the TAP experiments shed light on the nature of the active oxygen species and the important reaction pathways.

TRANSIENT AND IN-SITU REACTION STUDIES

Transient TAP Reactor

Temporal Analysis of Products (TAP) is a new device for study of reaction dynamics of solid-catalyzed vapor phase reactions (4,35). Only a brief description of the system will be given here. Key features of the TAP reactor are the following: i) the reactor is a micro-scale temperature controlled fixed bed holding 0.5 cc of 250-500 micron bulk form catalyst; ii) the reactor is connected to two high speed pulsed valves for introducing transients (pulse width 150 microseconds); iii) the entire reactor/pulse system is contained in a vacuum system which is linked through a differential pumping chamber to a QMS; iv) the QMS provides identification of products and reactants and allows analysis of their real time elution from the reactor with submillisecond time resolution; v) the number of molecules injected into the reactor through independently controlled pulsed valves ranges from 10^{14} to 10^{19} molecules per pulse. (A 10^{15} molecule per pulse intensity will address only 1/10000 of the available surface area of a 10 m^2/g catalyst.) An inert gas is used to characterize the flow through the reactor. The data can be collected in a number of modes. The scan mode measures mass intensity versus mass number and is used to identify products and estimate conversions and yields. The single pulse TAP mode measures mass intensity of a single mass peak characterizing a particular species versus time, and is used to evaluate adsorption/desorption properties and the real time elution of intermediates, end-products and reactants. The multipulse TAP mode measures mass intensity of a single mass component as a function of a fast multiple pulse train of reactant and is used to examine slower time processes, such as catalyst deactivation or depletion of surface species. A double pulse experiment yields another data format. In this experiment two separate pulses are admitted to the reactor at different and variable times, and an individual mass component is studied versus time and nature of the reactant injected. Importantly, through variations in the time between the reactant pulses, it is possible to examine effects of surface lifetimes of adsorbed species on product formation.

Detection of Reaction Intermediates

The selective oxidation of butane to maleic anhydride may proceed via a single site mechanism with no desorption of reaction intermediates or a multisite mechanism in which partially oxidized species such as butene or butadiene desorb and then react at another site. TAP reactor experiments show that reaction intermediates can desorb from an equilibrated vanadyl pyrophosphate surface. Figure 3 shows a composite spectrum of the reaction products when a 4:1 oxygen/butane mixture was pulsed through a sample of equilibriated vanadyl pyrophosphate heated to 420°C. To facilitate a comparison of the time dependence of the different products the peak maxima are normalized. Actual butane conversion was approximately 1% so the product peaks were about 100 times less intense then shown. The observed products are butene, butadiene, furan, carbon dioxide, and water. Maleic anhydride was not observed in this experiment. The water peak is not displayed since it is very broad and appears to be a straight line on the time scale of this experiment, and the carbon dioxide peak shape will be discussed later.

In addition to showing that intermediates can desorb in an oxidizing atmosphere, the 4:1 oxygen/butane data indicates the reaction sequence. The butane curve resembles an inert gas curve indicating that butane chemisorption is slow relative to the rate of escape from the reactor. The curve shape as well as the low butane conversion are consistent with butane chemisorption being an activated process. The peak maxima of the different intermediates are shifted to later times relative to the butane maximum. The order follows that for stepwise butane oxidation. Interestingly, the butene curve decays more rapidly than the butadiene curve, which decays more rapidly than butane. This is the result of the rapid conversion of butene and butadiene to other products. The shape of the furan decay indicates that very little furan is converted to products. This is consistent with the fact that no maleic anhydride is observed.

If the equilibrated catalyst is preoxidized prior to feeding the 4:1 oxygen/butane mixture intermediate products are initially not observed. In one preoxidation experiment O_2 was pulsed over the equilibrated catalyst and CO_2 production was monitored. When the CO_2 production

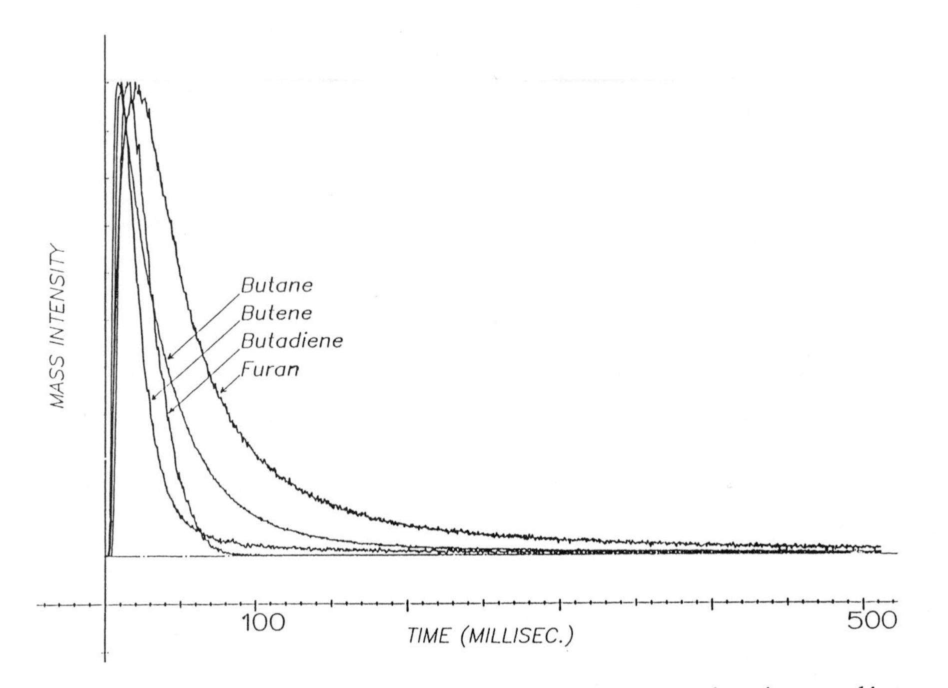

Fig. 3. Normalized temporal curves of desorbing reaction intermediates from butane oxidation over equilibrated $(VO)_2P_2O_7$ catalyst.

ceased the feed was switched to the 4:1 oxygen/butane mixture and the products were monitored as before. In this case maleic anhydride was the observed product. Further pulsing of the 4:1 feed led to diminished maleic anhydride yield and the eventual appearance of intermediate products. Reoxidizing the catalyst resulted in the production of more CO_2 and the re-establishment of the initial maleic yield upon further pulsing with the 4:1 feed mixture.

If higher oxygen to butane ratios are employed, such as an 8:1 mixture, then maleic anhydride is the only product observed during the course of a prolonged period of pulsing. If butane is pulsed without oxygen then products are formed for only a short period of time. In a typical anerobic experiment butane produces maleic anhydride for less than 50 pulses (pulse intensities = 10^{15} molecules/pulse). The amount of maleic produced depends on the initial oxidation state of the catalyst. For example, a preoxidized catalyst will produce maleic for 40 - 50 pulses while an equilibrated catalyst will produce maleic from 0 - 20 pulses. The latter result is due to the variability in the equilibrated catalysts surface oxidation state.

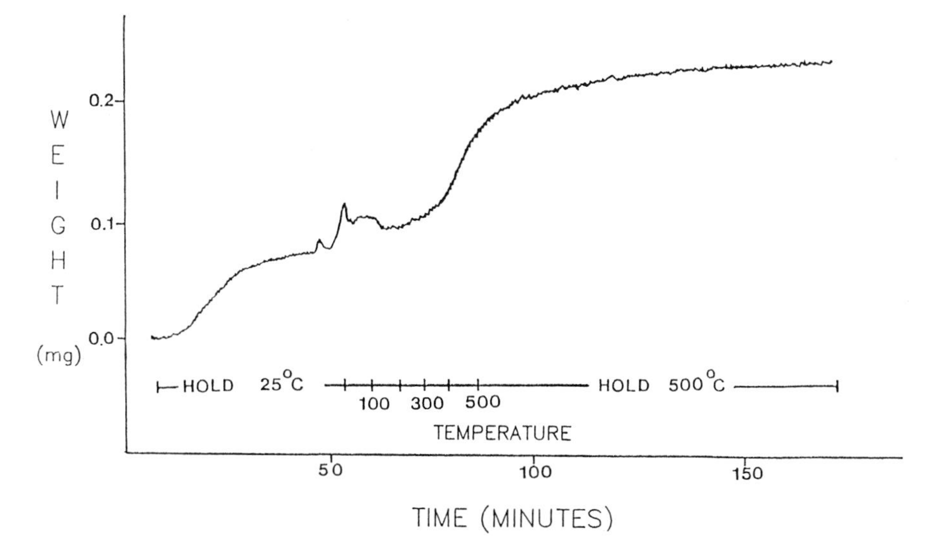

Fig. 4. Thermogravimetric study of oxygen chemisorption on equilibrated vanadium phosphorous oxide catalyst.

Types of Active Oxygen

Because of the importance of oxygen concentration in the activation of butane as indicated by kinetic studies, as well as its significance in the avoidance of intermediate desorption, we studied the chemisorption of oxygen and conducted TAP studies to ascertain the sources of selective and non-selective oxygen. We examined the chemisorption of oxygen as a function of temperature by thermal gravimetric analysis using an instrument previously described in the literature (4). The sample was 100 milligrams of equilibrated $(VO)_2P_2O_7$ catalyst (oxidation state 4.01) obtained from an operating catalytic reactor cooled down in a nitrogen stream. First, the sample was pretreated in the TGA system by purging with helium at 500°C for an hour to remove adsorbed water. A small weight loss of .04 mg was observed at the intermediate hold temperature of 250°C, and no further significant weight loss was seen upon heating the sample to 500°C. The sample was cooled to room temperature and exposed to a dry 20% O_2/He gas mixture. The results of the oxygen chemisorption experiment are depicted in Figure 4. An increase in weight of .07 mg was observed at room temperature in this experiment, but the amount of this type of weakly adsorbed oxygen was found to be somewhat variable from sample to sample. The oxygen adsorption at room temperature ranged from .01 - .08 mg.

Upon raising the temperature at 15°C/min from room temperature to 500°C, a significant chemisorption of oxygen was observed, and the onset of this chemisorption occurred at approximately 300°C and the maximum rate occurred at about 450°C. A total weight gain of .175 mg was measured, corresponding to the adsorption of 5.5 micromoles of O_2. This is approximately 3.6 micromoles/m^2 of surface area. Based on the crystal structure of $(VO)_2P_2O_7$, an estimate of the surface vanadium concentration is 11 micromoles of vanadium per m^2 of surface area. In this experiment, one oxygen molecule was adsorbed per three surface vanadium. This estimation suggests that much of the surface vanadium is available for oxygen chemisorption. After completing the adsorption experiment, the gas stream was switched back to helium and the weight loss monitored over time at 500°C. The weight loss observed was .04 mg over several hours. Cooling back to room temperature and repeating the oxygen adsorption as before replenished the .04 mg lost. Thus, roughly 75% of the oxygen remained on the surface at 500°C. In these experiments, the oxygen chemisorption did not become significant until temperatures near butane reaction temperature were reached. Further, the adsorption was largely irreversible, which implies a strong bond to the surface vanadium. As would be expected, the amount of surface oxidation was insufficient to be detected by bulk vanadium oxidation state determinations. These results suggest to us that the activated chemisorption of oxygen on $(VO)_2P_2O_7$ involves oxidation of surface V^{+4} to surface V^{+5}. The exact chemical form of this adsorbed oxygen is undetermined, and we will designate it O* for the remainder of the paper. However, it is tempting to suggest that the oxygen chemisorption occurs via a dissociative pathway on vanadium dimers (recall the V - V dimer distance is 3.15 - 3.20 angstroms in the bulk structure) leading to a V^{+5} oxo surface species capable of activating the alkane.

In order to examine the chemistry of the surface activated oxygen species, TAP experiments were conducted using intermediates as molecular probes of the surface. The results of TAP studies on butane showed that partially oxidized products such as butene, butadiene and furan desorb when the $(VO)_2P_2O_7$ surface is not fully oxidized. They also indicated that while butene and butadiene are readily oxidized to furan, furan is not readily converted to maleic anhydride on a partially oxidized surface. The difference between furan reactivity and butene or butadiene reactivity on a partially reduced surface suggests that there is more than one type of active oxygen involved. Multipulse TAP experiments (a train of pulses is admitted to the reactor and the intensities of the individual pulses in the train are monitored as a function of time) with furan and butene clearly indicate that this is the case. First, the catalyst was pulsed extensively with oxygen to form the O* species on the surface at 450°C. The oxygen pulse train gradually increased in intensity as less and less oxygen was adsorbed per pulse until each pulse was the same intensity. Then, furan was pulsed under <u>anaerobic</u> conditions at 420°C and the maleic anhydride was monitored. The maleic anhydride yield in the furan reaction started at a maximum and decreased to zero within 80 pulses. When the same experiment was done on an oxidized surface with cis-2-butene, and the furan peak intensity monitored, the furan yield started at zero and rose to a maximum and remained there for more than 10000 pulses. The initial zero furan yield was observed because of its further conversion to maleic anhydride. Figure 5 depicts the results for these two experiments. Importantly, it was further observed a $(VO)_2P_2O_7$ surface that had O* removed by the reaction of furan to maleic anhydride was no longer able to activate butane. These experiments indicate two types of

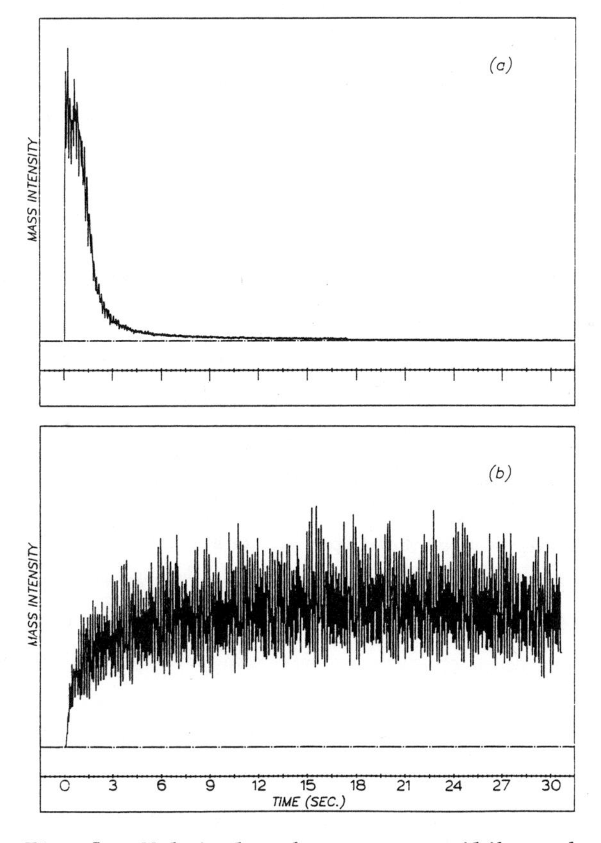

Fig. 5. Multipulse data over equilibrated $(VO)_2P_2O_7$ under anaerobic conditions showing (a) maleic anhydride production from furan and (b) furan production from cis-2-butene.

selective oxygen are present at the $(VO)_2P_2O_7$ surface. One type is the O* species which is present in low concentration and is rapidly depleted in the conversion of furan to maleic anhydride. Experiments pulsing oxygen-18 and butene or butadiene and monitoring the furan-O16 and furan-O18 as a function of pulse number indicate the second type of oxygen is associated with surface lattice oxygen. At the start of the pulsing, the initial furan formed with both hydrocarbons contained only oxygen-16. With continued pulsing the furan-O16 yield dropped and the furan-O18 yield increased until after about 30000 pulses only furan-O18 was detected. An estimate of the amount of oxygen inserted from the catalyst in forming furan-O16 indicates subsurface lattice oxygen does not participate. The lack of participation by bulk lattice oxide has also been observed in experiments by Sundareson et al. (24). Thus, a Mars van Krevelyn mechanism (28), in which the oxide of the surface lattice performs oxidation, is also operative with this catalyst system. The surface oxide layer of the $(VO)_2P_2O_7$ can be used for oxydehydrogenation of olefins and oxygen insertion to form furan.

Results of double pulse experiments indicate that O* may involve two chemisorbed species with different surface lifetimes. In double pulse experiments utilizing two pulsed valves connected to the microreactor, independent molecular injections of reactants separated by a controllable time interval from tenths of a millisecond to seconds are made to the microreactor. This type of experiment permits examination of the surface chemistry as a function of the surface lifetimes of adsorbed reactants. In these experiments the equilibrated $(VO)_2P_2O_7$ was fed alternating pulses of oxygen and furan. The pulse intensities were adjusted so the relative oxygen to furan ratio was 8:1. The interval between the valve pulses was initially set at 600 milliseconds. The maleic anhydride product curve was recorded and two peaks were observed. The major peak corresponded to the point when furan was pulsed into the reactor. The other peak, which appeared at the time of the oxygen injection was approximately 50 times smaller then the major peak. The pulse separation was then set at 0.1 milliseconds and the maleic product curve was recorded again. When the maleic product peaks corresponding to the furan injection times for the two different intervals were compared (Figure 6), it was found that the maleic yield was significantly smaller when the interval between the pulses was 600 milliseconds. This indicates that some of the chemisorbed oxygen had a relatively short surface lifetime and that other reaction channels were competing with the furan for the activated oxygen. It is important to note that the maleic yield did not continuously decrease as the

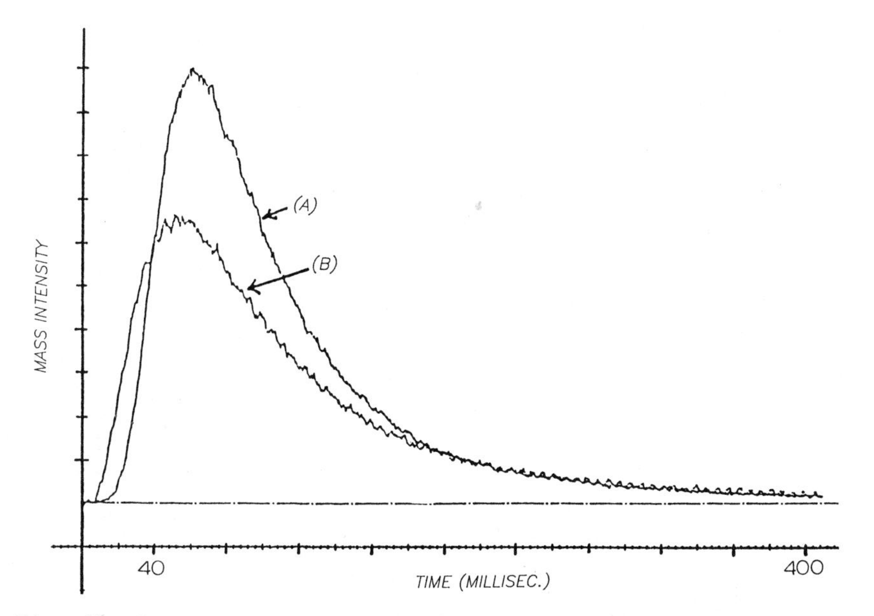

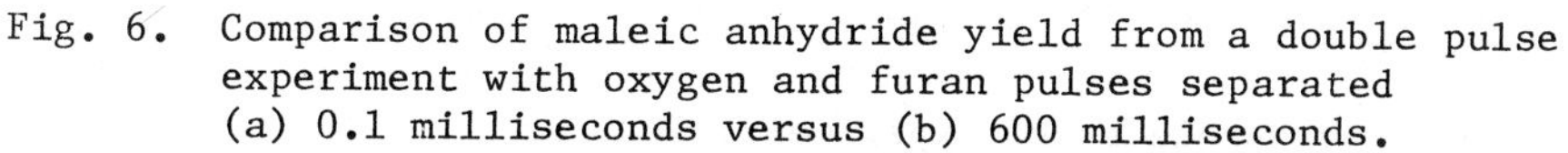

Fig. 6. Comparison of maleic anhydride yield from a double pulse experiment with oxygen and furan pulses separated (a) 0.1 milliseconds versus (b) 600 milliseconds.

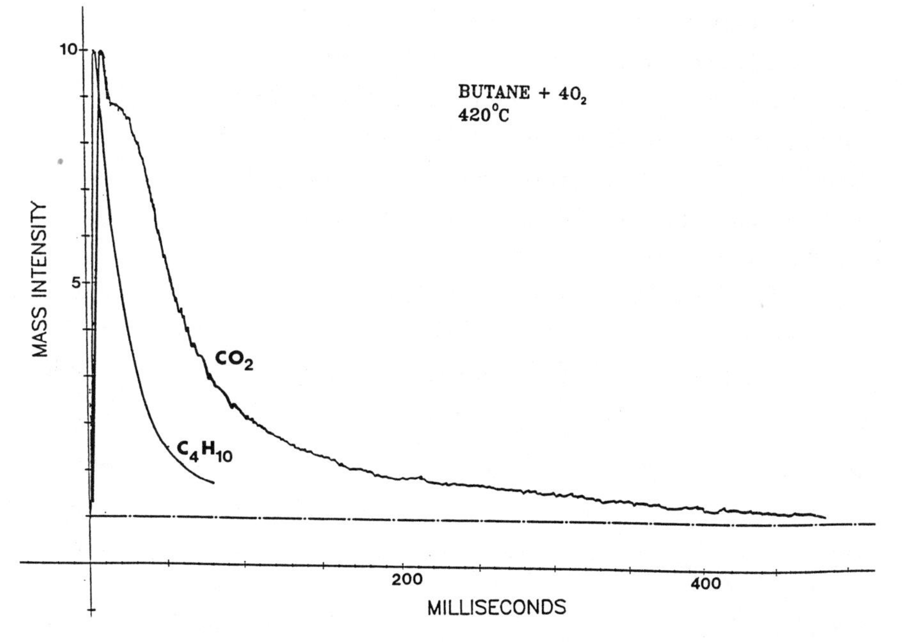

Fig. 7. Normalized temporal curves of butane and carbon dioxide from butane oxidation over equilibrated $(VO)_2P_2O_7$ catalyst.

interval between the pulses was extended beyond 600 milliseconds. This result is in agreement with the oxygen chemisorption experiments which demonstrated an appreciable lifetime for one form of O*. However, these TAP results suggest that another form of chemisorbed oxygen may be important. As we were tempted to suggest the long lived O* species resulted from dissociative chemisorption of oxygen, we speculate here that the short lived species is chemisorbed dioxygen. Molecular structure identification of these clearly observed oxygen types is a remaining experimental challenge.

Production of Carbon Oxides

The TAP curve shape of the carbon dioxide product peak from the reaction of a 4:1 oxygen/butane mixture at 420°C over equilibrated $(VO)_2P_2O_7$ is presented in Figure 7. The shape is suggestive of more than one pathway to CO_2 because it appears to be the convolution of a fast peak and a broad slower process. A similar pulse profile is seen when furan and cis-2-butene are the hydrocarbon reactants, but the portion of the peak attributed to a slower process is more predominant with these reactants. In this section we will focus only on these latter two reactants. To help shed light on the nature of the reactions responsible for carbon dioxide formation, single and double pulsed TAP experiments were conducted using oxygen-16 or oxygen-18, and adsorbed species were examined by in-situ FTIR. First, while conducting the furan/oxygen-16 double pulsed experiments described in the previous section, the production of carbon dioxide was monitored as well. When the two independent pulses of oxygen and furan separated by 400 milliseconds in time were admitted to the microreactor, two CO_2 pulses were observed. Opposite to the maleic product formation previously described, the CO_2 peak coinciding with the oxygen injection was larger than the peak corresponding to the furan pulse by a factor of about 2.5. This result clearly points to a longer lived surface hydrocarbon species which can produce a substantial amount of carbon dioxide.

The formation of surface hydrocarbon species was investigated in our laboratory using in-situ FTIR (36). In this experiment, reactant gas mixtures are fed directly through a thin catalyst pellet contained in the cavity of the in-situ cell which was silicon windows for obtaining the FTIR spectra (37). Figure 8 shows the spectra of surface deposits formed in reactions of oxygen mixtures of cis-2-butene, 1,3-butadiene and furan. The adsorbed species remained on the surface after evacuation and purging of the cell with argon at 450°C. They could only

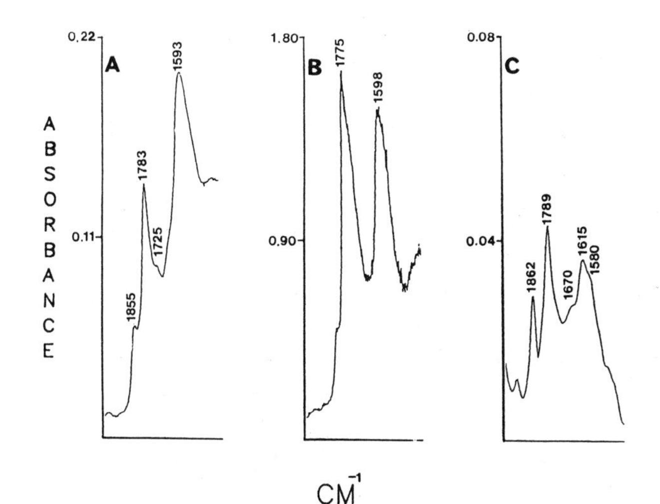

Fig. 8. FTIR of surface deposits formed by reaction of oxygen/hydrocarbon mixtures with equilibrated $(VO)_2P_2O_7$ catalyst: (A) 9:1 oxygen/cis-2-butene at 425°C; (B) 9:1 oxygen/1,3-butadiene at 450°C; and (C) 15:1 oxygen/furan at 450°C.

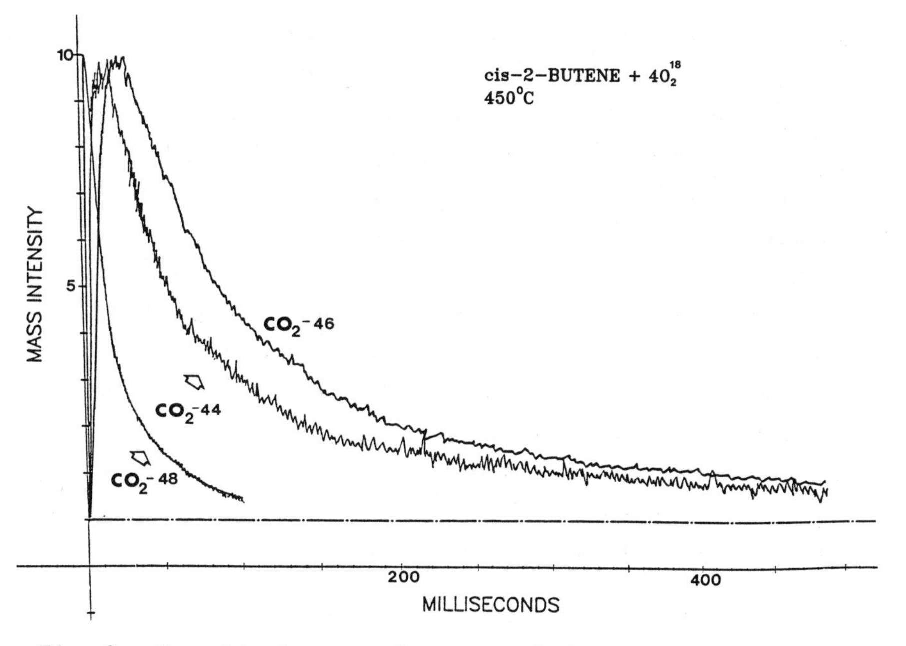

Fig. 9. Normalized temporal curves of the isotopes of carbon dioxide from pulsing cis-2-butene/oxygen-18 over equilibrated $(VO)_2P_2O_7$ catalyst.

be slowly removed from the surface by oxygen, and their oxidation produced primarily carbon oxides. The surface hydrocarbon species contains carbonyl groups and unsaturation as evidenced by the bands in the 1700-1870 cm^{-1} and 1500-1600 cm^{-1} regions, respectively. Studies of adsorbed maleic anhydride and lactone (butenolide) on this catalyst indicated the observed surface hydrocarbon species could not be attributed to either of these molecules. The olefins exhibited a much greater tendency to form the surface hydrocarbon species than did furan, especially 1,3-butadiene. Further, the amount of surface hydrocarbon species increased with increasing oxygen concentration.

The source of oxygen for the oxidation of the surface hydrocarbon species and the hydrocarbon reactant was investigated using oxygen-18 in a double pulsed experiment. The catalyst was first oxidized with oxygen-16 and then double pulsed with oxygen-18 and furan as before. Interestingly, the largest CO_2 peak corresponding to the oxygen-18 injection was from $CO^{16}O^{16}$ and it was approximately 1.5 times larger than $CO^{18}O^{16}$ and an order of magnitude greater than the $CO^{18}O^{18}$ isotope. In order to better understand the time dependencies of this process, single pulse studies were conducted with cis-2-butene and furan.

The cis-2-butene experiment was performed by pulsing a 4/1 mixture of O_2^{18}/butene through an equilibrated $(VO)_2P_2O_7$ catalyst and monitoring the CO_2 temporal curves of the various isotopes. Figure 9 shows the normalized curve shapes of the three isotopes ($CO^{16}O^{16}$, $CO^{16}O^{18}$, $CO^{18}O^{18}$) collected during the first 100 pulses. The initial ratio of peak intensities of $CO^{16}O^{16}$:$CO^{16}O^{18}$:$CO^{18}O^{18}$ is 16:3:1. Further, each peak has a unique curve shape and the various peak maxima occur at different times. The fastest process was $CO^{18}O^{18}$ formation using gas phase or chemisorbed oxygen-18, and the slower and dominant process involved lattice oxygen. Monitoring the amounts of the various isotopes as a function of pulse number showed the expected gradual increase in the intensity of the $CO^{18}O^{18}$ peak at the expense of $CO^{16}O^{16}$ and $CO^{16}O^{18}$. Further, the peak shape of $CO^{18}O^{18}$ changed with continued pulsing, and eventually contained the slow broad peak as well. Qualitatively similar results were obtained using furan as the reactant.

These results show that at least two types of oxygen are involved in the nonselective pathway to CO_2. A small amount of chemisorbed oxygen reacts with the hydrocarbon directly to form CO_2 by a fast process. The surface lattice oxygen plays a significant role in producing CO_2 by a much slower process. In this latter slower mechanism, the oxygen from the gas phase is first dissociatively adsorbed on the surface and incorporated into the surface lattice. Further, there must be high surface mobility of the surface lattice oxygen because the oxygen activation is accompanied by the oxidation of surface hydrocarbon deposits and the release of carbon dioxide with predominantly lattice oxygen. This mechanism is particularly important with the intermediate olefins and furan because of their propensity to form surface deposits as compared to butane.

FINAL REMARKS

Our results show the active catalyst for butane oxidation to maleic anhydride is $(VO)_2P_2O_7$. Butane oxidation may involve either a single site (Figure 10A-D) or multiple sites depending on oxygen availability. If oxygen cannot be channeled to the reaction quickly enough, then an intermediate product may desorb which can react at a different site (Figure 10D-F). This latter stepwise process has lower selectivity. At least two, and possibly three, types of oxygen play a role in butane oxidation. Activated oxygen [O*], which we suggest is formed by the irreversible dissociative chemisorption of dioxygen via oxidation of V^{+4} to surface V^{+5}, is responsible for oxidation of the CH bonds of butane (Figure 10A) and furan. Surface lattice oxygen is responsible for allyl oxidation and ring insertion (Figure 10B-C). Our results showing a dependence of maleic product concentration on the surface residence time of oxygen suggests that a shorter lifetime [O*], perhaps a partially reduced dioxygen species (Figure 10C), increases maleic yield from furan. Non-selective reactions lead to the formation of partially oxidized hydrocarbon surface species which can react with activated oxygen, primarily through the surface lattice to produce carbon oxides (Figure 10F-H). Since this latter process channels active oxygen away from the active site, it decreases selectivity.

ACKNOWLEDGEMENTS

We wish to express our appreciation to those who participated in the experimental work, including Bill Andrews and Fred Strauser (TAP), Sylvia Nadean and Henry Yuen (TGA), Frank May and Mike Thompson (X-ray), Cathy Ragland and Jim Wrobleski (PVO analysis), Isaam Hammadeh and Bob Friedman (FTIR), John Freeman (IR Microprobe) and Victoria Franchetti for her support of this fundamental work.

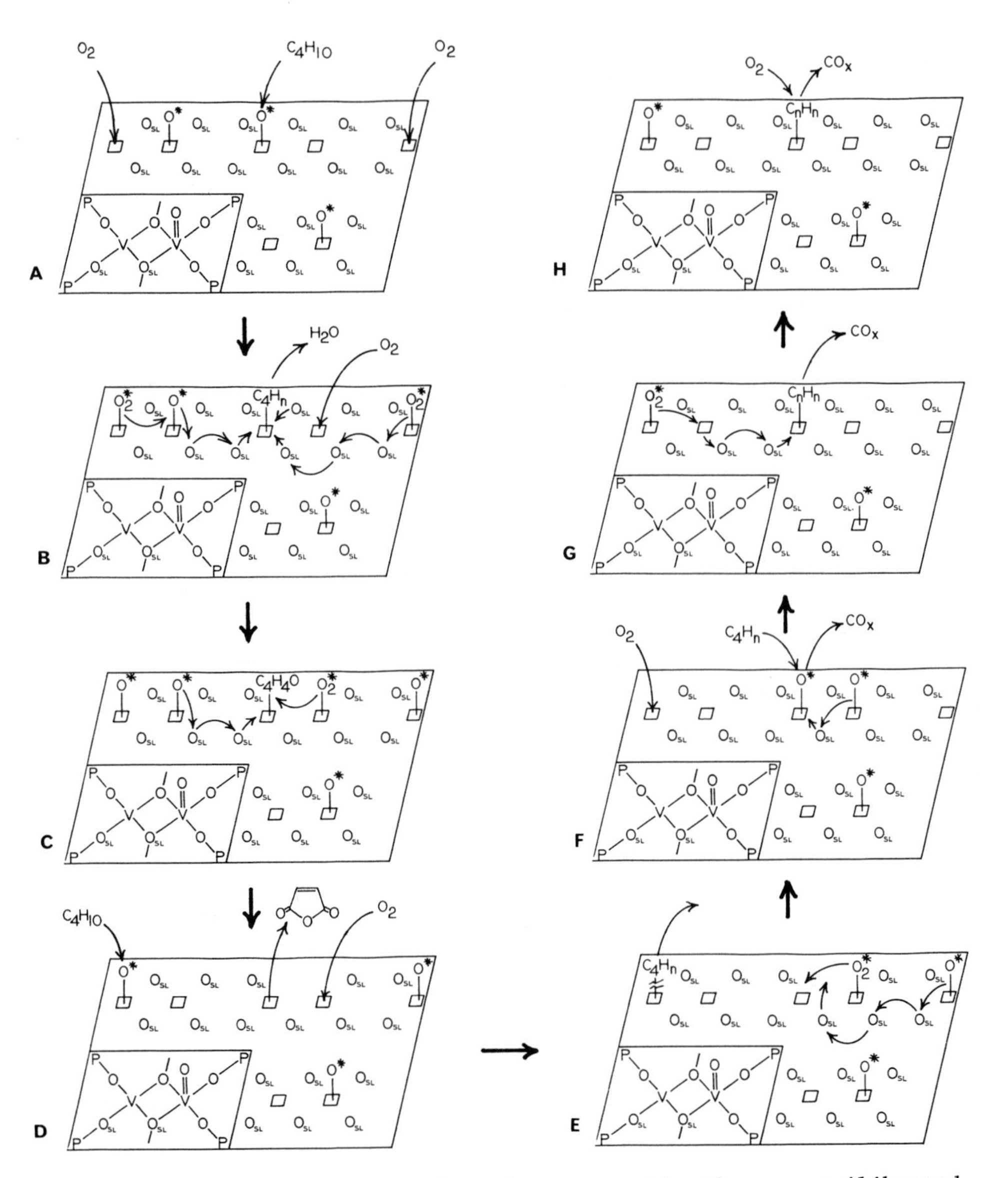

Fig. 10. Schematic representation of oxygen activation on equilibrated $(VO)_2P_2O_7$.

REFERENCES

1. J. B. Moffat, In "Topics in Phosphorous Chemistry", M. Grayson, E. J. Griffith, eds., John Wiley and Sons, New York, 10:285 (1980).

2. J. R. Ebner, U. S. Patent 4457905 (to Monsanto Co.) 1984.

3. J. R. Ebner, J. T. Gleaves, T. C. Kuechler, and T. P. Li, Div. of Petroleum Chemistry, Preprints, 31(1):54 (1986).

4. J. R. Ebner, J. T. Gleaves, T. C. Kuechler, T. P. Li, "Industrial Chemicals via C1 Processes", D. R. Fahey, ed., ACS Symposium Series No. 328, American Chemical Society, Washington, D.C., p. 189-205 (1986).

5. E. Bordes and P. Courtine, J. Chem. Soc. Chem. Comm., 294 (1985).

6. E. Bordes and P. Courtine, J. Catal., 57:236 (1979).

7. G. Centi, I. Manenti, A. Riva, and F. Trifiro; Appl. Catal. 9:177 (1984).

8. G. Bergeret, J. P. Broyer, M. David, P. Gallezot, J. C. Volta and G. Hecquet, J. Chem. Soc. Chem. Comm., 825 (1986).

9. J. S. Buchanan, J. Apostolakis and S. Sundaresan, Appl. Catal., 19:65 (1985).

10. B. K. Hodnett and B. Delmon, J. Catal., 88:43 (1984).

11. R. W. Wenig and G. L. Schrader, Ind. Eng. Chem. Fundam., 25:612 (1986).

12. J. W. Johnson, D. C. Johnston, A. J. Jacobson and J. F. Brady, J. Am. Chem. Soc., 106:8123 (1984).

13. E. Bordes, P. Courtine and J. W. Johnson, J. Solid State Chem., 55:270 (1984).

14. M. A. Pepera, J. L. Callahan, M. J. Desmond, E. C. Milberger, P. R. Blum and N. J. Bremer, J. Am. Chem. Soc., 107:4883 (1985).

15. G. Busca, G. Centi and F. Trifiro', J. Am. Chem. Soc., 107:7757 (1985).

16. G. Centi and F. Trifiro', J. Molec. Catal., 35, 255 (1986).

17. T. P. Moser and G. L. Schrader, J. Catal., 92:43 (1985).

18. B. K. Hodnett and B. Delmon, Ind. Eng. Chem. Fundam., 23: 465 (1984).

19. R. M. Contractor, H. E. Bergna, H. S. Horowitz, C. M. Blackstone, B. Malone, C. C. Torardi, B. Griffiths, U. Chowdhry and A. W. Sleight, Catal. Today, 1:49 (1987).

20. A. Escardino, C. Sola and F. Ruiz, F. An. Quim., 69:1157 (1973).

21. K. Wohlfahrt and H. Hofmann, Chem. Ing. Techn., 52:811 (1980).

22. R. K. Sharma and D. L. Cresswell, Paper from AICHE Meeting, San Francisco, Nov. 1984.

23. J. J. Lerou, Paper from Pittsburgh/Cleveland Catalysis Society Meeting, May 1986.

24. J. S. Buchanan and S. Sundareson, Appl. Catal. 26:211 (1986).

25. G. Centi, G. Fornasari and F. Trifiro', Ind. Eng. Chem. Prod. Res. Dev., 24:32 (1985).

26. G. K. Boreskov in "Catalysis Science and Technology"; J. R. Anderson, M. Boudart, eds, Springer-Verlag, New York, 3:39 (1982).

27. M. Che and A. J. Tench, Adv. in Catal. 31:78 (1982).

28. P. Mars and D. W. van Krevelen, Chem. Eng. Sci. (Special Suppl.) 3:41 (1954).

29. R. A. Keppel and V. M. Franchetti, U. S. Patent 4632915 (to Monsanto), 1986.

30. J. T. Wrobleski, J. W. Edwards, C. R. Graham, R. A. Keppel and H. Raffelson, U. S. Patent 4562268 (to Monsanto), 1985.

31. J. T. Wrobleski, R. F. Brooks, W. S. Ellis and R. A. Mount, Monsanto Internal Report, (1982).

32. T. P. Moser and G. L. Schrader, J. Catal., 92:216 (1985).

33. Yu. E. Gorbunova and S. A. Linde, Sov. Phys. Dokl., 24:138 (1979).

34. G. Centi, G. Fornasari and F. Trifiro', J. Catal., 89:44 (1984).

35. J. T. Gleaves and J. R. Ebner, U. S. Patent 4626412 (to Monsanto Co.) 1986.

36. R. M. Friedman and H. C. Dannhardt, Rev. Sci. Instrum., 56(8):1589 (1985).

37. R. M. Friedman and I. M. Hammadeh, Monsanto Internal Report, (1986).

THE OXIDATION OF ORGANIC COMPOUNDS BY METAL COMPLEXES IN ZEOLITES

Chadwick A. Tolman and Norman Herron

Central Research and Development Department
E. I. DuPont de Nemours & Co., Experimental Station
Wilmington, DE 19898

INTRODUCTION

The selective oxidation of organic compounds to desired products has long been a challenge to chemists. The industrial oxidation of hydrocarbons to useful oxygenated compounds is commercially important and is carried out on a very large scale - on the order of several billions of pounds per year. The reactions are usually carried out at high temperatures(>150°C) and pressures,[1] and often leave much to be desired in terms of selectivity. The difficulty lies in the fact that the desired products are often themselves easily oxidizable, so that a certain percentage of the carbon is inevitably lost to CO and CO_2, and other byproducts.

Biological systems are well known for their ability to carry out chemical reactions with high selectivity under ambient conditions. The enzymes responsible for hydrocarbon oxidations are the cytochromes P-450.[2] Some of them, the omega-hydroxylases, have the amazing abiliity to hydroxylate the terminal methyl groups of long chain hydrocarbons.[3] In this paper we describe some of our work aimed at mimicking the P-450 enzymes. We have studied reversible O_2 binding to cobalt complexes in zeolites, but have emphasized oxidizing saturated hydrocarbons at ambient temperature using iron complexes in zeolites as catalysts. The idea here was to use the inorganic aluminosilicate framework of the zeolite to direct the approach of substrates to the oxidation site - filling the role of the tertiary structure of the enzyme's protein. The molecular sieving action of the zeolites was also expected to give substrate selectivity based on molecular size. We were encouraged in this respect by earlier work on olefin hydrogenation by Rh in zeolites, where we were able to get a selectivity of over 40:1 favoring hydrogenation of cyclopentene over methylcyclohexene.[4]

BACKGROUND

Table I shows heats of combustion for a variety of organic compounds, along with the number of moles of O_2 required for complete combustion. The quotient of the two, in the last column, is the heat evolved per mole of O_2 consumed. The values are remarkably constant (an average of about 105 kcal/mol of O_2 or 52 kcal/g atom of O), even for the carbohydrate sucrose. Individual steps of partial oxidation are close: 47 kcal from cyclohexane to cyclohexanol and 50 from cyclohexanol to cyclohexanone. The high heat of combustion of hydrocarbons largely accounts for their use as engine fuels. This serves to point out the difficulty of stopping an oxidation at an intermediate point short of CO_2 and water.

In spite of the thermodynamic instability of hydrocarbons and other organic compounds in the presence of oxygen, the kinetics of the reactions are usually very slow, except at elevated temperatures. The bond dissociation energies in Table II help to explain why. The very weak O-H bond in $\cdot$OO-H means that hydrogen abstraction from an alkane by O_2 as in equation (1) is endothermic by 40 to 50 kcal/mol.

$$O_2 + R\text{-}H \longrightarrow \cdot O_2\text{-}H + R\cdot \quad (1)$$

Since the Arrhenius activation energy is at least as large as the endothermicity, the rate constant for (1) is extremely small below temperatures of about 300°C. The first step in the oxidation of a hydrocarbon instead usually involves hydrogen atom abstraction by an oxyradical that can form a stronger O-H bond. In the case of the

$$CyO_2\cdot + CyH \longrightarrow CyO_2H + Cy\cdot \quad (2)$$

$$Cy\cdot + O_2 \longrightarrow CyO_2\cdot \quad (3)$$

industrial oxidation of cyclohexane, the oxyradicals are generated by metal ion catalysed decomposition of cyclohexylhydroperoxide.[5] As seen in Table II, hydrogen abstractions from CyH by CyO$\cdot$ and $CyO_2\cdot$ are exothermic by 8 or endothermic by 4 kcal/mol, respectively. The subsequent reaction of Cy$\cdot$ with O_2 is extremely rapid and highly exothermic. Cycling through reactions (2) and (3) provides a radical

Table I. Heats of Combustion[a]

Compound	$-\Delta H$ (kcal/mol)	nO_2[b]	$-\Delta H/n$
Methane	211	2.0	105
Propane	526	5.0	105
Cyclohexane	938	9.0	104
Cyclohexanol	891	8.5	105
Cyclohexanone	841[c]	8.0	105
Adipic acid	669	6.5	103
Benzene	782	7.5	104
Sucrose	1350	12.0	112

[a] To burn the compound to CO_2 and H_2O (liq.), taken from the Handbook of Chemistry and Physics, 41st Edition, Chemical Rubber Publishing Co., Cleveland, 1959-60, pp. 1913-1920, unless noted otherwise.

[b] n is the number of moles of O_2 required per mole of compound.

[c] Acta. Chem. Scand., 16, 46 (1962).

Table II. Typical Bond Energies[a]

$$A\text{-}B \longrightarrow A\cdot + B\cdot$$

O-H	C-H	O-O	D(kcal/mole)
HO-H			119
$CH_3CH_2CH_2C(=O)O\text{-}H$			108[b]
	$C_6H_5\text{-}H$		104
$CH_3CH_2O\text{-}H$; CyO-H[c]			102
	$C_2H_5\text{-}H$		98
	Cy-H		94
	$(CH_3)_3C\text{-}H$		91
$HO_2\text{-}H$; CyOO-H[c]			90
	$CH_3C(=O)\text{-}H$; cyclohexanone (α C-H)[c]; cyclohexanol (HO, H)[c]		88
	$C_6H_5CH_2\text{-}H$		85
•OO-H			47[d]
		$CH_3O\text{-}OH$	43
		$CH_3O\text{-}OCH_3$	36

a Taken from J. A. Kerr, Chem. Rev., 66, 465, (1966) except as noted otherwise.
b Calculated from ΔH°_f for H•, $CH_3CO_2\cdot$, and CH_3CO_2H from S. W. Benson "Thermochemical Kinetics", John Wiley & Sons, New York, 1976.
c Estimated by analogy with compounds of similar structure.
d Calculated from ΔH°_f for H• and $HO_2\cdot$ from S. W. Benson.

chain mechanism in which many molecules of substrate can be oxidized for each initiation event. Eventually the chains are terminated by radical-radical reactions like (4), where K and A represent the ketone and alcohol, respectively.

$$CyO_2\cdot + CyO_2\cdot \longrightarrow K + A + O_2 \qquad (4)$$

Because the propagation reaction (2) has an activation energy of about 18 kcal/mol,[6] whereas termination (4) has nearly zero,[7] the oxidation is still very slow below temperatures of about 150°C.

The P-450 enzymes work quite differently. Figure 1 shows the currently accepted catalytic cycle. Substrate binding is followed by a

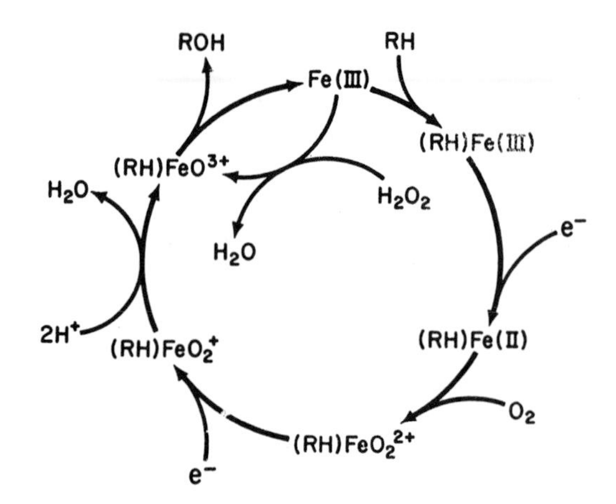

Fig. 1 The cytochrome P-450 catalytic cycle. The porphyrin dianion is not shown.

one-electron reduction of the Fe(III) to Fe(II), which can bind O_2 much like hemoglobin. A further one-electron reduction is followed by removal of one oxygen atom as water, leaving a reactive ferryl (FeO) which is capable of attacking the hydrocarbon. The ferryl can also be produced directly by what is known as the peroxide shunt, supplying the two electrons and two protons with the oxygen as H_2O_2. Model studies with model iron porphyrin complexes have often used iodosobenzene (PhIO) as an O atom transfer reagent, following Groves.[8] While the FeO group is believed to have some oxyradical character, any alkyl radical intermediates formed by hydrogen abstraction must be extremely short lived, since hydroxylation of optically active 1-H,D,T-octane gave 1-D,T-octanol with retention of configuration.[9]

The P-450 enzymes are monooxygenases - that is they incorporate only one oxygen atom of O_2 into the substrate, while the other is reduced to water. A number of nonbiological systems which can oxidize hydrocarbons with oxygen at room temperature, including Barton's Gif and Gif-Orsay systems,[10] also employ either reducing agents as cofactors or their electrochemical equivalent as electrons. It seemed very puzzling at first that oxygen could be made into a more active oxidizing agent by supplying a reducing agent. The reason can be understood, however, by reference to Table III. It takes 59 kcal to produce a g-atom of O atoms - powerful oxidizing agents which will react rapidly with hydrocarbons even below room temperature.[11] Ozone is also a powerful oxidant, though it is 25 kcal/mol less energetic than atomic oxygen; N_2O is a still weaker oxidant. Entries with a positive ΔH are stronger oxidants than O_2 itself. $OPPh_3$ is the weakest on the list, as PPh_3 is the strongest reducing agent. Thermodynamically, the best reducing agents on the list are capable of reducing one atom of O_2 to give a free O atom. For example (5) is exothermic by 9 kcal/mol.

$$H_2 + O_2 \longrightarrow H_2O + O \qquad (5)$$

We don't know where in the list to place the ferryl of P-450, or the iodosobenzene often used in model studies, but it is reasonable to put PhIO below HClO (+8) and a ferryl below that but above about -50, the value for partial oxidation of an organic substrate.

Table III. Heats of Oxidation[a]

Reaction	ΔH (kcal/mol)
$1/2\ O_2 \longrightarrow O$	59
$O_2 + 1/2\ O_2 \longrightarrow O_3$	34
$CH_3CO_2H + 1/2\ O_2 \longrightarrow CH_3CO_3H$	34
$H_2O + 1/2\ O_2 \longrightarrow H_2O_2$	24
$N_2 + 1/2\ O_2 \longrightarrow N_2O$	17
$HCl + 1/2\ O_2 \longrightarrow HClO$	8
$Me_2S + 1/2\ O_2 \longrightarrow Me_2SO$	-32[b]
$H_2 + 1/2O_2 \longrightarrow H_2O$	-68
$CO + 1/2\ O_2 \longrightarrow CO_2$	-68
$PPh_3 + 1/2\ O_2 \longrightarrow OPPh_3$	-70[b]

[a] For compounds in their standard state at 25°, taken from the Handbook of Chemistry and Physics, 41st Edition, Chemical Rubber Company, Cleveland, 1959-60, pp. 1800-1808, unless noted otherwise.
[b] From D. R. Stully, "The Chemical Thermydynamics of Organic Compounds," John Wiley & Sons, New York, 1969.

SHIP-IN-A-BOTTLE O_2 COMPLEXES

While many dioxygen complexes of transition metals are known which can be considered models for O_2 binding in biological systems, most suffer from an instability in solution at room temperature resulting from dimer formation, with CoOOCo bridges in the case of cobalt, or ultimately FeOFe bridges in the case of iron.[12] The problem can be circumvented by enclosing the metal center with organic bulk, as in the case of Collman's picket fence Fe porphyrin.[13] The protein must prevent dimerization in the enzymes. We reasoned that a metal complex capable of binding O_2, encapsulated inside a zeolite, would be prevented from dimerizing by the inorganic framework.

Co(Salen) (Salen = the diimine of salicylaldehyde with ethylenediammine) was prepared in a 13X zeolite by metal ion exchange followed by heating in molten H_2Salen.[14] The initially blue [Co(II)] zeolite turned orange as expected. An initial washing gave a colored solution, but further washing or even extended Soxhlet extraction was unable to remove any further complex, indicating that the Co(Salen) was trapped inside, like a ship-in-a-bottle. The 13 Å diameter supercages of the zeolite can accommodate the 12 Å long complex, but it can't pass out through the 7 Å windows. A prep with the smaller 5 Å window zeolite 5A also gave an orange product, but washing gave back a blue zeolite. In this case the 6 Å aromatic rings couldn't get in. Addition of pyridine to the 13X Co(Salen) gave the expected color change, but the material did not give a detectable esr signal for an O_2 complex. A reversible O_2 complex could be prepared, however, by changing the order of synthesis - adding the pyridine before the H_2Salen. Apparently the preformed Co(Salen) so blocked the channels of the zeolite that the pyridine had access only to the 1 or 2% of the complexes near the surface. Measurements of the equilibrium constant for O_2 binding to the Co(Salen)py show that binding is less favorable in the zeolite than in solution, possibly because of steric crowding in the supercage. Typical esr spectral

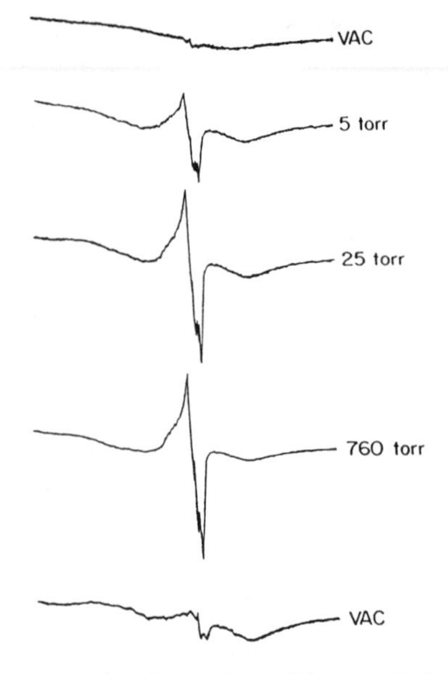

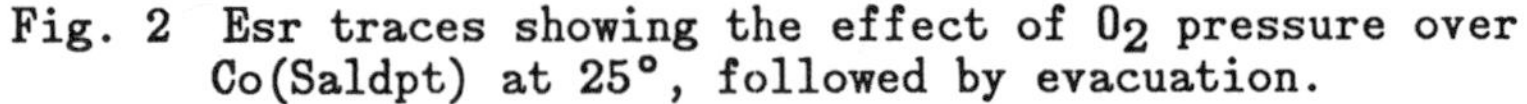
Fig. 2 Esr traces showing the effect of O_2 pressure over Co(Saldpt) at 25°, followed by evacuation.

traces at various O_2 pressures are shown in Figure 2*; the last shows that O_2 can be removed under vacuum. The encapsulated O_2 complex, unlike its solution analog, is indefinitely stable at room temperature under an O_2 atmosphere.

IRON PHTHALOCYANINE WITH IODOSOBENZENE

The Fe phthalocyanine (FePc) in zeolite catalysts[15] were prepared by ion exchanging Fe(II) into X or Y zeolites to various extents (20% exchange corresponds to one Fe per supercage), followed by heating in molten o-dicyanobenzene for four hours.[16] The catalyst was then washed with acetone and Soxhlet extracted with pyridine and then chloronaphthalene for seven days, followed by oven drying at 100°C. Pains were taken to remove all catalyst from the surface. The formation of FePc inside the zeolite is supported by the elemental analysis, the diffuse relectance spectrum, and by the fact that we could recover FePc by dissolving the zeolite in aqueous H_2SO_4. Oxidations were carried out at room temperature for 20 hours, using ratios of substrate:PhIO:Fe of 50:10:1. The reactions desired are (6) and (7).

$$\text{PhIO} + \text{FePc} \longrightarrow \text{PhI} + \text{OFePc} \quad (6)$$

$$\text{RH} + \text{OFePc} \longrightarrow \text{ROH} + \text{FePc} \quad (7)$$

---

* The traces shown are actually those of the O_2 complex of Co(Saldpt) (the ligand from salicylaldehyde and dipropyltriamine), but those with Co(Salen)py are similar.

Table IV. Products of Methylcyclohexane Oxidation Using PhIO and FePc Catalysts[a]

Catalyst		Turnover	K/A
FePc[b]		1.1	0.32
Fe(II)/13X	20%[c]	<0.01	-
FePc/Y20	1%	5.6	0.33
FePc/X	2%	4.1	0.48
FePc/X	20%	0.5	0.56

[a] From ref. 16
[b] Turned brown
[c] The percentage of cations exchanged by Fe(II).

Table IV shows some of the results with methylcyclohexane as the substrate. The unsupported FePc catalyst solution turned dark brown and lost activity, presumably because of attack upon itself.* The iron exchanged zeolite without the Pc ligand showed no signs of activity. About five cycles of oxidation were observed in the most favorable cases, with 1 or 2% Fe exchange. The encapsulated catalysts did not change color, but lost activity because of channel blockage by iodoso- or iodoxybenzene (as shown by gc after heating), but could be restored to their original activity by heating in a vacuum oven. The iodoxybenzene can be formed in side-reaction (8).

$$PhIO + OFePc \rightarrow PhIO_2 + FePc \qquad (8)$$

The 20% exchanged catalyst, with one FePc per supercage, gave the lowest turnover, presumably because access to the interior of the particles was blocked and only the molecules of complex near the surface were active. The higher ketone/alcohol ratio found with the zeolite catalysts suggests that some of the molecules of alcohol were oxidized to ketone before they could escape into solution.

Some stereochemical selectivity could be seen in the products, for example in the ratio of trans/cis-4-methylcyclohexanol, which was slightly larger in the zeolite. Similarly small selectivities were seen in the norborneols, with higher endo/exo ratios. The zeolite must have some orienting effect on the substrates.

In our early work[16] we saw only a small rate preference (factors of about 1.2) for cyclohexane in competitions with the larger cyclododecane, when comparing the encapsulated and unencapsulated FePc catalysts, presumably because the channels of the faujasite structure are relatively large, and the substrates are quite flexible. Subsequently, however, we found that selectivities of 10:1 could be achieved by ion exchanging various large cations into the zeolites.[18] The effect was reminiscent of the large increases in hydrogenation selectivity observed earlier[4] by the addition of water.

* A second order disappearance of Fe(TTP) (TTP = tetraphenylporphyrin)[17] in the presence of PhIO supports this idea.

Fe(III)/Pd(0) in 5A WITH O_2/H_2 AS OXIDANT

While the catalysts with FePc in X or Y showed a small regio-selectivity (slightly increasing the ratio of 4-/2-octanol) with n-octane,[16] it is clear that the windows to the supercages are too large to have much effect in orienting the substrate to the metal center. A 5A zeolite with 5 Å windows has the advantage of a nice tight fit, as shown in Figure 3. However, an aromatic compound like PhIO has a 6 Å diameter and is too large to get in, so that a smaller oxidant is required. Also the 11 Å supercage limits the size of ligand one may use. We thought, however, that it might be possible to make a mixed function catalyst containing Fe(III) and Pd(0), and use a mixture of O_2 and H_2 as the oxidizing agent. The idea was that the Pd(0) might use the H_2 to reduce off one of the oxygen atoms, leaving the other on the iron as a ferryl. We were encouraged by reports that Pd(0)/H_2 in an acidic environment could reduce O_2 to H_2O_2.[19] It worked![20] All of the components of the system - Fe(III)*, Pd(0), O_2, and H_2 must be present in the zeolite for the system to work, as shown by control experiments. No selectivity is observed with the same system on amorphous aluminosilicate.

Reactions were typically carried out for four hours at 25° under 45 psig O_2 and 15 psig H_2, with 200 mg catalyst (about 5 mg Fe and 2 mg Pd) in 9 ml methylene chloride and 1 ml substrate. Figure 4 shows a gc trace of the products after a competition of a 1:1 mixture of n-octane and cyclohexane. The large peak in the center is the chlorobenzene internal standard. The peaks to the right, from right to left, are 1-, 2-, 3-, and 4-octanol, respectively. A trace peak between these and the chlorobenzene is cyclohexanol. The selectivity for n-octane over cyclohexane is over 150:1! The ratio of 1-/2-octanol (0.7 on a per hydrogen basis) is the largest yet observed for an omega-hydroxylase mimic - including Suslick's tetra(o,o'-diphenyl)phenylporphyrin system.[21] One difference between our system and Suslik's is that our 1-/2-alkanol ratio is quite insensitive to chain length, as shown in Figure 5, while his (and the natural enzyme's) become much less selective with shorter chains.

Fig 3. A computer drawn space-filling model of a normal alkane entering a channel of zeolite 5A.

* The Fe(II) used for ion exchange is rapidly converted to Fe(III) in the zeolite in the presence of O_2.

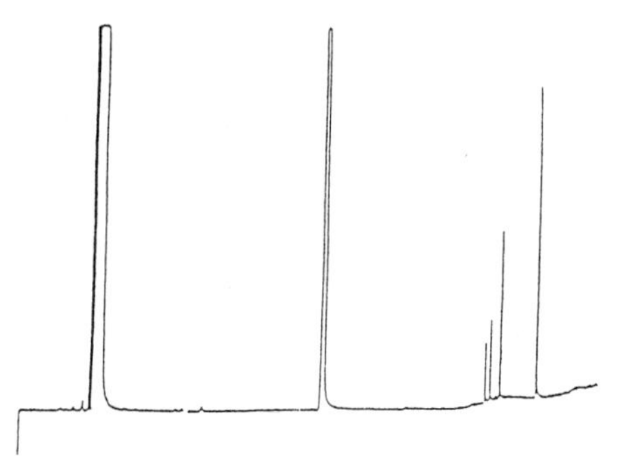

Fig. 4 A gc trace of the product mixture obtained by oxidizing a 1:1 mixture of cyclohexane and n-octane with O_2/H_2 over a Fe(III)Pd(0) in 5A catalyst at 25°. The strong central peak is the chlorobenzene internal standard.

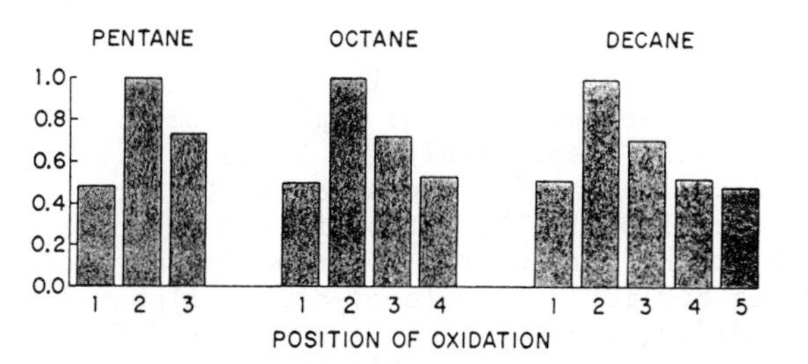

Fig. 5 Regioselectivity of oxidation of various normal alkanes over a Fe(III)/Pd(0) in 5A catalyst, on a per hydrogen basis, normalized to 1.0 for C2. From ref. 20.

The trace shown in Figure 4 was run after washing the catalyst with water and dissolving the zeolite with aqueous sulfuric acid. 5A zeolite has such small channels that the oxidation products formed inside get stuck. We are currently exploring catalysts with channels which are large enough to allow the products to escape while still giving good selectivity.

SUMMARY

Most commercial oxidations using O_2 as the oxidant are operated at high temperatures, where selectivities are often not good, because of the need for high temperatures to get adequate rates. Biological enzymes of the P-450 type circumvent this difficulty by reducing one coxygen atom of O_2 to water while using the other to produce a reactive ferryl iron porphyrin whose protected position in an envelope of protein prevents dimerization and restricts access of substrates, and in favorable cases permits selective hydroxylation of the terminal methyl groups of linear alkanes. We have been able to mimic a number of features of the enzymes, including reversible O_2 binding and omega-hydroxylation of linear alkanes, using the inorganic structure of the zeolite to replace the protein. The most selective catalysts we have found have both Fe(III) and Pd(0) in 5A and use an O_2/H_2 mixture as the oxidant.*

The advantages of our metal-in-zeolite catalysts, compared with their soluble counterparts, include: ease of preparation, increased catalyst life, ease of separation of catalyst for reuse, selectivities based on substrate size and shape, and tunability. Disadvantages include: slow rates of catalyst turnover, size limitations on complexes that can be built inside a zeolite, and pore blockage in some cases by the catalyst complex or products of reaction.

REFERENCES

1. K. Weissermel and H.-J. Arpe, "Industrial Organic Chemistry," Verlag Chemie, New York (1978).
2. R. E. White and M. J. Coon, Ann. Rev. Biochem., 49, 315 (1980); I. C. Gunsales and S. C. Sligar, Adv. Enzymology, 47, 1 (1978).
3. M. Hamburg, B. Samuelsson, I. Bjorkhem, and H. Danielsson, p. 29 in "Molecular Mechanisms of Oxygen ACtivation," O. Hayaishi, ed., Academic Press, New York (1974).
4. D. R. Corbin, W. C. Seidel, L. Abrams, N. Herron, G. D. Stucky, and C. A. Tolman, Inorg. Chem. 24, 1800 (1985)).
5. C. A. Tolman, N. Herron, M. J. Nappa, and J. D. Druliner, 'Hydrocarbon Oxidation Studies in Du Pont's Central Research' in "Activation and Functionalization of Alkanes," C. L. Hill, ed., in preparation.
6. S. Korcek, J. H. B. Chenier, J. A. Howard, and K. U. Ingold, Can. J. Chem. 50, 2285 (1972).
7. R. L. McCarthy, and A. MacLachlan, Trans. Farad. Soc., 57, 1107 (1961).
8. J. T. Groves, T. E. Nemo, and R. S. Meyers, J. Am. Chem. Soc., 101, 1032 (1979).
9. E. Casper, S. Shapiro, and J. Piper, J. Chem. Soc. Chem. Commun., 1198 (1981).
10. D. H. R. Barton, M. J. Gastiger, and W. B. Motherwell, J. Chem. Soc. Chem. Commun., 41 and 731 (1983); G. Balavoine, D. H. K. Barton, J. Boivin, A. Gref, N. Ozbalik, and H. Riviere, J. Chem. Soc. Chem. Commun., 1727 (1986).
11. T. H. Varkony, S. Pass, and Y. Mazur, J. Chem. Soc. Chem. Commun., 457 (1975).
12. R. D. Jones, D. A. Summerville, F. Basolo, Chem. Revs., 79, 139 (1979).
13. J. P. Collman, R. R. Gagne, C. A. Reed, T. R. Halbert, G. Lang, and W. T. Robinson, J. Am. Chem. Soc., 97, 1427 (1975).
14. N. Herron, Inorg. Chem., 25, 4714 (1986).
15. Romanovsky, B. V., Proc. Int. Symp. on Zeolite Catalysis, Siofok, May 13-16, 1985, p. 215.
16. N. Herron, C. A. Tolman, and G. D. Stucky, J. Chem. Soc. Chem. Commun., 1521 (1986).
17. M. J. Nappa, and C. A. Tolman, Inorg. Chem., 24, 4711 (1985).
18. N. Herron and C. A. Tolman, unpublished results.

---

* More recent experiments have shown that selectivities can be further improved using other zeolites containing Fe(III), using H_2O_2 as the oxidant.[18]

19. G. W. Schoenthal, Ger. Patent 2,615,625 (1976); Y. Izumi, H. Miazaki, and S. Kawahara, U. S. Patent 4,009,252 (1977); F. Mosley, U. S. Patent 4,128,627 (1978).
20. N. Herron, and C. A. Tolman, J. Am. Chem. Soc., accepted for publication.
21. B. R. Cook, T. J. Reinert, and K. S. Suslick, J. Am. Chem. Soc., 108, 7281 (1986).

ABSTRACTS OF POSTERS

Further information on the subject matter of these posters may be obtained from the senior authors, indicated by *.

KINETICS AND MECHANISMS OF DEGRADATION OF BINUCLEAR COBALT DIOXYGEN COMPLEXES

Arup K. Basak and Arthur E. Martell*

Department of Chemistry
Texas A&M University
College Station, Texas 77843

The stability constants and oxygenation constants of the cobalt(II) dioxygen complex, **1**, of the polyamine 1,6-bis(2-pyridyl)-2,5-diazahexane, PYEN, have been determined by potentiometric methods. The degradation of the binuclear dibridged (μ-peroxo-μ-hydroxo)dioxygen complex formed by Co(II)-PYEN is compared with those of the monobridged cobalt(II)-dioxygen complexes, **2** and **3**, formed by the cobalt(II) complexes of the pentadentate polyamines 1,9-bis(2-pyridyl)2,5,8,-triazanonane (PYDIEN)[1] and 2,6-bis(2-(3,6-diazahexyl)pyridine) (EPYDEN)[2], respectively. The equilibrium constants obtained determine the concentrations of the complexes formed by PYEN in aqueous solution and distribution curves are presented, which show the pH range in which the dioxygen complex **1** is the predominant species.

All of the dioxygen complexes, **1**, **2**, and **3**, are stable in aqueous solution at 25°C, but undergo degradation to inert complexes at moderately elevated temperatures (35°C for **2** and **3** and 50°C for **1**). Dioxygen complexes **1** and **2** undergo base-catalyzed oxidative dehydrogenation of the ligands by coordinated dioxygen to give the Co(II) complexes of the corresponding monoimines. Dioxygen complex **3**, on the other hand, undergoes base-catalyzed displacement of hydrogen peroxide to form the inert Co(III) complex of the unchanged polyamine. Rate laws are presented for these degradation reactions and second-order rate constants for **1** are reported. The oxidative dehydrogenation reactions of **1** and **2** and the lack of such a reaction for **3** shows that oxidative dehydrogenation will not occur without a conformation of the coordinated ligand that allows the direct transfer of a proton from the α-CH_2 group to coordinated dioxygen, the generation of a trigonal nitrogen, and the formation of an imine group conjugated with the aromatic ring. These factors and the existence of a large kinetic deuterium isotope effect (8.5) for the second order oxidative dehydrogenation of the PYEN ligands in **1** leads to the suggestion of a concerted reaction mechanism and a transition state is proposed in which direct proton transfer to coordinated dioxygen is accompanied by electron transfer from the ligand to the dioxygen through the coordinated metal ion.

REFERENCES

1. C. J. Raleigh and A. E. Martell, Inorg. Chem., 24:142 (1985).
2. C. J. Raleigh and A. E. Martell, J. Coord. Chem., 14:113 (1985).

METHANE ACTIVATION OVER LANTHANIDE OXIDES

K. D. Campbell, H. Zhang, and J. H. Lunsford*

Department of Chemistry
Texas A&M University
College Station, Texas 77843

In order to better understand the characteristics of catalysts which are capable of generating methyl radicals from methane and to gain insight into the reactions these radicals undergo once formed on the surface, the activities of the lanthanide oxides for gas-phase methyl radical production were examined using a matrix isolation electron spin resonance technique.[1,2] Oxides of the metals with stable multiple oxidation states (Cr, Pr, Tb) exhibited very low activities. Thus, for this series of oxides, the existence of an active metal center with multiple oxidation states is not a requirement. This agrees with the results reported by Otsuka et al.[3] However our results show that the greatest activities occur for La_2O_3, Sm_2O_3 and Nd_2O_3, whereas Otsuka reported Sm_2O_3 has much greater activity.

Conventional flow reactor studies were also carried out on selected lanthanide oxides to determine if gas phase methyl radical production could be correlated to methane conversion or C_2 selectivity. The results show qualitative agreement, except for Nd_2O_3, which supports the role of gas phase radical coupling in the catalytic conversion of CH_4 to C_2H_6 and C_2H_4.

REFERENCES

1. W. Martir and J. H. Lunsford, J. Am. Chem. Soc., 103:3728 (1981).
2. D. J. Driscoll, W. Martir, J. X. Wang, and J. H. Lunsford, J. Am. Chem. Soc., 107:58 (1985).
3. K. Otsuka, K. Jinna, and A. Morikawa, J. Catal., 100:353 (1986).

DIOXYGEN AFFINITIES OF SOME SYNTHETIC COBALT SCHIFF BASE COMPLEXES

Dian Chen and A. E. Martell*

Department of Chemistry
Texas A&M University
College Station, Texas 77843

The synthesis and dioxygen affinities of six cobalt(II) Schiff base complexes in diglyme solution in the presence of excess aromatic bases are reported. The cobalt(II) chelates investigated are bis-salicylaldehyde-tetramethylethylenediiminocobalt(II) (CoSALTMEN), bis-salicylaldehyde-*o*-phenylenediiminocobalt(II) (CoSALOPHEN), bis-(2-hydroxyacetophenone)-ethylenediiminocobalt(II) (Co CH_3SALEN), bis-(3,5-dichlorosalicylaldehyde)-*o*-phenylenediiminocobalt(II) (Co35ClSALOPHEN), bis-(3-methoxysalicylaldehyde-)*o*-phenylenediiminocobalt(II) (Co3MeOSALOPHEN), and the parent compound, bis-salicylaldehydeethylenediiminocobalt(II) (SALCOMINE). Axial bases employed are pyridine, 4-methylpyridine, 4-dimethylaminopyridine, and 4-cyanopyridine. In the solvent medium employed the Schiff base chelates combine with only one axial base; no 2:1 adducts were observed. Electron-withdrawing substituents on the Schiff bases were found to decrease the affinity of the cobalt Schiff base for dioxygen. Equilibrium dioxygen uptake measurements over a range of temperatures provide values of ΔH^o and ΔS^o of oxygenation which fall in the range -6 to -13 kcal $mole^{-1}$ for ΔH^o and -30 to -52 cal $degree^{-1}$ $mole^{-1}$ for ΔS^o, and are in line with values reported for analogous dioxygen complexes in the literature.

TEMPORAL ANALYSIS OF PRODUCTS (TAP): A UNIQUE CATALYST EVALUATION SYSTEM WITH SUB-MILLISECOND TIME RESOLUTION

J. T. Gleaves, J. R. Ebner*, and P. L. Mills

Monsanto Company
St. Louis, Missouri 63167

A new real-time/in-situ technique called TAP (Temporal Analysis of Products) used for investigation of gas-solid interactions is presented along with examples of its application to a variety of catalyst problems. Key features of TAP include: i) The reactor configuration is a micro-scale fixed bed that accepts a bulk form of the catalyst, ii) The reactants are introduced to the reactor using a pulse input so that the transient response of the catalyst is observed, iii) The time-scale of the experiment is minimally 10 microseconds so that the potential of observing reaction intermediates is increased when compared to previous methods, iv) The reaction products are sensed by mass spectroscopy for positive identification, v) The catalyst-bed temperature can be either isothermal or programmed for adsorption or desorption studies, vi) The number of molecules introduced to th catalyst bed can be varied over a wide range, and vii) Two separate gas pulses can be introduced to the reactor through independently controlled input pulse valves. Key applications of TAP include the determination of adsorption-desorption energies of activation, identification of reaction intermediates, and the deducement of reaction mechanisms and reaction networks. TAP is a general technique and can be applied to the study of most gas/solid reaction systems including selective oxidation reactions, hydrodesulfurization, and zeolite catalysis.

DIOXYGEN INSERTION INTO METAL-CARBON BONDS OF METALLOPORPHYRINS: FORMATION AND CHARACTERIZATION OF ALKYLPEROXY METALLOPORPHYRINS

Maureen K. Geno* and Walid Al-Akhdar

Department of Chemistry
University of Alabama
Tuscaloosa, Alabama 35487-9671

Under photolytic and/or thermal-reaction conditions in organic solvents, dioxygen inserts into the metal-carbon bond of a series of metalloporphyrins, (P)M(R) and (P)M(R)(L), where P = octaethylporphyrin (OEP), tetraphenylporphyrin (TPP), or tetratolylporphyrin (TTP), M = Co(III), R = $-CH_3$, $-CH_2CH_3$, $-CH_2C_6H_5$, or $-CH_2C(CH_3)_3$, and L = PR_3, pyridine, or 1-methylimidazole to form alkylperoxy metalloporphyrins, (P)M(OOR) and (P)M(OOR)(L). The alkylperoxy metalloporphyrns isolated have been characterized by ^1H NMR, FT-IR, UV-visible spectrophotometry, cyclic voltammetry, and elemental analysis. Activation parameters for dioxygen insertion into the Co-R bond have been determined from kinetic experiments. The insertion reaction proceeds via a Co(II) intermediate, which can be detected spectrophotometrically. The rate of formation of the $(P)Co^{III}(OOR)(L)$ complexes is strongly dependent upon the Co-R bond dissociation energies of the parent $(P)Co^{III}(R)$ complexes, which have been determined previously[1]. The generality of this reaction for other metalloporphyrins and the utility of these complexes in understanding heterolytic and homolytic O-O bond cleavage reactions common in the biological manipulation of dioxygen by metalloporphyrins will be discussed.

REFERENCE

1. M. K. Geno and J. Halpern, *J. Am. Chem. Soc.*, 109:1238 (1987).

IRON PORPHYRIN CATALYZED AIR OXIDATION OF ALDEHYDES AND ALKENES

Isam M. Arafa, Kenton R. Rodgers, and Harold M. Goff*

University of Iowa
Iowa City, Iowa 52242

Propionaldehyde undergoes autoxidation to propionic acid in benzene solution in the presence of millimolar quantitites of iron(III) tetraphenylporphyrin complexes. The iron prophyrin catalyst promotes oxidation of some 200 equivalents of aldehyde prior to prophyrin ring destruction that takes place over a period of hours. Addition of cyclohexene to the initial reaction mixture serves to diminish the yield of propionic acid, and produce cyclohexene oxide, 2-cyclohexene-1-ol, and 2-cyclohexen-1-one as major oxidation products. Yields of the oxide approximate those of propionic acid, and are higher than the allylic cyclohexene oxidation products. A mechanism is suggested that involves significant metal-oxo group transfer, rather than total radical chain oxidation. The manganese(III) tetraphenylporphyrin analogue also serves as an efficient catalyst.

OXIDATIVE DIMERIZATION OF METHANE OVER SODIUM-PROMOTED CALCIUM OXIDES

Chiu-Hsun Lin, Ji-Xiang, and Jack H. Lunsford*

Department of Chemistry
Texas A&M University
College Station, Texas 77843-3255

Sodium-promoted calcium oxides are active and selective catalysts for the partial oxidation of methane to ethane and ethylene using molecular oxygen as an oxidant. In a conventional fixed-bed flow reactor, operating at atmospheric pressure, a 45% C_2 (sum of ethane and ethylene) selectivity was achieved to a 33% methane conversion over 2.0 g of the catalyst at 725°C with a gas mixture of CH_4/O_2 = 2. The other products were CO and CO_2. EPR results indicate that $[Na^+O^{\cdot-}]$ centers in Na/CaO are responsible for the catalytic production of $CH_3^{\cdot}$ from methane through hydrogen-atom abstraction. These $CH_3^{\cdot}$ radicals dimerize, primarily in the gas-phase, to form C_2H_6 which further oxidizes to C_2H_4. Increasing temperatures reverse the gas-phase equilibrium $CH_3^{\cdot} + O_2 \rightleftharpoons CH_3O_2^{\cdot}$ to produce more $CH_3^{\cdot}$ and increase the C_2 selectivity. The $CH_3O_2^{\cdot}$ eventually is converted to carbon oxides under the reaction conditions employed, therefore increasing O_2 pressures decrease the C_2 selectivity. There is evidence that $CH_3O_2^{\cdot}$ in the presence of C_2H_6 initiates a chain reaction which enhances the methane conversion. The addition of Na to CaO also reduces the surface area of the catalysts, thus minimizing a nonselective oxidation pathway via surface methoxide intermediates.

SYNTHESIS AND METAL ION AFFINITIES OF A BINUCLEATING POLYAMINE: REVERSIBLE FORMATION OF A COBALT DIOYXGEN COMPLEX

Rached Menif and Arthur E. Martell*

Department of Chemistry
Texas A&M University
College Station, Texas 77843-3255

The synthesis of m-xylyltrien(1,3-bis(2,5,8,11-tetraazaundecyl)-benzene) is described. Potentiometric equilibrium studies of the stability constants of this binucleating ligand with Cu(II), Co(II) and Ni(II) are reported.[1] Equilibrium data are determined for the formation of mononuclear and dinuclear chelates of these metal ions, as well as several protonated and hydroxo chelates. The oxygenation constant of the binuclear Co(II) complex is reported. The cobalt dioxygen complex was found[2,3] to undergo first order degradation at 50°C and 0.1 M ionic strength. Rate constants are reported as a function of p[H].

REFERENCES

1. A. K. Basak and A. E. Martell, in: "Frontiers in Bioinorganic Chemistry", A. V. Xavier, Ed., VCH Publishers, Weinheim, Germany, (1986), p.345-352.
2. A. K. Basak and A. E. Martell, Inorg. Chem., 25:1182 (1986).
3. C. J. Raleigh and A. E. Martell, Inorg. Chem., 25:1190 (1986).

POTENTIOMETRIC DETERMINATION OF STABILITIES OF COBALT(II) COMPLEXES OF POLYAMINE SCHIFF BASES AND THEIR DIOXYGEN ADDUCTS

Ramunas J. Motekaitis and Arthur E. Martell*

Department of Chemistry
Texas A&M University
College Station, Texas 77843

SALCOMINE, ethylenebis(salicylideneiminato)cobalt(II) and 3FSALCOPHEN, _o_-phenylenebis(3-fluorosalicylideneiminato)cobalt(II) combine with dioxygen to form adducts in the presence of 4-methylpyridine

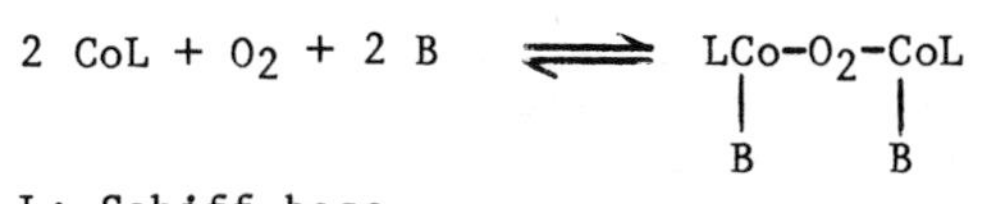

L: Schiff base

B: 4-Methylpyridine

In aqueous dioxane (30% water) solution SALCOMINE and 3-FSALCOPHEN are only partially formed. Depending on pH, under nitrogen the solution will contain various concentrations of Co^{2+}, free neutral and deprotonated salicylaldehyde, uncomplexed neutral, protonated and deprotonated diamine, uncomplexed bis(salicylaldehyde) and mono(salicylaldehyde) Schiff bases together with their several deprotonated forms, and the various possible 1:1 and 2:1 ligand to Co^{2+} complexes. The presence of dioxygen and 4-methylpyridine serves to organize the various components and provide a considerable equilibrium concentration of the binuclear dioxygen-bridged bis-Schiff base complex possessing axial pyridines.

These exceedingly complex solutions were first systematically analyzed anaerobically by equilibrium potentiometry in aqueous dioxane in order to obtain the requisite equilibrium constants relating the various above species before the final equilibrium measurements could be made under dioxygen.

CATALYSIS OF COBALT SCHIFF BASE COMPLEXES FOR THE OXYGENATION OF OLEFINS.

MECHANISMS FOR THE KETONIZATION REACTION

A. Nishinaga*[+], T. Yamada, H. Fujisawa, K. Ishizaki,
H. Ihara, and T. Matsuura

Department of Synthetic Chemistry
Kyoto University, Kyoto 606, Japan

Cobalt Schiff base complexes are interesting because in aprotic solvents they exhibit dioxygenase-like activities as well as the reversible formation of dioxygen complexes.[1] Our recent findings on oxidation-reduction reactions of cobalt Schiff base complexes in alcohols[2] prompted us to investigate oxygenations of organic molecules in protic solvents. Four coordinate cobalt(II) Schiff base complexes, $Co(L^1)$-$Co(L^6)$, are now found to catalyze efficiently, in primary or secondary alcohols under an atmospheric pessure of O_2 at 60°C, the oxygenations of olefin substrates **1**, **5**, and **6**, which result in ketonization without carbon-carbon bond cleavage (Table 1). In the $Co(L^1)$-catalyzed oxygenation of **1a** in $PhCH_2OH$, a mixture composed of **2a**, **3a**, and PhCHO was obtained. Time course of the reaction showed that the reaction proceeded as a co-oxidation of **1a** and $PhCH_2OH$ with a 1:1 stoichiometry (98% selectivity). The rate of the

Table 1. $Co(L^1)$ Catalyzed Oxygenation of Olefins[a]

Olefin	Solvent	Reaction Time (h)	Conversion (%)	Product (%) 2	3	4
1a	MeOH	22	100	91	9	-
1a	EtOH	12	100	83	17	-
1a	i-PrOH[b]	8	100	85	15	-
1a	t-BuOH[b]	48	0	-	-	-
1b	MeOH	22	100	87	13	-
1c	MeOH	14	100	89	11	-
1d	MeOH	12	100	17	70	-
1e	MeOH	10	100	10	60	10
1f	MeOH	24	97	100	0	-
5	EtOH	24	40		90	**7**
6	EtOH	40	60		65	**8**

[a] Olefin (2 mmol), $Co(L^1)$ (0.2 mmol), ROH (10 ml) O_2 (1 atm), 60°C. [b] CH_2ClCH_2Cl (20 ml) was added to dissolve $Co(L^1)$.

[+] Present address: Department of Applied Chemistry, Osaka Institute of Technology, Ohmiya 5, Asahi-ku, Osaka 535, Japan.

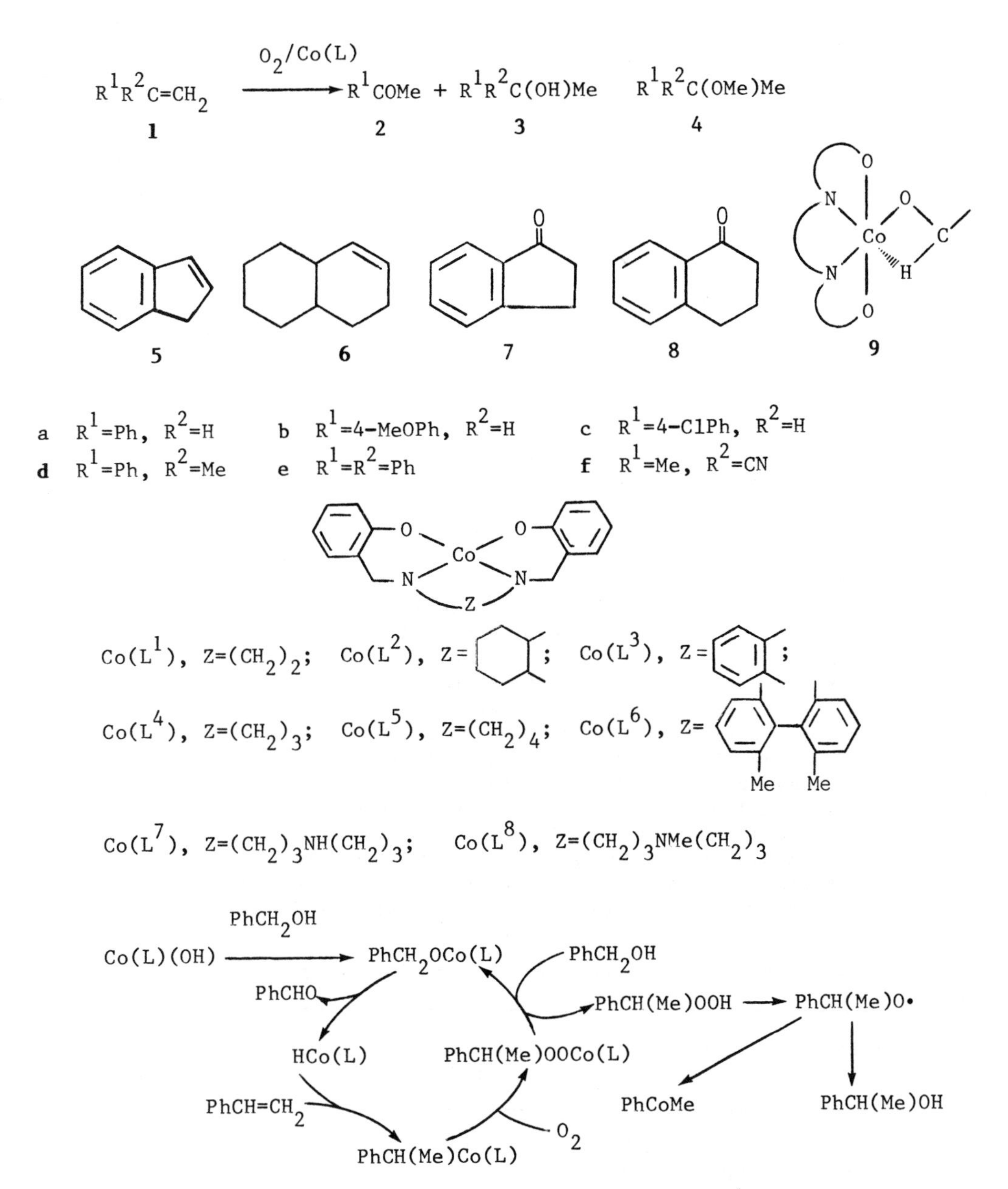

reaction depended on concentrations of $PhCH_2OH$ and Co(L^1), whereas was independent of those of **1a** and O_2. The reaction was retarded entirely by the addition of 1-methylimidazole but accelerated by PPh_3 with increasing amount of **3a**. These results suggest the above mechanistic diagram, where the rate determining step is the decomposition of $PhCH_2OCo(L)$ to PhCHO and a hydridocobalt species. The addition of the hydrido complex to **1a** was supported by the formation of $PhCOCH_2D$ in the oxygenation of **1a** using $PhCD_2OH$. The effect of PPh_3 may be reasonably understood by the reduction of PhCH(Me)OOCo(L) to PhCH(Me)OCo(L). Interestingly, nonplanar complexes, Co(L^4)-Co(L^6), were highly reactive compared to planar ones, Co(L^1)-Co(L^3), among which Co(L^3) was less reactive than Co(L^1). The reactivity of five coordinate complexes was quite low (Table 2). These results suggest that a twist of formula **9**, is important for the transition state. Increase in the amount of **3a** in cases with Co(L^4)-Co(L^6) (Table 2) may be due to prolongation of the life time of 1-phenylethoxy radical (shown in the diagram), which results from a paramagnetic interaction between the radical and the catalysts.

The recently reported Drago conclusion: involvement of addition of a hydroperoxocobalt(III) species to **1a** as a key reaction step,[3] is in conflict with the present observations.

Table 2. Influence of the Structure of Co(L) on the oxygenation of **1a**[a]

	$Co(L^1)$	$Co(L^2)$	$Co(L^3)$	$Co(L^4)$	$Co(L^5)$	$Co(L^6)$	$Co(L^7)$	$Co(L^8)$
$k \times 10^5/s$[b]	1.67	5.41	0.15	19.44	22.00	22.22	0.13	0.14
2a/3a	4.2	5.0	2.8	0.3	1.0	1.3	0.9	0.8

[a] **1a** (43.6 mM), BzOH (0.48 M), Co(L) (0.5 mM), CH_2Cl_2CCl (10 ml), O_2 (1 atm) at 60°C. [b] Initial rate constant.

REFERENCES

1. A. Nishinaga and H. Tomita, J. Mol. Catal., 7:179 (1980).
2. A. Nishinaga, T. Kondo and T. Natsuura, Chem. Lett., 905:1319 (1985).
3. D. E. Hamilton, R. S. Drago and A. Zombeck, J. Am. Chem. Soc., 109:374 (1987).

SELECTIVE OXIDATION OF SATURATED HYDROCARBONS BY THE GIF AND GIF-ORSAY SYSTEMS

Derek H. R. Barton*, Frank Halley, Nubar Ozbalik, and Esme Young

Department of Chemistry
Texas A&M University
College Station, Texas 77843

The Gif system[1,2] for the selective oxidation of saturated hydrocarbons consists of pyridine (as solvent), acetic acid (or other carboxylic acid), zinc powder (as source of electrons), oxygen and an iron salt as catalyst. The iron salt is rapidly complexed with *ortho*-dipyridyl (produced by reduction of pyridine). This system is interesting because it oxidizes hydrocarbons selectively at the secondary positions giving mainly ketones. Primary and tertiary positions give minor amounts of products. This substitution pattern excludes a radical attack (as in the Fenton reagent). However, adamantane shows exceptional behavior in that tertiary oxidation products arise, at least in part, from carbon radical. This radical partitions itself between oxygen and pyridine. Reduction of oxygen pressure increases the apparent selectivity of the reaction and leads to increased formation of *tertiary* pyridine coupled products. No such behavior is seen at the secondary position which is, therefore, oxidized without involving carbon radicals.

As normally run, the Gif system is inefficient in its use of electrons. More recent studies, in collaboration with the electrochemical group of Professor Balavoine (Orsay), has shown[3] that the Gif system can be applied electrochemically to hydrocarbon oxidation and that in a unicellular mode it is possible to obtain high electronic yields (Gif-Orsay system).

REFERENCES

1. D.H.R. Barton, J. Boivin, M. Gastiger, J. Morzycki, R. S. Hay-Motherwell, W. B. Motherwell, N. Ozbalik, and K. M. Scwartzentruber, *J. Chem. Soc. Perkins Trans.*, 1:947 (1986).
2. D.H.R. Barton, J. Boivin, W. B. Motherwell, N. Ozbalik, and K. M. Schwarzentruber, *Nouveau J. Chimie*, 10:387 (1986).
3. G. Balavoine, D.H.R. Barton, J. Boivin, A. Gref, N. Ozbalik, and H. Riviere, *Tetrahedron Letts.*, 2849 (1986); *J. Chem. Soc. Chem. Commun.*, 1727 (1986).

THE FORMATION, CHARACTERIZATION, AND REACTIVITY OF THE OXENE ADDUCT OF TETRAKIS(2,6-DICHLOROPHENYL)PORPHINATO-IRON(III) PERCHLORATE IN ACETONITRILE

Hiroshi Sugimoto, Hui-Chan Tung and Donald T. Sawyer*

Department of Chemistry
Texas A&M University
College Station, Texas 77843-3255

Combination of tetrakis(2,6-dichlorophenyl)porphinato-iron(III) perchlorate with pentafluoroiodosobenzene or _m_-chloroperbenzoic acid in acetonitrile at -35°C yields a green porphyrin-oxene adduct. This species, which has been characterized by spectroscopic, magnetic and electrochemical methods, cleanly and directly epoxidizes olefins. The reaction chemistry and electronic characterization of the adduct are consistent with an oxygen atom covalently bound to an iron(II)-porphyrin radical center $[(P^{\overline{\cdot}})Fe^{II}(O)^{+}]$. The latter has a reactivity and spectral characteristics that are closely similar to Compound I of horseradish peroxidase, and the selective stereospecific oxygenase character of the reactive intermediate for cytochrome P-450. The iron(III) porphyrin is an efficient catalyst for the stereospecific epoxidation of olefins by F_5PhIO and _m_-ClPhC(O)OOH; with H_2O_2 the reaction is much less efficient and there is extensive attack of the porphyrin ring.

PREPARATION AND CHARACTERIZATION OF A BINUCLEAR IRON(III)-HYDROXO-μ-HYDROXYPEROXY COMPLEX $[(PH_3PO)_4(HO)Fe^{III}(HOOH)Fe^{III}(OH)(OPPh_3)_4](ClO_4)_4$

M. Steven McDowell, Lee Spencer, Paul K.-S. Tsang and
Donald T. Sawyer*

Department of Chemistry
Texas A&M University
College Station, Texas 77843-3255

The combination of $[Fe^{II}(OPPh_3)_4](ClO_4)_2$ with H_2O_2, m-$ClC_6H_4C(O)OOH$, Me_3COOH, PhIO, $PhI(OAc)_2$, $Bu_4N(IO_4)$, O_3, or NaOCl in anhydrous acetonitrile results in the formation of the binuclear complex $[(Ph_3PO)_4(HO)Fe^{III}(HOOH)Fe^{III}(OH)(OPPh_3)_4](ClO_4)_4$, **1**. The same material is produced from the addition of H_2O_2 to $[Fe^{III}(Ph_3PO)_4](ClO_4)_3$ in acetonitrile. The complex has been charactrized by elemental analysis; electronic, vibrational, and ESR spectroscopy; solid- and solution-phase magnetic susceptibility measurements; and electrochemistry. A mechanism is proposed for the formation of **1**, and for its reactivity with halide ions.

AUTOXIDATION OF Fe^{II}(DIHYDROXYPHENANTHROLINE)$_3$

David M. Stanbury

Department of Chemistry
Rice University
Houston, Texas 77251

$Fe^{II}(OHP)_3$ (OHP = 4,7-dihydroxy-1,10-phenanthroline) reacts with oxygen rapidly in alkaline aqueous solution. A major product of the reaction is HO_2^-; $O_2^-\cdot$ is a feasible intermediate in the reaction. $Fe^{II}(OHP)_3$ reacts with HO_2^- to produce a 1:1 mixture of $Fe^{III}(OHP)_3$ and "$Fe^{III}(OHP)_2(OH)$"; this reaction has a rate law that is first order in [$Fe^{II}(OHP)_3$] but zero order in [peroxide], with a rate constant of 1.8×10^{-2} s^{-1} at 22°C, pH 13. The substitution reaction of $Fe^{II}(OHP)_3$ with CN^- has the same rate law as the redox reaction with HO_2^- and a rate constant of 8×10^{-3} s^{-1}. Both the redox reaction of HO_2^- and the substitution reaction of CN^- are interpreted as having loss of the ligand OHP as the rate-limiting step. A sensitive and specific HPLC method has been developed for determination of HO_2^- in complex mixtures.

PHOSPHINE-RUTHENIUM(II)-AQUO REDOX CHEMISTRY: THE AEROBIC CATALYTIC OXIDATION OF CYCLOHEXENE

R. A. Leising+, M. E. Marmion+, J. J. Gryzbowski‡, and K. J. Takeuchi+*

Department of Chemistry
+ SUNY at Buffalo, Buffalo, New York 14214
‡ Gettysburg College, Gettysburg, Pennsylvania 17325

Recently, our laboratory synthesized ruthenium(IV)-oxo complexes containing tertiary phosphine ligands, which act as stoichiometric oxidation reagents toward a variety of organic and inorganic substrates (alcohols, olefins, aldehydes, phosphines, sulfides, and sulfoxides). Currently, we are investigating the aerobic oxidation of cyclohexene, phosphine-ruthenium(II)-aquo complexes. The activation of molecular oxygen occurs at room pressure and temperature, without the need for a coreductant. The active oxidant in the catalytic cycle appears to be the phosphine-ruthenium(IV)-oxo species, for the product distribution from the catalytic oxidation of cyclohexene is identical to the product distribution from the stoichiometric oxidation of cyclohexene by a phosphine-ruthenium(IV)-oxo complex. The rate of product formation over a 24 hour period is constant, with the catalytic reaction sampled after 1, 2, 4, 8, and 24 hours. The initial catalyst was isolated intact at the end of the reaction. The oxidation of cyclohexene produces 2-cyclohexene-1-one, 2-cyclohexene-1-ol, and cyclohexene oxide in a product ratio of 16:8:1. After twenty-four hours, the oxidation of a 2.2 M solution of cyclohexene using a 5.0×10^{-4} M solution of catalyst yields a turnover number of 1560, where the production of 2-cyclohexen-1-one from cyclohexene requires two turnovers of catalyst per molecule. From experimental observations, a catalytic cycle can be suggested for the aerobic oxidation of cyclohexene. Initially, a five-coordinate phosphine-ruthenium(II) complex is generated, by the the loss of an aquo ligand. The combination of a molecule of dioxygen with two of these five-coordinate complexes forms a dinuclear, oxygen-bridged intermediate. Homolytic cleavage of the O-O bond yields two phosphine-ruthenium(IV)-oxo molecules, the proposed active oxidant. The phosphine-ruthenium(IV)-oxo complex oxidizes the target organic substrate, forming a phosphine-ruthenium(II)-oxidized substrate complex. Dissociation of the oxidized organic substrate by this complex yields the five-coordinate phosphine-ruthenium(II) complex, which continues in the catalytic cycle.

THE OXIDATION OF ORGANIC SUBSTRATES BY MOLECULAR OXYGEN; CATALYSIS BY Ru(III) AND Ru(III)-EDTA

M. M. Taqui Khan

Division of Coordination Chemistry & Homogeneous Catalysis
Central Salt & Marine Chemicals Research Institute
Bhavnagar 364 002 India

Oxidation reactions of organic substrates by molecular oxygen that are catalyzed by Ru(III) ion and Ru(III)-EDTA are described. These reactions proceed either through an electron-transfer (oxidase) or oxygen-atom insertion (oxygenase) routes.

In the pH range 1.5-2.5 Ru(III) ion primarily exists[1] as $RuCl_2(H_2O)_4^+$ and catalyzes the oxidation[2,3] of ascorbic acid to dehydroascorbic acid. The oxidase type of reactions of $RuCl_2(H_2O)_4^+$ also include the oxidation of allyl alcohol to acrolien[4] and cyclohexanol to cyclohexanone.[5]

Ru(III)-EDTA is an effective catalyst in the homodioxygenation of organic substrates, where both oxygen atoms of molecular oxygen are inserted in two molecules of the substrate. The reaction proceeds through the formation of a μ-peroxo-Ru(IV)-EDTA complex.[6] Such reactions include the oxygenation of triphenylphosphine[7] to phosphine oxide and cyclohexene and allyl alcohol to the epoxides.[8] For saturated substrates such as triethylamine[9] or diethylamine,[10] the Ru(III)-EDTA catalyzed oxidation by molecular oxygen leads to N-dealkylation via N-H or C-H hydride abstraction, respectively. The nature of the oxidized substrate depends on its thermodynamic stability.

The system Ru(III)-EDTA-ascorbic acid-molecular oxygen, the Ru(III) analog of Udenfriend's system[11], is a much better oxidant than the Udenfriend's Fe(III)-EDTA-ascorbic acid-O_2 system. The system catalyzes the oxidation of saturated hydrocarbons such as cyclohexane to cyclohexanol and cyclohexanaone. Cyclohexanol is hydroxylated to cis-cyclohexane-1,3 diol. Cyclohexene and cyclooctene are oxidized to the pure epoxides. The reactions proceed by an ionic pathway and the system is an excellent model for cytochrome P-450.

REFERENCES

1. M. M. Taqui Khan, G. Ramachandriah, and A. Prakash Rao, Inorg. Chem., 25:665 (1986).
2. M. M. Taqui Khan and R. S. Shukla, J. Mol. Catal., 34:19 (1986).
3. M. M. Taqui Khan and R. S. Shukla, J. Mol. Catal., 37:269 (1986)
4. M. M. Taqui Khan and A. Prakash Rao, J. Mol. Catal., 35:237 (1986).
5. M. M. Taqui Khan and H. C. Bajaj, React. Kinet. & Catal. Letters, 28:339 (1985).

6. M. M. Taqui Khan, A. Hussain, G. Ramachandriah and M. A. Moiz, *Inorg. Chem.*, 25:3023 (1986).
7. M. M. Taqui Khan, M. R. H. Siddiqui, A. Hussain and M. A. Moiz, *Inorg. Chem.*, 25:2765 (1986).
8. M. M. Taqui Khan, *Proc. Ind. Natl. Sci. Acad.* (1986).
9. M. M. Taqui Khan, S. A. Mirza and H. C. Bajaj, *J. Mol. Catal.*, 37:253 (1986).
10. M. M. Taqui Khan, S. A. Mirza and H. C. Bajaj, *J. Mol. Catal.*, in press.
11. M. M. Taqui Khan and A. E. Martell, "Homogeneous Catalysis by Metal Complexes", Vol.I, Academic Press, New York (1974).

OXYGENATION OF TRYPTOPHANE CATALYZED BY POLYAMINE COBALT DIOYXGEN COMPLEXES

Kyte H. Terhune and Arthur E. Martell*

Department of Chemistry
Texas A&M University
College Station, Texas 77843

The catalytic conversion of tryptophane to an oxygenated product has been accomplished with a polyamine cobalt dioxygen complex, $[Co_2(TETREN)O_2]^{+4}$ (TETREN = tetraethylenepentaamine), as the catalyst. The apparent reaction rates measured by molecular oxygen consumption are the same in pure methanol and methanol that contains 1M tetramethylammonium chloride at 40°C reaction temperature. The reaction is first order in tryptophan, and the primary product has been tentatively identified as a dioxygenated substrate.

RESONANCE RAMAN SPECTROSCOPY OF THE Fe(IV)=O GROUP IN PEROXIDASE INTERMEDIATES

J. Terner*, Catherine M. Reczek, and Andrew J. Sitter

Department of Chemistry
Virginia Commonwealth University
Richmond, Virginia 23284

The heme proteins known as peroxidases and catalases catalyze reactions of peroxides. The reaction sequences of peroxidases and catalases involve two colored intermediates, known as compounds I and II. Horseradish peroxidase is one of the most studied peroxidases because of its universal availability from commercial sources, and because its reactions are typical of a large number of peroxidases. Upon reaction with hydrogen peroxide, horseradish peroxidase forms a green colored intermediate known as compound I, which is two oxidation equivalents above the resting enzyme. A one electron reduction of compound I results in a red colored intermediate known as compound II. Compound II is an Fe(IV) heme, one oxidation equivalent above the resting enzyme. While compound I is formally an Fe(V) heme, it is believed to contain an Fe(IV) with another electron removed from the highest occupied molecular orbital of the porphyrin group, resulting in a porphyrin π-radical cation.

For many years the coordination of the heme iron in the oxidized forms was not understood. Several structures had been proposed, which included Fe(IV)-OH, Fe(IV)-OOH, and Fe(IV)=O among others. Through the use of isotopic ^{18}O substitution we were able to locate bands which we could assign to the heme resonance Raman Fe(IV)=O stretching vibrations of horseradish peroxidase compound II[1] and several other peroxidases.

We subsequently observed that the Fe(IV)=O frequencies were especially sensitive to the environmental effects in the heme pocket.[2] It was suggested that, in order for a peroxidase to have enzymatic activity, it was required that the Fe(IV)=O group be hydrogen bonded to a distal amino acid group. The low peroxidative activity of ferryl myoglobin could be rationalized by minimal hydrogen bonding. Our recent observations indicate that differing reactivities among the peroxidases and catalases are reflected in differences in Fe(IV)=O frequencies.[3]

REFERENCES

1. J. Terner, A. J. Sitter, and C. M. Reczek, Biochim. Biophys. Acta., 828:73 (1985).
2. A. J. Sitter, C. M. Reczek, and J. Terner, J. Biol. Chem., 260:7515 (1985).
3. C. M. Reczek, A. J. Sitter, and J. Terner, (1987), in preparation.

REACTION OF DIOXYGEN WITH SYNTHETIC COPPER(I) COMPOUNDS OF BIOLOGICAL RELEVANCE

J. A. Goodwin+, D. M. Stanbury+, L. J. Wilson+*,
G. A. Bodager+, and W. R. Scheidt‡

Department of Chemistry
+ Rice University, Houston, Texas 77251
‡ The University of Notre Dame, Notra Dame, Indiana 46556

The copper(I) complex of the imidazole-bearing pentadentate Schiff base ligand, (bis[2,6-(2-imidazol-4-ylethylimino)ethyl]pyridine), $[Cu^{I}(imidH)_2DAP]^+$ has been previously presented as a reversible dioxygen carrier at room temperature in nonaqueous media. Spectroscopic, electrochemical, and manometric results were used to propose that dioxygen reversibly binds to Cu(I) in a fashion similar to hemocyanain, to form an antiferromagnetically-coupled binuclear Cu(II) species with a peroxo bridge between the metal centers by the reaction stoichiometry:[1]

$$2[Cu^{I}(imidH)_2DAP]^+ + O_2 \rightarrow [(O_2^{2-})(Cu^{II}(imidH)_2DAP)_2]^{2+}$$

Prompt removal of O_2 from the product solution apparently allows regeneration of the parent $[Cu^{I}(imidH)_2DAP]^+$ compound.

The present work documents the pentacoordinate structure of the parent copper(I) compound and the nature of the reaction that regenerates a Cu(I) species from the proposed binuclear dixoygen adduct.

The dioxygen adduct of the previously reported "reversible" dioxygen-carrier, $[Cu^{I}(imidH)_2DAP]^+$, has been shown to regenerate a Cu(I) species without release of O_2 under the conditions of the Toepler-pump experiment. A new working hypotheses which accounts for this observation by way of a disproportionation pathway is now under consideration. Indications that the intermediate $[Cu_2O_2]$ adduct is stable only at low temperatures will probably require low-temperature spectroscopic probes for its detection. Crystal structure determinations reveal that the parent copper(I) species can exist in both pentacoordinate and tetracoordinate geometries, with the latter produced from the pentacoordinate species upon prolonged standing in solution. The reactivity of O_2 with the newly-discovered tetracoordinate Cu(I) species is, as of yet, completely unknown.

REFERENCE

1. M. G. Simmons, C. L. Merrill, L. J. Wilson, L. A. Bottomly, and K. M. Kadish, J. Chem. Soc., Dalton Trans., 1827 (1980).

THOROUGH ELUCIDATION OF OXYGENATION- AND OXIDATION-MECHANISMS OF IRON(II) PORPHYRIN ON THE BASIS OF A NEW HYPOTHESIS FOR AN ELECTRON-TRANSFER PATHWAY

Yasuo Yamamoto

Department of Chemistry
Shimane University
Matsue 690 Japan

It is widely believed that the hydrophobicity of the heme pocket and the protection against dimerization of oxymyoglobin stabilizes the heme-iron(II) from oxidation. However, a number of model compounds designed on the basis of this consideration have not always been stable. Recently, some investigators[1-4] have pointed out the significance of the polar interactions at the distal side. In this presentation, the discussion will be focused on the titled mechanisms and will be shown that these are thoroughly elucidated by the hypothesis that has been proposed by the author[5].

The hypothesis is described as follows: "In an electron transfer reaction, each of the electron-donor and the electron-acceptor orbitals must have the nodal plane, both sides of which the lobes of the orbital must have reflection symmetry. An electron can be transferred only when these nodal planes are co-planar." An application of the hypothesis to the $Fe-O_2$ system is illustrated in the figure.

The dioxo-iron complex has traditionally been formulated as $Fe(III)-O_2^-$. However, if the net charge of an electron were transferred from iron to dioxygen, it is quite unfavorable that the superoxide ion leave an electron behind in the deoxygenation. According to X-ray crystallography, Fe-O-O angles of oxygenated picket fence and oxymyoglobin are 129°[6] and 116°[7], respectively. These results indicate that the atomic orbitals of both atoms of coordinated oxygen molecule must be those of sp^2 hybridization. In this configuration there is no unpaired electron, all the electrons are paired in the bonding (O=O, σ and π) and nonbonding orbitals, being only one $2p^*$ orbital (electron-acceptor orbital) empty. It should be noted that the charge transfer is a reversible movement of electron while the electron transfer (net charge is transferred) is irreversible. When the electron density in the π^*-orbital of the oxygen molecule is increased by such a reversible charge transfer from the d_{xz} and/or d_{yz} of iron(II), the O-O bond strength must be weakened, and result in an elongation of the bond length and a lower-energy shift of its stretching vibration. $Fe(II)^+-O_2^-$ is the suitable formulation.

The X-ray crystallography also shows that the Fe-O-O conformation of oxygenated picket fence porphyrin is identical with the type (a) in the figure. This conformation is due to the hydrogen bonds with the four amide groups in the picket linkages.[2] The model compounds which are relatively stable to oxidation must have the same conformation. Oxymyoglobin is stabilized by the type (c), being responsible to the distal histidine which imide proton electrophilically attacks in the π-electrons of dioxygen. When the imide proton is replaced by -CN, it has been reported,[8] myoglobin becomes quite unstable and is oxidized very rapidly. The electronic effect, such as lone pair[3,4] electron, might exert on the empty 2p*-oribtal to give distortion of the lobe from symmetry.

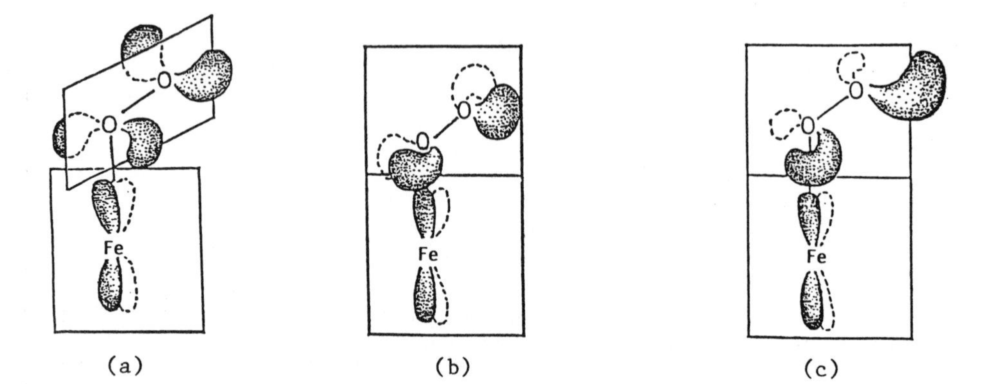

(a) (b) (c)

The application of the hypothesis to Fe(II)-O_2 system

(a) Irreversible electron transfer does not occur, because the nodal planes are not coplanar, i.e., stable oxygenation. Reversible charge transfer may occur.
(b) Irreversible electron transfer occurs, i.e., oxidation.
(c) Electron transfer does not occur, because lobes of electron-acceptor orbital are not reflection symmetry, i.e., stable oxygenation.

REFERENCES

1. M. Momenteau, M., J. Mispelter, B. Loock, and J-M. Lhoste, J. Chem. Soc. 221 (1985).
2. G. B. Jameson and R. S. Drago, J. Am. Chem. Soc. 107:3017 (1985).
3. T. G. Traylor, N. Koga, and L. A. Deardurff, J. Am. Chem. Soc. 107:6504 (1985).
4. S. Tsuchiya, Inorg. Chem. 24:4450 (1985).
5. a) Y. Yamamoto, K. Ishizu, and Y. Shimizu, Chem. Lett. 1977:735 (1977).
   b) Y. Yamomoto, T. Sasaki, T. Izumitani, and T. Kurata, Inorg. Nucl. Chem. Lett. 14:239 (1978).
6. G. B. Jameson, G. A. Rodley, W. T. Robinson, R. R. Gagne, C. A. Reed, and J. P. Collman, Inorg. Chem., 17:850 (1978).
7. S.E.V. Philips, J. Mol. Biol. 142:531 (1979).
8. Y. Shiro and I. Morishima, Biochem. 23:4879 (1984).

DESCRIPTION OF THE INDUSTRY-UNIVERSITY CHEMISTRY COOPERATIVE PROGRAM (IUCCP)

ADMINISTRATIVE ORGANIZATION OF THE IUCCP

Advisory Committee

An advisory board has been established to serve as the governing board of the IUCCP. It controls the scope and functions of the organization and is responsible for future planning activities.

Each member company has one seat on the advisory committee and the Texas A&M Chemistry Department has three seats.

Industrial-Academic Liaison

One of the three members appointed to the advisory committee by the Department functions as the Coordinator of the IUCCP. He is responsible for administering all IUCCP programs and for communicating with the member companies. He also supervises the publication of bulletins describing IUCCP activities, Departmental and company news items, research summaries, etc. In order to properly execute these functions, a half-time staff assistant is appointed to assist the Coordinator.

Each member company also appoints an individual, not necessarily the advisory board member, to serve as a contact person. This individual is responsible for transmitting news items and activities for publication in the bulletin. In turn, he receives all communications from the Chemistry Department dealing with IUCCP business and is responsible for disseminating this information within the company.

ACTIVITIES

The activities of the IUCCP are intended to serve two basic purposes:

1. Promote scientific interaction between the member companies and the Texas A&M Chemistry Department.
2. Accelerate the rate of Departmental Development.

Joint Symposium

The department organizes and hosts an annual symposium focussing on scientific topics of current interest to IUCCP members. Topics and speakers are determined by the advisory committee. Speakers are drawn from both industry and university sources worldwide. The symposia are structured in such a way as to maximize communication among the participants. For this purpose, a Gordon Research Conference format has been employed as the most appropriate. The following is a list of the international symposia that have been held, and the title of the meeting being organized for 1988.

1.	Organometallic Compounds: Synthesis, Structure and Theory	April 17-20, 1983
2.	Heterogeneous Catalysis	April 1- 4, 1984
3.	New Directions in Chemical Analysis	March 31-April 3, 1985
4.	Design of New Materials	March 24-26, 1986
5.	Oxygen Complexes and Oxygen Activation by Transition Metals	March 23-26, 1987
6.	Functional Polymers	March 21-24, 1988

The proceedings of these symposia are published in book form, and three free copies are supplied to member companies. The first three volumes were published by the Texas A&M University Press, and also may be obtained by contacting the IUCCP Coordinator (A. E. Martell). Volumes 4-6 are being (or will be) published by Plenum Press.

Information Exchange

a. An information exchange is organized and maintained by the program coordinator. Member companies and the Department provide up-to-date information on their major areas of research emphasis and their more active current projects. In addition, the Department keeps members apprised of student thesis projects and progress toward graduation.

b. News Bulletin. Departmental activities are reported to member companies on a regular basis.

c. Research Bulletins. Bimonthly lists of papers and abstracts submitted for publication are distributed to the Liaison officer of each company as a means of supplying prepublication information on research activities.

Colloquium Programs

a. Speaker Exchange. A colloquium speaker exchange has been organized between the member companies and the department. Company scientists are invited to present both scientific talks and to discuss the industrial research environment with our students.

b. Speaker Sharing. The sharing of outside colloquium speakers who visit the University is encouraged. Lists of desired speakers are exchanged. Joint invitations are issued where feasible and costs are shared.

Fellowship Program

a. A joint graduate fellowship program has been established. These fellowships consist of departmentally financed teaching assistantships combined with company sponsored research assistantships. Ideally, each member company sponsors one fellowship. These fellowships, named for the sponsoring company, are awarded to entering graduate students on a competitive basis. During the first year, such students would receive nine months of fellowship support at $1,000 per month, and an additional $333 per month for a 1/3 teaching assignment during the academic year. He would then be supported under his IUCCP fellowship during the summer at $1,000 per month, making the one-year package $15,000. Support in subsequent years at $1,000 per month would come from research grants.

b. Undergraduate scholarships are offered on a competitive basis to outstanding entering freshmen. These fellowships are designed to attract larger numbers of high quality undergraduate students into chemistry programs.

Program Extensions

The advisory committee is charged with the responsibility for recommending additional mechanisms for cooperative activities. Such activities might include collaborative research, sponsored research, student co-op programs, short courses, instrument sharing programs, consulting arrangements, etc. The funds provided under IUCCP are for the purposes stated and do not impact on or restrict in any way the establishment of research grants and contracts by the member companies with individual research groups in the Department.

ADVANTAGES OF MEMBERSHIP

The program is designed to be one of mutual support and benefit. The success of such a program largely depends upon the attitude of the participants, and we want to assure you that our Department is wholeheartedly behind the program. If the member companies also approach this program with the same commitment, then we can envisage many benefits to be derived from membership. Those that readily come to mind are as follows:

1. It provides close contact with resources of a truly major chemistry department. Advance notice of new research developments are provided on a regular basis. Abstracts of papers and theses are provided to member companies long before publication, and preprints are available on request.

2. It provides an opportunity to present the industrial point of view to a large faculty and student body.

3. It provides access to a large body of high quality students, both graduate and undergraduate, for potential employment.

4. It helps to attract a greater number of quality speakers to the yearly international symposia.

5. It provides a means of developing educational programs of interest to member companies.

6. Arrangements can be made to provide graduate educational opportunities for employees of member companies.

7. It encourages and assists personnel of member companies in participating in collaborative research projects with faculty members.

8. It aids in the establishment of sponsored research projects of interest to member companies.

9. It offers the satisfaction to be derived from participating in the development of this Department into one of the great Chemistry Departments in the Country.

FINANCING OF THE IUCCP

In our drive towards excellence, it is imperative that we continue to recruit the highest quality faculty and graduate students. However, this is increasingly difficult to do, given the current level of competition for such personnel and the constraints on our University budget. Most of our funds are earmarked for teaching functions, defined as classroom teaching and the advising of thesis research. Furthermore, the bulk of our operating budget is restricted to instructional support and non-academic staff. Thus, funds are very limited for faculty and graduate student recruiting, fellowships, equipment maintenance, moving expenses, speakers programs, and other needs. This situation requires us to find other sources to supply the supplementary funds necessary to bring us into the front rank of chemistry departments. The operating costs of the IUCCP are supported by annual dues paid by the member companies, set at $15,000 per year. Because this amount is obviously not sufficient to support the activities described above, and provide one graduate fellowship per year per company at $15,000, each member company is invited to provide the graduate fellowship funds over and above the $15,000 base.

Advisory Board

Dr. R. J. H. Voorhoeve
Technical Director
Celanese Specialty Operations
P.O. Box 9077
Corpus Christi, Texas 78469-9077

Dr. Doug Hunter
Dow Chemical U.S.A.
Urethanes Department B-4810
Freeport, Texas 77541

Dr. William M. Haynes, Director
Physical Sciences Center
Monsanto Company
800 N Lindbergh Boulevard
St. Louis, Missouri 63166

Dr. Harvey Klein
Corporate Research & Development Science
Shell Development Company
P.O. Box 1380
Houston, Texas 77001

Dr. Daniel R. Herrington
The Standard Oil Company
Research Center
4440 Warrensville Center Road
Cleveland, Ohio 44128

Liaison Officers

Dr. Edward G. Zey
Celanese Chemical Company
P.O. Box 9077
Corpus Christi, Texas 78469-9077

Dr. Doug Hunter
Dow Chemical U.S.A.
Urethanes Department B-4810
Freeport, Texas 77541

Mr. Burnell Curtis
Monsanto Company
P.O. Box 711
Alvin, Texas 77512-9888

Dr. J. L. Laity
Chemical Research Applications Department
Shell Development Company
P.O. Box 1380
Houston, Texas 77001

Dr. Daniel R. Herrington
The Standard Oil Company
Research Center
4440 Warrensville Center Road
Cleveland Ohio 44128

INDEX